THE LOEB CLASSICAL LIBRARY

FOUNDED BY JAMES LOEB 1911

EDITED BY

JEFFREY HENDERSON

EDITOR EMERITUS

G. P. GOOLD

GALEN

ON THE NATURAL FACULTIES

LCL 71

GALEN

ON THE NATURAL FACULTIES

WITH AN ENGLISH TRANSLATION BY

ARTHUR JOHN BROCK

HARVARD UNIVERSITY PRESS
CAMBRIDGE, MASSACHUSETTS
LONDON, ENGLAND

First published 1916
Reprinted 1928, 1947, 1952, 1963, 1979,
1991, 2000

LOEB CLASSICAL LIBRARY® is a registered trademark
of the President and Fellows of Harvard College

ISBN 0-674-99078-1

Printed in Great Britain by St Edmundsbury Press Ltd,
Bury St Edmunds, Suffolk, on acid-free paper.
Bound by Hunter & Foulis Ltd, Edinburgh, Scotland.

CONTENTS

PREFACE

THE text used is (with a few unimportant modifications) that of Kühn (Vol. II), as edited by Georg Helmreich; Teubner, Leipzig, 1893. The numbers of the pages of Kühn's edition are printed at the side of the Greek text, a parallel mark (‖) in the line indicating the exact point of division between Kühn's pages.

Words in the English text which are enclosed in square brackets are supplementary or explanatory; practically all explanations, however, are relegated to the footnotes or introduction. In the footnotes, also, attention is drawn to words which are of particular philological interest from the point of view of modern medicine.

I have made the translation directly from the Greek; where passages of special difficulty occurred, I have been able to compare my own version with Linacre's Latin translation (1523) and the French rendering of Charles Daremberg (1854-56); in this respect I am also peculiarly fortunate in having had the help of Mr. A. W. Pickard Cambridge of Balliol College, Oxford, who most kindly went through the

PREFACE

proofs and made many valuable suggestions from the point of view of exact scholarship.

My best thanks are due to the Editors for their courtesy and for the kindly interest they have taken in the work. I have also gratefully to acknowledge the receipt of much assistance and encouragement from Sir William Osler, Regius Professor of Medicine at Oxford, and from Dr. J. D. Comrie, first lecturer on the History of Medicine at Edinburgh University. Professor D'Arcy W. Thompson of University College, Dundee, and Sir W. T. Thiselton-Dyer, late director of the Royal Botanic Gardens at Kew, have very kindly helped me to identify several animals and plants mentioned by Galen.

I cannot conclude without expressing a word of gratitude to my former biological teachers, Professors Patrick Geddes and J. Arthur Thomson. The experience reared on the foundation of their teaching has gone far to help me in interpreting the great medical biologist of Greece.

I should be glad to think that the present work might help, however little, to hasten the coming reunion between the "humanities" and modern biological science ; their present separation I believe to be against the best interest of both.

A. J. B.

22ND STATIONARY HOSPITAL, ALDERSHOT.
    March, 1916.

INTRODUCTION

IF the work of Hippocrates be taken as repre- senting the foundation upon which the edifice of historical Greek medicine was reared, then the work of Galen, who lived some six hundred years later, may be looked upon as the summit or apex of the same edifice. Galen's merit is to have crystallised or brought to a focus all the best work of the Greek medical schools which had preceded his own time. It is essentially in the form of Galenism that Greek medicine was transmitted to after ages.

The ancient Greeks referred the origins of medicine to a god Asklepios (called in Latin Aesculapius), thereby testifying to their appreciation of the truly divine function of the healing art. The emblem of Aesculapius, familiar in medical symbolism at the present day, was a staff with a serpent coiled round it, the animal typifying wisdom in general, and more particularly the wisdom of the medicine-man, with his semi-miraculous powers over life and death.

" Be ye therefore wise as serpents and harmless as doves."

ix

INTRODUCTION

The Ascle-
piea or
Health-
temples.
The temples of Aesculapius were scattered over
the ancient Hellenic world. To them the sick and
ailing resorted in crowds. The treatment, which was
in the hands of an hereditary priesthood, combined
the best of the methods carried on at our present-day
health-resorts, our hydropathics, sanatoriums, and
nursing-homes. Fresh air, water-cures, massage,
gymnastics, psychotherapy, and natural methods in
general were chiefly relied on.

Hippocrates
and the
Unity of the
Organism.
Hippocrates, the " Father of Medicine " (5th to 4th
centuries, B.C.) was associated with the Asclepieum
of Cos, an island off the south-west coast of Asia
Minor, near Rhodes. He apparently revitalized the
work of the health-temples, which had before his
time been showing a certain decline in vigour, coupled
with a corresponding excessive tendency towards
sophistry and priestcraft.

Celsus says : " *Hippocrates Cous primus quidem ex
omnibus memoria dignis ab studio sapientiae disciplinam
hanc separavit.*" He means that Hippocrates first gave
the physician an independent standing, separating
him from the cosmological speculator. Hippocrates
confined the medical man to medicine. He did with
medical thought what Socrates did with thought in
general—he " brought it down from heaven to earth."
His watchword was " Back to Nature ! "

At the same time, while assigning the physician
his post, Hippocrates would not let him regard that
post as sacrosanct. He set his face against any

tendency to mystery-mongering, to exclusiveness, to sacerdotalism. He was, in fact, opposed to the spirit of trade-unionism in medicine. His concern was rather with the physician's duties than his " rights."

At the dawn of recorded medical history Hippocrates stands for the fundamental and primary importance of *seeing clearly*—that is of *clinical observation*. And what he observed was that the human organism, when exposed to certain abnormal conditions—certain stresses—tends to behave in a certain way : that in other words, each "disease" tends to run a certain definite course. To him a disease was essentially a process, one and indivisible, and thus his practical problem was essentially one of *prognosis*— " what will be the natural course of this disease, if left to itself?" Here he found himself to no small extent in opposition with the teaching of the neighbouring medical school of Cnidus, where a more static view-point laid special emphasis upon the minutiae of *diagnosis*.

Observation taught Hippocrates to place unbounded faith in the recuperative powers of the living organism—in what we sometimes call nowadays the *vis medicatrix Naturae*. His observation was that even with a very considerable " abnormality" of environmental stress the organism, in the large majority of cases, manages eventually by its own inherent powers to adjust itself to the new conditions. " Merely give Nature a chance," said the father of medicine in effect, " and most diseases will cure themselves." And accordingly his treatment was mainly directed towards "giving Nature a chance."

His keen sense of the solidarity (or rather, of the constant interplay) between the organism and its environment (the "conditions" to which it is exposed) is instanced in his book, "Airs, Waters, and Places." As we recognise, in our popular everyday psychology, that "it takes two to make a quarrel," so Hippocrates recognised that in pathology, it takes two (organism and environment) to make a disease.

As an outstanding example of his power of clinical observation we may recall the *facies Hippocratica*, an accurate study of the countenance of a dying man.

His ideals for the profession are embodied in the "Hippocratic oath."

Anatomy. Impressed by this view of the organism as a unity, the Hippocratic school tended in some degree to overlook the importance of its constituent *parts*. The balance was re-adjusted later on by the labours of the anatomical school of Alexandria, which, under the aegis of the enlightened Ptolemies, arose in the 3rd century B.C. Two prominent exponents of anatomy belonging to this school were Herophilus and Erasistratus, the latter of whom we shall frequently meet with in the following pages (*v.* p. 95 *et seq.*).

INTRODUCTION

After the death of the Master, the Hippocratic school tended, as so often happens with the best of cultural movements, to show signs itself of diminishing vitality: the letter began to obscure and hamper the spirit. The comparatively small element of theory which existed in the Hippocratic physiology was made the groundwork of a somewhat over-elaborated "system." Against this tendency on the part of the "Dogmatic" or "Rationalist" school there arose, also at Alexandria, the sect of the Empiricists. "It is not," they said, "the cause but the cure of diseases that concerns us; not how we digest, but what is digestible."

Horace said "*Graecia capta ferum victorem* *cepit*." Political domination, the occupation of territory by armies, does not necessarily mean real conquest. Horace's statement applied to medicine as to other branches of culture.

The introducer of Greek medicine into Rome was Asclepiades (1st century B.C.). A man of forceful personality, and equipped with a fully developed philosophic system of health and disease which commended itself to the Roman *savants* of the day, he soon attained to the pinnacle of professional success in the Latin capital: he is indeed to all time the type of the fashionable (and somewhat "faddy") West-end physician. His system was a purely mechanistic one, being based upon

xiii

the atomic doctrine of Leucippus and Democritus, which had been completed by Epicurus and recently introduced to the Roman public in Lucretius's great poem " *De Rerum Natura.*" The disbelief of Asclepiades in the self-maintaining powers of the living organism are exposed and refuted at considerable length by Galen in the volume before us.

The Methodists.

Out of the teaching of Asclepiades that physiological processes depend upon the particular way in which the ultimate indivisible molecules come together (ἐν τῇ ποίᾳ συνόδῳ τῶν πρώτων ἐκείνων σωμάτων τῶν ἀπαθῶν) there was developed by his pupil, Themison of Laodicea, a system of medicine characterised by the most engaging simplicity both of diagnosis and treatment. This so-called " Methodic " system was intended to strike a balance between the excessive leaning to apriorism shown by the Rationalist (Hippocratic) school and the opposite tendency of the Empiricists. " A pathological theory we must have," said the Methodists in effect, " but let it be simple." They held that the molecular groups constituting the tissues were traversed by minute channels (πόροι, " pores ") ; all diseases belonged to one or other of two classes ; if the channels were constricted the disease was one of *stasis* (στέγνωσις), and if they were dilated the disease was one of *flux* (ῥύσις). Flux and stasis were indicated respectively by increase and diminution of the natural secretions ;

xiv

treatment was of opposites by opposites—of stasis by methods causing dilatation of the channels, and conversely.

Wild as it may seem, this pathological theory of the Methodists contained an element of truth; in various guises it has cropped up once and again at different epochs of medical history; even to-day there are pathologists who tend to describe certain classes of disease in terms of vaso-constriction and vaso-dilatation. The vice of the Methodist teaching was that it looked on a disease too much as something fixed and finite, an independent *entity*, to be considered entirely apart from its particular setting. The Methodists illustrate for us the tyranny of *names*. In its defects as in its virtues this school has analogues at the present day; we are all acquainted with the medical man to whom a name (such, let us say, as " tuberculosis," " gout," or "intestinal auto-intoxication ") stands for an entity, one and indivisible, to be treated by a definite and unvarying formula.

To such an individual the old German saying " *Jedermann hat am Ende ein Bischen Tuberkulose* " is simply—incomprehensible.

* * * * * *

All the medical schools which I have mentioned Galen. were still holding their ground in the 2nd century A.D., with more or less popular acceptance, when the great Galen made his entry into the world of Graeco-Roman medicine.

INTRODUCTION

His Nature and Nurture.

Claudius Galenus was born at Pergamos in Asia Minor in the year 131 A.D. His father was one Nicon, a well-to-do architect of that city. "I had the great good fortune," says Galen,[1] "to have as a father a highly amiable, just, good, and benevolent man. My mother, on the other hand, possessed a very bad temper; she used sometimes to bite her serving-maids, and she was perpetually shouting at my father and quarrelling with him—worse than Xanthippe with Socrates. When, therefore, I compared the excellence of my father's disposition with the disgraceful passions of my mother, I resolved to embrace and love the former qualities, and to avoid and hate the latter."

Nicon called his son Γαληνός, which means *quiet, peaceable*, and although the physician eventually turned out to be a man of elevated character, it is possible that his somewhat excessive leaning towards controversy (exemplified in the following pages) may have resulted from the fact that he was never quite able to throw off the worst side of the maternal inheritance.

His father, a man well schooled in mathematics and philosophy, saw to it that his son should not lack a liberal education. Pergamos itself was an ancient centre of civilisation, containing, among other culture-institutions, a library only second in importance to that of Alexandria itself; it also contained an Asclepieum.

[1] *On the Affections of the Mind*, p. 41 (Kühn's ed.),

INTRODUCTION

Galen's training was essentially eclectic: he studied all the chief philosophical systems of the time— Platonic, Aristotelian, Stoic, and Epicurean—and then, at the age of seventeen, entered on a course of medical studies; these he pursued under the best teachers at his own city, and afterwards, during a period of *Wanderjahre*, at Smyrna, Alexandria, and other leading medical centres.

Returning to Pergamos, he received his first professional appointment—that of surgeon to the gladiators. After four years here he was drawn by ambition to Rome, being at that time about thirty-one years of age. At Rome the young Pergamene attained a brilliant reputation both as a practitioner and as a public demonstrator of anatomy; among his patients he finally numbered even the Emperor Marcus Aurelius himself.

Medical practice in Rome at this time was at a low ebb, and Galen took no pains to conceal his contempt for the ignorance, charlatanism, and venality of his fellow-practitioners. Eventually, in spite of his social popularity, he raised up such odium against himself in medical circles, that he was forced to flee the city. This he did hurriedly and secretly in the year 168 A.D., when thirty-six years of age. He betook himself to his old home at Pergamos, where he settled down once more to a literary life.

His respite was short, however, for within a year he was summoned back to Italy by imperial mandate. Marcus Aurelius was about to undertake an expedition against the Germans, who at that time were threatening the northern frontiers of the Empire, and he was anxious that his consulting physician should accompany him to the front. " Patriotism " in this sense, however, seems to have had no charms for the Pergamene, and he pleaded vigorously to be excused. Eventually, the Emperor gave him permission to remain at home, entrusting to his care the young prince Commodus.

Thereafter we know little of Galen's history, beyond the fact that he now entered upon a period of great literary activity. Probably he died about the end of the century.

Subsequent History of Galen's Works.

Galen wrote extensively, not only on anatomy, physiology, and medicine in general, but also on logic ; his logical proclivities, as will be shown later, are well exemplified in his medical writings. A considerable number of undoubtedly genuine works of his have come down to us. The full importance of his contributions to medicine does not appear to have been recognized till some time after his death, but eventually, as already pointed out, the terms Galenism and Greek medicine became practically synonymous.

A few words may be devoted to the subsequent history of his writings.

Byzantine Medicine.

During and after the final break-up of the Roman Empire came times of confusion and of social reconstruction, which left little opportunity for scientific thought and research. The Byzantine Empire, from the 4th century onwards, was the scene of much internal turmoil, in which the militant activities of the now State-established Christian church played a not inconsiderable part. The Byzantine medical scholars were at best compilers, and a typical compiler was Oribasius, body-physician to the Emperor Julian (4th century, A.D.); his excellent *Synopsis* was written in order to make the huge mass of the Galenic writings available for the ordinary practitioner.

Greek medicine spread, with general Greek culture, throughout Syria, and from thence was carried by the Nestorians, a persecuted heretical sect, into Persia; here it became implanted, and hence eventually spread to the Mohammedan world. Several of the Prophet's successors (such as the Caliphs Harun-al-Rashid and Abdul-Rahman III) were great patrons of Greek learning, and especially of medicine. The Arabian scholars imbibed Aristotle and Galen with avidity. A partial assimilation, however, was the farthest stage to which they could attain; with the exception of pharmacology, the Arabians made practically no independent additions to medicine. They were essentially systematizers and commentators. " *Averrois che il gran comento feo* " [1]

Arabian Medicine.

---

[1] " Averrhoës who made the great Commentary " (Dante). It was Averrhoës (Ebn Roshd) who, in the 12th century, introduced Aristotle to the Mohammedan world, and the " Commentary " referred to was on Aristotle.

may stand as the type *par excellence* of the Moslem sage.

Avicenna (Ebn Sina), (10th to 11th century) is the foremost name in Arabian medicine: his "Book of the Canon in Medicine," when translated into Latin, even overshadowed the authority of Galen himself for some four centuries. Of this work the medical historian Max Neuburger says: "Avicenna, according to his lights, imparted to contemporary medical science the appearance of almost mathematical accuracy, whilst the art of therapeutics, although empiricism did not wholly lack recognition, was deduced as a logical sequence from theoretical (Galenic and Aristotelian) premises."

Introduction of Arabian Medicine to the West. Arabo-Scholastic Period.

Having arrived at such a condition in the hands of the Mohammedans, Galenism was now destined to pass once more to the West. From the 11th century onwards Latin translations of this "Arabian" Medicine (being Greek medicine in oriental trappings) began to make their way into Europe; here they helped to undermine the authority of the one medical school of native growth which the West produced during the Middle Ages—namely the School of Salerno.

Blending with the Scholastic philosophy at the universities of Naples and Montpellier, the teachings of Aristotle and Galen now assumed a position of supreme authority: from their word, in matters scientific and medical, there was no appeal. In reference to this period the Pergamene was referred to in later times as the "Medical Pope of the Middle Ages."

It was of course the logical side of Galenism which chiefly commended it to the mediaeval School-men, as to the essentially speculative Moslems.

The year 1453, when Constantinople fell into the hands of the Turks, is often taken as marking the commencement of the Renascence. Among the many factors which tended to stimulate and awaken men's minds during these spacious times was the re-discovery of the Greek classics, which were brought to Europe by, among others, the scholars who fled from Byzantium. The Arabo-Scholastic versions of Aristotle and Galen were now confronted by their Greek originals. A passion for Greek learning was aroused. The freshness and truth of these old writings helped to awaken men to a renewed sense of their own dignity and worth, and to brace them in their own struggle for self-expression. *The Renascence.*

Prominent in this " Humanist " movement was the English physician, Thomas Linacre (*c.* 1460–1524) who, having gained in Italy an extraordinary zeal for the New Learning, devoted the rest of his life, after returning to England, to the promotion of the *litterae humaniores,* and especially to making Galen accessible to readers of Latin. Thus the " *De Naturalibus Facultatibus* " appeared in London in

1523, and was preceded and followed by several other translations, all marked by minute accuracy and elegant Latinity.

Two new parties now arose in the medical world—the so-called " Greeks " and the more conservative " Arabists."

Paracelsus. But the swing of the pendulum did not cease with the creation of the liberal " Greek " party ; the dazzling vision of freedom was to drive some to a yet more anarchical position. Paracelsus, who flourished in the first half of the 16th century, may be taken as typifying this extremist tendency. His one cry was, " Let us away with all authority whatsoever, and get back to Nature ! " At his first lecture as professor at the medical school of Basle he symbolically burned the works of Galen and of his chief Arabian exponent, Avicenna.

The Renascence Anatomists. But the final collapse of authority in medicine could not be brought about by mere negativism. It was the constructive work of the Renascence anatomists, particularly those of the Italian school, which finally brought Galenism to the ground.

Vesalius (1514–64), the modern " Father of Anatomy," for dissecting human bodies, was fiercely assailed by the hosts of orthodoxy, including that stout Galenist, his old teacher Jacques Dubois (Jacobus Sylvius). Vesalius held on his way, however, proving, *inter alia*, that Galen had been wrong

xxii

in saying that the interventricular septum of the heart was permeable (*cf.* present volume, p. 321).

Michael Servetus (1509–53) suggested that the blood, in order to get from the right to the left side of the heart, might have to pass through the lungs. For his heterodox opinions he was burned at the stake.

Another 16th-century anatomist, Andrea Cesalpino, is considered by the Italians to have been a discoverer of the circulation of the blood before Harvey; he certainly had a more or less clear idea of the circulation, but, as in the case of the " organic evolutionists before Darwin," he failed to prove his point by conclusive demonstration.

William Harvey, the great Englishman who founded modern experimental physiology and was the first to establish not only the fact of the circulation but also the physical laws governing it, is commonly reckoned the Father of Modern Medicine. He owed his interest in the movements of the blood to Fabricio of Acquapendente, his tutor at Padua, who drew his attention to the valves in the veins, thus suggesting the idea of a circular as opposed to a to-and-fro motion. Harvey's great generalisation, based upon a long series of experiments *in vivo*, was considered to have given the *coup de grâce* to the Galenic physiology, and hence threw temporary discredit upon the whole system of medicine associated therewith. William Harvey (1578–1657).

Modern medicine, based upon a painstaking

xxiii

research into the details of physiological function, had begun.

Back to Galen !

While we cannot sufficiently commend the results of the long modern period of research-work to which the labours of the Renascence anatomists from Vesalius to Harvey form a fitting prelude, we yet by no means allow that Galen's general medical outlook was so entirely invalidated as many imagine by the conclusive demonstration of his anatomical errors. It is time for us now to turn to Galen again after three hundred years of virtual neglect : it may be that he will help us to see something fundamentally important for medical practice which is beyond the power even of our microscopes and X-rays to reveal. While the value of his work undoubtedly lies mainly in its enabling us to envisage one of the greatest of the early steps attained by man in medical knowledge, it also has a very definite intrinsic value of its own.

Galen's Debt to his Precursors.

No attempt can be made here to determine how much of Galen's work is, in the true sense of the word, original, and how much is drawn from the labours of his predecessors. In any case, there is no doubt that he was much more than a mere compiler and systematizer of other men's work : he was great enough to be able not merely to collect, to digest, and to assimilate all the best of the work done before his time, but, adding to this the outcome of his own observations, experiments, and reflections, to present

xxiv

the whole in an articulated "system" showing that perfect balance of parts which is the essential criterion of a work of art. Constantly, however, in his writings we shall come across traces of the influence of, among others, Plato, Aristotle, and writers of the Stoic school.

Although Galen is an eclectic in the best sense of the term, there is one name to which he pays a very special tribute—that of his illustrious forerunner Hippocrates. Him on quite a number of occasions he actually calls "divine" (*cf.* p. 293). *Influence of Hippocrates on Galen.*

"Hippocrates," he says, "was the first known to us of all who have been both physicians and philosophers, in that *he was the first to recognise what nature does.*" Here is struck the keynote of the teaching of both Hippocrates and Galen ; this is shown in the volume before us, which deals with " the *natural* faculties "—that is with the faculties of this same " Nature " or vital principle referred to in the quotation.

If Galen be looked on as a crystallisation of Greek medicine, then this book may be looked on as a crystallisation of Galen. Within its comparatively short compass we meet with instances illustrating perhaps most of the sides of this many-sided writer. The " Natural Faculties " therefore forms an excellent prelude to the study of his larger and more specialised works. *" The Natural Faculties.*

INTRODUCTION

What, now, is this "Nature" or biological principle upon which Galen, like Hippocrates, bases the whole of his medical teaching, and which, we may add, is constantly overlooked—if indeed ever properly apprehended—by many physiologists of the present day? By using this term Galen meant simply that, when we deal with a living thing, we are dealing primarily with a unity, which, *quâ* living, is not further divisible; all its parts can only be understood and dealt with as being *in relation to* this principle of unity. Galen was thus led to criticise with considerable severity many of the medical and surgical specialists of his time, who acted on the assumption (implicit if not explicit) that the whole was merely the sum of its parts, and that if, in an ailing organism, these parts were treated each in and for itself, the health of the whole organism could in this way be eventually restored.

Galen expressed this idea of the unity of the organism by saying that it was governed by a *Physis* or Nature (ἡ φύσις ἥπερ διοικεῖ τὸ ζῷον), with whose "faculties" or powers it was the province of φυσιολογία (physi-ology, Nature-lore) to deal. It was because Hippocrates had a clear sense of this principle that Galen called him master. "Greatest," say the Moslems, "is Allah, and Mohammed is his prophet." "Greatest," said Galen, "is the Physis, and Hippocrates is its prophet." Never did Mohammed more zealously maintain the unity of the Godhead than Hippocrates and Galen the unity of the organism.

xxvi

INTRODUCTION

But we shall not have read far before we discover that the term *Physiology*, as used by Galen, stands not merely for what we understand by it nowadays, but also for a large part of *Physics* as well. This is one of the chief sources of confusion in his writings. Having grasped, for example, the uniqueness of the process of *specific selection* (ὁλκὴ τοῦ οἰκείου), by which the tissues nourish themselves, he proceeds to apply this principle in explanation of entirely different classes of phenomena; thus he mixes it up with the physical phenomenon of the attraction of the lode-stone for iron, of dry grain for moisture, etc. It is noteworthy, however, in these latter instances, that he does not venture to follow out his comparison to its logical conclusion; he certainly stops short of hinting that the lodestone (like a living organ or tissue) *assimilates* the metal which it has attracted!

Setting aside, however, these occasional half-hearted attempts to apply his principle of a φύσις in regions where it has no natural standing, we shall find that in the field of biology Galen moves with an assurance bred of first-hand experience.

Against his attempt to "biologize" physics may be set the converse attempt of the mechanical Atomist school. Thus in Asclepiades he found a doughty defender of the view that physiology was "merely" physics. Galen's ire being roused, he is not content with driving the enemy out of the biological camp, but must needs attempt also to

INTRODUCTION

dislodge him from that of physics, in which he has every right to be.

In defence of the universal validity of his principle, Galen also tends to excessive disparagement of morphological factors; witness his objection to the view of the anatomist Erasistratus that the calibre of vessels played a part in determining the secretion of fluids (p. 123), that digestion was caused by the mechanical action of the stomach walls (p. 243), and dropsy by induration of the liver (p. 171).

While combating the atomic explanation of physical processes, Galen of course realised that there were many of these which could only be explained according to what we should now call "mechanical laws." For example, non-living things could be subjected to φορά (passive motion), they answered to the laws of gravity (ταῖς τῶν ὑλῶν οἰακιζόμενα ῥοπαῖς, p. 126). Furthermore, Galen did not fail to see that living things also were not entirely exempted from the operation of these laws; they too may be at least partly subject to gravity (loc. cit.); a hollow organ exerts, by virtue of its cavity, an attraction similar to that of dilating bellows, as well as, by virtue of the living tissue of its walls, a specifically "vital" or selective kind of attraction (p. 325).

As a type of characteristically vital action we may take *nutrition*, in which occurs a phenomenon

xxviii

which Galen calls *active motion* (δραστικὴ κίνησις) or, more technically, *alteration* (ἀλλοίωσις). This active type of motion cannot be adequately stated in terms of the passive movements (groupings and re-groupings) of its constituent parts according to certain empirical "laws." Alteration involves *self-movement*, a self-determination of the organism or organic part. Galen does not attempt to explain this fundamental characteristic of *alteration* any further; he contents himself with referring his opponents to Aristotle's work on the "Complete Alteration of Substance" (p. 9).

The most important characteristic of the Physis or Nature is its τέχνη—its artistic creativeness. In other words, the living organism is a creative artist. This feature may be observed typically in its primary functions of *growth* and *nutrition*; these are dependent on the characteristic *faculties* or powers, by virtue of which each part draws to itself what is proper or appropriate to it (τὸ οἰκεῖον) and rejects what is foreign (τὸ ἀλλότριον), thereafter appropriating or assimilating the attracted material; this assimilation is an example of the *alteration* (or qualitative change) already alluded to; thus the food eaten is "altered" into the various tissues of the body, each of these having been provided by "Nature" with its own specific faculties of attraction and repulsion.

Any of the operations of the living part may be looked on in three ways, either (*a*) as a δύναμις, The Three Categories.

faculty, potentiality; (*b*) as an ἐνέργεια, which is this δύναμις in operation; or (*c*) as an ἔργον, the product or effect of the ἐνέργεια.[1]

[1] What appear to me to be certain resemblances between the Galenical and the modern vitalistic views of Henri Bergson may perhaps be alluded to here. Galen's vital principle, ἡ τεχνικὴ φύσις ("creative growth"), presents analogies with *l'Evolution créatrice*: both manifest their activity in producing qualitative change (ἀλλοίωσις, *changement*): in both, the creative change cannot be analysed into a series of static states, but is one and continuous. In Galen, however, it comes to an end with the *development of the individual*, whereas in Bergson it continues indefinitely as the *evolution of life*. The three aspects of organic life may be tabulated thus:—

δύναμις	ἐνέργεια	ἔργον
Work to be done. Future aspect.	Work being done. Present aspect. Function. The *élan vital*. A changing which cannot be understood as a sum of static parts; a constant becoming, never stopping — at least till the ἔργον is reached.	Work done, finished. Past aspect. Structure. A "thing."
Bergson's "teleological" aspect.	Bergson's "philosophical" aspect.	Bergson's "outlook of physical science."

Galen recognized "creativeness" (τέχνη) in the *development* of the individual and its parts (ontogeny) and in the maintenance of these, but he failed to appreciate the creative *evolution* of species (phylogeny), which is, of course, part of the same process. To the teleologist the possibilities (δυνάμεις) of the Physis are limited, to Bergson they are un-

INTRODUCTION

Like his master Hippocrates, Galen attached Galen's Method.
fundamental importance to clinical observation—
to the evidence of the senses as the indispensable
groundwork of all medical knowledge. He had
also, however, a forte for rapid generalisation from
observations, and his logical proclivities disposed him

limited. Galen and Bergson agree in attaching most practical
importance to the middle category—that of Function.

While it must be conceded that Galen, following Aristotle,
had never seriously questioned the fixity of species, the
following quotation from his work *On Habits* (chap. ii.) will
show that he must have at least had occasional glimmerings
of our modern point of view on the matter. Referring to
assimilation, he says: "Just as everything we eat or drink
becomes *altered in quality*, so of course also does the altering
factor itself become altered. . . . A clear proof of the assimilation of things which are being nourished to that which
is nourishing them is the change which occurs in plants and
seeds; this often goes so far that what is highly noxious in
one soil becomes, when transplanted into another soil, not
merely harmless, but actually useful. This has been largely
put to the test by those who compose memoirs on farming
and on plants, as also by zoological authors who have
written on the changes which occur according to the
countries in which animals live. Since, therefore, not only
is the nourishment altered by the creature nourished, but
the latter itself also undergoes some slight alteration, *this
slight alteration must necessarily become considerable in the
course of time*, and thus properties resulting from prolonged
habit must come to be on a par with natural properties."

Galen fails to see the possibility that the "natural" properties themselves originated in this way, as activities which
gradually became habitual—that is to say, that the effects of
nurture may become a "second nature," and so eventually
nature itself.

The whole passage, however, may be commended to modern
biologists—particularly, might one say, to those bacterio-
logists who have not yet realised how extraordinarily *relative* is the term "specificity" when applied to the subject-
matter of their science.

INTRODUCTION

particularly to deductive reasoning. Examples of an almost Euclidean method of argument may be found in the *Natural Faculties* (*e.g.* Book III. chap. i.). While this method undoubtedly gave him much help in his search for truth, it also not unfrequently led him astray. This is evidenced by his attempt, already noted, to apply the biological principle of the φύσις in physics. Characteristic examples of attempts to force facts to fit premises will be found in Book II. chap. ix., where our author demonstrates that yellow bile is "virtually" dry, and also, by a process of exclusion, assigns to the spleen the function of clearing away black bile. Strangest of all is his attempt to prove that the same principle of specific attraction by which the ultimate tissues nourish themselves (and the lodestone attracts iron!) accounts for the reception of food into the stomach, of urine into the kidneys, of bile into the gall-bladder, and of semen into the uterus.

These instances are given, however, without prejudice to the system of generalisation and deduction which, in Galen's hands, often proved exceedingly fruitful. He is said to have tried "to unite professional and scientific medicine with a philosophic link." He objected, however, to such extreme attempts at simplification of medical science as that of the Methodists, to whom diseases were isolated entities, without any relationships in time or space (*v.* p. xv. *supra*).

He based much of his pathological reasoning upon

the "humoral theory" of Hippocrates, according to which certain diseases were caused by one or more of the four humours (blood, phlegm, black and yellow bile) being in excess—that is, by various *dyscrasiae*. Our modern conception of "hormone" action shows certain resemblances with this theory.

Besides observation and reasoning, Galen took his stand on *experiment*; he was one of the first of experimental physiologists, as is illustrated in the present book by his researches into the function of the kidneys (p. 59 *et seq.*). He also conducted a long series of experiments into the physiology of the spinal cord, to determine what parts controlled movement and what sensibility.

As a practitioner he modelled his work largely on the broad and simple lines laid down by Hippocrates. He had also at his disposal all the acquisitions of biological science dating from the time of Aristotle five hundred years earlier, and reinforced by the discoveries in anatomy made by the Alexandrian school. To these he added a large series of researches of his own.

Galen never confined himself to what one might call the academic or strictly orthodox sources of information; he roamed the world over for answers to his queries. For example, we find him on his journeys between Pergamos and Rome twice visiting the island of Lemnos in order to procure some of the *terra sigillata*, a kind of earth which had a reputation for healing the bites of serpents and other wounds. At other times he visited the copper-mines of Cyprus in search for copper, and Palestine for the resin called Balm of Gilead.

By inclination and training Galen was the reverse of a " party-man." In the *Natural Faculties* (p. 55) he speaks of the bane of sectarian partizanship, " harder to heal than any itch." He pours scorn upon the ignorant " Erasistrateans " and " Asclepiadeans," who attempted to hide their own incompetence under the shield of some great man's name (*cf.* p. 141).

Of the two chief objects of his censure in the *Natural Faculties*, Galen deals perhaps less rigorously with Erasistratus than with Asclepiades. Erasistratus did at least recognize the existence of a vital principle in the organism, albeit, with his eye on the structures which the scalpel displayed he tended frequently to forget it. The researches of the anatomical school of Alexandria had been naturally of the greatest service to surgery, but in medicine they sometimes had a tendency to check progress by diverting attention from the whole to the part.

The Pneuma or Spirit.
Another novel conception frequently occurring in Galen's writings is that of the *Pneuma* (*i.e.* the breath, *spiritus*). This word is used in two senses, as meaning (1) the inspired air, which was drawn into the left side of the heart and thence carried all over the body by the arteries; this has not a few analogies with oxygen, particularly as its action in the tissues

xxxiv

is attended with the appearance of the so-called "innate heat." (2) A vital principle, conceived as being made up of matter in the most subtle imaginable state (*i.e.* air). This vital principle became resolved into three kinds: (*a*) πνεῦμα φυσικόν or *spiritus naturalis*, carried by the veins, and presiding over the subconscious vegetative life; this "natural spirit" is therefore practically equivalent to the φῦσις or "nature" itself. (*b*) The πνεῦμα ζωτικόν or *spiritus vitalis*; here particularly is a source of error, since the air already alluded to as being carried by the arteries tends to be confused with this principle of "individuality" or relative autonomy in the circulatory (including, perhaps, the vasomotor) system. (*c*) The πνεῦμα ψυχικόν or *spiritus animalis* (anima = ψυχή), carried by longitudinal canals in the nerves; this corresponds to the ψυχή.

This view of a "vital principle" as necessarily consisting of matter in a finely divided, fluid, or "etheric" state is not unknown even in our day. Belief in the fundamental importance of the Pneuma formed the basis of the teaching of another vitalist school in ancient Greece, that of the Pneumatists.

It is unnecessary to detail here the various ways in which Galen's physiological views differ from those of the Moderns, as most of these are noticed in footnotes to the text of the present translation. His ignorance of the circulation of the blood does not lessen the force of his general physiological conclu-

Galen and the Circulation of the Blood.

sions to the extent that might be anticipated. In his opinion, the great bulk of the blood travelled with a to-and-fro motion in the veins, while a little of it, mixed with inspired air, moved in the same way along the arteries ; whereas we now know that all the blood goes outward by the arteries and returns by the veins ; in either case blood is carried to the tissues by blood-vessels, and Galen's ideas of tissue-nutrition were wonderfully sound. The ingenious method by which (in ignorance of the pulmonary circulation) he makes blood pass from the right to the left ventricle, may be read in the present work (p. 321). As will be seen, he was conversant with the " anastomoses " between the ultimate branches of arteries and veins, although he imagined that they were not used under " normal " conditions.

Galen's Character.

Galen was not only a man of great intellectual gifts, but one also of strong moral fibre. In his short treatise " That the best Physician is also a Philosopher " he outlines his professional ideals. It is necessary for the efficient healer to be versed in the three branches of " philosophy," viz. : (*a*) *logic*, the science of how to think ; (*b*) *physics*, the science of what is—*i.e.* of " Nature " in the widest sense ; (*c*) *ethics*, the science of what to do. The amount of toil which he who wishes to be a physician must undergo—firstly, in mastering the work of his pre-decessors and afterwards in studying disease at first hand—makes it absolutely necessary that he should possess perfect self-control, that he should scorn money and the weak pleasures of the senses, and should live laborious days.

Readers of the following pages will notice that Galen uses what we should call distinctly immoderate language towards those who ventured to differ from the views of his master Hippocrates (which were also his own). The employment of such language was one of the few weaknesses of his age which he did not transcend. Possibly also his mother's choleric temper may have predisposed him to it.

The fact, too, that his vivisection experiments (*e.g.* pp. 59, 273) were carried out apparently without any kind of anaesthetisation being even thought of is abhorrent to the feelings of to-day, but must be excused also on the ground that callousness towards animals was then customary, men having probably never thought much about the subject.

Galen is a master of language, using a highly polished variety of Attic prose with a precision which can be only very imperfectly reproduced in another tongue. Every word he uses has an exact and definite meaning attached to it. Translation is particularly difficult when a word stands for a physiological conception which is not now held; instances are the words *anadosis, prosthesis,* and *prosphysis,* indicating certain steps in the process by which nutriment is conveyed from the alimentary canal to the tissues.

Galen's Greek Style.

INTRODUCTION

Readers will be surprised to find how many words are used by Galen which they would have thought had been expressly coined to fit modern conceptions; thus our author employs not merely such terms as *physiology, phthisis, atrophy, anastomosis,* but also *haematopoietic, anaesthesia,* and even *aseptic*! It is only fair, however, to remark that these terms, particularly the last, were not used by Galen in quite their modern significance.

Summary.   To resume, then: What contribution can Galen bring to the art of healing at the present day? It was not, surely, for nothing that the great Pergamene gave laws to the medical world for over a thousand years!

Let us draw attention once more to:

(1) The high ideal which he set before the profession.

(2) His insistence on immediate contact with nature as the primary condition for arriving at an understanding of disease; on the need for due consideration of previous authorities; on the need also for reflection—for employment of the mind's eye (ἡ λογικὴ θεωρία) as an aid to the physical eye.

(3) His essentially broad outlook, which often helped him in the comprehension of a phenomenon through his knowledge of an analogous phenomenon in another field of nature.

INTRODUCTION

(4) His keen appreciation of the unity of the organism, and of the inter-dependence of its parts ; his realisation that the vital phenomena (physiological and pathological) in a living organism can only be understood when considered in relation to the *environment* of that organism or part. This is the foundation for the war that Galen waged *à outrance* on the Methodists, to whom diseases were things without relation to anything. This dispute is, unfortunately, not touched upon in the present volume. What Galen combated was the tendency, familiar enough in our own day, to reduce medicine to the science of finding a label for each patient, and then treating not the patient, but the label. (This tendency, we may remark in parenthesis, is one which is obviously well suited for the *standardising* purposes of a State medical service, and is therefore one which all who have the weal of the profession at heart must most jealously watch in the difficult days that lie ahead.)

(5) His realisation of the inappropriateness and inadequacy of physical formulae in explaining physiological activities. Galen's disputes with Asclepiades over τὰ πρῶτα ἐκεῖνα σώματα τὰ ἀπαθῆ, over the ἄναρμα στοιχεῖα καὶ ληρώδεις ὄγκοι, is but another aspect of his quarrel with the Methodists regarding their pathological "units," whose primary characteristic was just this same ἀπάθεια (impassiveness to environment, "unimpressionability"). We have of course

our Physiatric or Iatromechanical school at the present day, to whom such processes as absorption from the alimentary canal, the respiratory interchange of gases, and the action of the renal epithelium are susceptible of a purely physical explanation.[1]

(6) His quarrel with the Anatomists, which was in essence the same as that with the Atomists, and which arose from his clear realisation that that primary and indispensable desideratum, a view of the whole, could never be obtained by a mere summation of partial views; hence, also, his sense of the dangers which would beset the medical art if it were allowed to fall into the hands of a mere crowd of competing specialists without any organising head to guide them.

[1] In terms of filtration, diffusion, and osmosis.

BIBLIOGRAPHY

Codices

Bibliothèque Nationale. Paris. No. 2267.
Library of St. Mark. Venice. No. 275.

Translations

Arabic translations by Honain in the Escurial Library, and in the Library at Leyden. Hebrew translation in the Library at Bonn. Latin translations in the Library of Gonville and Caius College (MSS.), No. 947; also by Linacre in editions published, London, 1523; Paris, 1528; Leyden, 1540, 1548, and 1550; also by C. G. Kühn, Leipzig, 1821.

Commentaries and Appreciations

Nic. de Anglia in Bib. Nat. Paris (MSS.), No. 7015; J. Rochon, *ibidem*, No. 7025; J. Segarra, 1528; J. Sylvius, 1550, 1560; L. Joubert, 1599; M. Sebitz, 1644, 1645; J. B. Pacuvius, 1554; J. C. G. Ackermann, 1821, in the introduction to Kühn's translation, p. lxxx; Ilberg in articles on "Die Schriftstellerei des Klaudios Galenos," in *Rhein. Mus.*, Nos. 44, 47, 51, and 52 (years 1889, 1892, 1896 and 1897); I. von Mueller in *Quæstiones Criticae de Galeni libris*, Erlangen, 1871; Steinschneider in Virchow's *Archiv*, No. cxxiv. for 1891; Wenrich in *De auctorum graecorum versionibus et commentariis syriacis, arabicis, armiacis, persisque*, Leipzig, 1842.

SYNOPSIS OF CHAPTERS

BOOK I

CHAPTER I

Distinction between the effects of (*a*) the organism's *psyche* or soul (*b*) its *physis* or nature. The author proposes to confine himself to a consideration of the latter—the vegetative—aspect of life.

CHAPTER II

Definition of terms. Different kinds of *motion*. *Alteration* or qualitative change. Refutation of the Sophists' objection that such change is only apparent, not real. The four fundamental qualities of **Hippocrates** (later **Aristotle**). Distinction between *faculty*, *activity* (function), and *effect* (work or product).

CHAPTER III

It is by virtue of the *four qualities* that each part functions. Some authorities subordinate the dry and the moist principles to the hot and the cold. Aristotle inconsistent here.

CHAPTER IV

We must suppose that there are *faculties* corresponding in number to the visible *effects* (or products) with which we are familiar.

CHAPTER V

Genesis, growth, and nutrition. Genesis (embryogeny) subdivided into histogenesis and organogenesis. Growth is a tridimensional expansion of the solid parts formed during genesis. Nutrition.

SYNOPSIS OF CHAPTERS

CHAPTER VI

The process of genesis (embryogeny) from insemination onwards. Each of the simple, elementary, homogeneous parts (tissues) is produced by a special blend of the four primary alterative faculties (such secondary alterative faculties being *ostopoietic*, *neuropoietic*, etc.). A special *function* and *use* also corresponds to each of these special tissues. The bringing of these tissues together into *organs* and the disposal of these organs is performed by another faculty called *diaplastic*, *moulding*, or *formative*.

CHAPTER VII

We now pass from genesis to *growth*. Growth essentially a post-natal process; it involves two factors, expansion and nutrition, explained by analogy of a familiar child's game.

CHAPTER VIII

Nutrition.

CHAPTER IX

These three primary faculties (genesis, growth, nutrition) have various others subservient to them.

CHAPTER X

Nutrition not a simple process. (1) Need of subsidiary organs for the various stages of alteration, *e.g.*, of bread into blood, of that into bone, etc. (2) Need also of organs for excreting the non-utilizable portions of the food, *e.g.*, much vegetable matter is superfluous. (3) Need of organs of a third kind, for distributing the pabulum through the body.

CHAPTER XI

Nutrition analysed into the stages of application (*prosthesis*), adhesion (*prosphysis*), and assimilation. The stages illustrated by certain pathological conditions. Different shades of meaning of the term *nutriment*.

xliv

SYNOPSIS OF CHAPTERS

CHAPTER XII

The two chief medico-philosophical schools—Atomist and Vitalist. Hippocrates an adherent of the latter school—his doctrine of an original principle or "nature" in every living thing (doctrine of the unity of the organism).

CHAPTER XIII

Failure of Asclepiades to understand the functions of kidneys and ureters. His hypothesis of vaporization of imbibed fluids is here refuted. A demonstration of urinary secretion in the living animal; the forethought and artistic skill of Nature vindicated. Refutation also of Asclepiades's disbelief in the special selective action of purgative drugs.

CHAPTER XIV

While Asclepiades denies *in toto* the obvious fact of specific attraction, Epicurus grants the fact, although his attempt to explain it by the atomic hypothesis breaks down. Refutation of the Epicurean theory of magnetic attraction. Instances of specific attraction of thorns and animal poisons by medicaments, of moisture by corn, etc.

CHAPTER XV

It now being granted that the urine is secreted by the kidneys, the *rationale* of this secretion is enquired into. The kidneys are not mechanical filters, but are by virtue of their *nature* possessed of a specific faculty of attraction.

CHAPTER XVI

Erasistratus, again, by his favourite principle of *horror vacui* could never explain the secretion of urine by the kidneys. While, however, he acknowledged that the kidneys do secrete urine, he makes no attempt to explain this; he ignores, but does not attempt to refute, the Hippocratic doctrine of specific *attraction*. "Servile" position taken up by Asclepiades and Erasistratus in regard to this function of urinary secretion.

SYNOPSIS OF CHAPTERS

CHAPTER XVII

Three other attempts (by adherents of the Erasistratean
school and by Lycus of Macedonia) to explain how the
kidneys come to separate out urine from the blood. All
these ignore the obvious principle of attraction.

BOOK II

CHAPTER I

In order to explain dispersal of food from alimentary canal
viâ the veins (*anadosis*) there is no need to invoke with
Erasistratus, the *horror vacui*, since here again the principle of specific attraction is operative ; moreover, blood
is also driven forward by the compressing action of the
stomach and the contractions of the veins. Possibility,
however, of Erasistratus's factor playing a certain minor
rôle.

CHAPTER II

The Erasistratean idea that bile becomes separated out from
the blood in the liver because, being the thinner fluid, it
alone can enter the narrow stomata of the bile-ducts,
while the thicker blood can only enter the wider mouths
of the hepatic venules.

CHAPTER III

The morphological factors suggested by Erasistratus are
quite inadequate to explain biological happenings.
Erasistratus inconsistent with his own statements. The
immanence of the *physis* or nature ; her shaping is not
merely external like that of a statuary, but involves the
entire substance. In genesis (embryogeny) the semen is
the active, and the menstrual blood the passive, principle. Attractive, alterative, and formative faculties of
the semen. Embryogeny is naturally followed by
growth ; these two functions distinguished.

SYNOPSIS OF CHAPTERS

SYNOPSIS OF CHAPTERS

CHAPTER VII

In the last resort, the ultimate living elements (Erasistratus's *simple vessels*) must draw in their food by virtue of an inherent attractive faculty like that which the lodestone exerts on iron. Thus the process of anadosis, from beginning to end, can be explained without assuming a *horror vacui*

CHAPTER VIII

Erasistratus's disregard for the humours. In respect to excessive formation of bile, however, prevention is better than cure ; accordingly we must consider its pathology. Does blood pre-exist in the food, or does it come into existence in the body ? Erasistratus's purely anatomical explanation of *dropsy*. He entirely avoids the question of the four qualities (*e.g.* the importance of innate heat) in the generation of the humours, etc. Yet the problem of blood-production is no less important than that of gastric digestion. Proof that bile does not pre-exist in the food. The four fundamental qualities of Hippocrates and Aristotle. How the humours are formed from food taken into the veins : when heat is in proportionate amount, blood results : when in excess, bile ; when deficient, phlegm. Various conditions determining cold or warm temperaments. The four primary diseases result each from excess of one of the four qualities. Erasistratus unwillingly acknowledges this when he ascribes the indigestion occurring in fever to *impaired function* of the stomach. For what causes this *functio laesa ?* Proof that it is the fever (excess of innate heat).

If, then, heat plays so important a part in abnormal functioning, so must it also in normal (*i.e.* causes of eucrasia involved in those of dyscrasia, of physiology in those of pathology). A like argument explains the *genesis of the humours*. Addition of warmth to things already warm makes them bitter ; thus honey turns to bile in people who are already warm ; where warmth deficient, as in old people, it turns to useful blood. This is a proof that bile does not pre-exist, as such, in the food.

SYNOPSIS OF CHAPTERS

SYNOPSIS OF CHAPTERS

BOOK III

CHAPTER I

A recapitulation of certain points previously demonstrated. Every part of the animal has an attractive and an alterative (assimilative) faculty; it attracts the nutrient juice which is proper to it. Assimilation is preceded by adhesion (*prosphysis*) and that again, by application (*prosthesis*). Application the goal of attraction. It would not, however, be followed by adhesion and assimilation if each part did not also possess a faculty for *retaining in position* the nutriment which has been applied. *A priori* necessity for this *retentive* faculty.

CHAPTER II

The same faculty to be proved *a posteriori*. Its corresponding *function* (*i.e.* the activation of this faculty or potentiality) well seen in the large hollow organs, notably the uterus and stomach.

CHAPTER III

Exercise of the retentive faculty particularly well seen in the uterus. Its object is to allow the embryo to attain full development; this being completed, a new faculty—the expulsive—hitherto quiescent, comes into play. Characteristic signs and symptoms of pregnancy. Tight grip of uterus on growing embryo, and accurate closure of os uteri during operation of the retentive faculty. Dilatation of os and expulsive activities of uterus at full term, or when foetus dies. Prolapse from undue exercise of this faculty. *Rôle* of the midwife. Accessory muscles in parturition.

CHAPTER IV

Same two faculties seen in stomach. *Gurglings* or *borborygmi* show that this organ is weak and is not gripping its contents tightly enough. Undue delay of food in a weak

1

stomach proved not to be due to narrowness of pylorus : length of stay depends on whether *digestion* (another instance of the characteristically vital process of *alteration*) has taken place or not. Erasistratus wrong in attributing digestion merely to the mechanical action of the stomach walls. When digestion completed, then pylorus opens and allows contents to pass downwards, just as os uteri when development of embyro completed.

CHAPTER V

If attraction and elimination always proceeded *pari passu*, the content of these hollow organs (including gall-bladder and urinary bladder) would never vary in amount. A *retentive* faculty, therefore, also logically needed. Its existence demonstrated. Expulsion determined by qualitative and quantitative changes of contents. "Diarrhoea" of stomach. Vomiting.

CHAPTER VI

Every organic part has an *appetite* and *aversion* for the qualities which are appropriate and foreign to it respectively. Attraction necessarily leads to a certain *benefit* received. This again necessitates *retention*.

CHAPTER VII

Interaction between two bodies ; the stronger masters the weaker ; a deleterious drug masters the forces of the body, whereas food is mastered by them ; this mastery is an *alteration*, and the amount of alteration varies with the different organs ; thus a partial alteration is effected in mouth by saliva, but much greater in stomach, where not only gastric juice, but also bile, pneuma, innate heat (*i.e.* oxidation ?), and other powerful factors are brought to bear on it ; need of considerable alteration in stomach

li

as a transition-stage between food and blood; appearance of faeces in intestine another proof of great alteration effected in stomach. Asclepiades's denial of real qualitative change in stomach rebutted. Erasistratus's denial that digestion in any way resembles a *boiling* process comes from his taking words too literally.

SYNOPSIS OF CHAPTERS

neighbour. Reversal of direction of flow occurs not merely on occasion but also constantly (as in arteries, lungs, heart, etc.). The various stages of normal nutrition described. Why the stomach sometimes draws back the nutriment it had passed on to portal veins and liver. A similar ebb and flow in relation to the spleen. Comparison of the parts of the body to a lot of animals at a feast. The valves of the heart are a provision of Nature to prevent this otherwise inevitable regurgitation, though even they are not quite efficient.

Chapter XIV

The superficial arteries, when they dilate, draw in air from the atmosphere, and the deeper ones a fine, vaporous blood from the veins and heart. Lighter matter such as air will always be drawn in in preference to heavier; this is why the arteries in the food-canal draw in practically none of the nutrient matter contained in it.

Chapter XV

The two kinds of attraction—the mechanical attraction of dilating bellows and the " physical " (vital) attraction by living tissue of nutrient matter which is specifically allied or appropriate to it. The former kind—that resulting from *horror vacui*—acts primarily on light matter, whereas vital attraction has no essential concern with such mechanical factors. A hollow organ exercises, by virtue of its cavity, the former kind of attraction, and by virtue of the living tissue of its walls, the second kind. Application of this to question of contents of arteries ; *anastomoses of arteries and veins. Foramina in interventricular septum of heart*, allowing some blood to pass from right to left ventricle. Large size of aorta probably due to fact that it not merely carries the pneuma received from the lungs, but also some of the blood which percolates through septum from right ventricle. Thus arteries carry not merely pneuma, but also some light vaporous blood, which certain parts need more than the ordinary thick blood of the veins. The organic parts must have their blood-supply sufficiently near to allow them to absorb it ; comparison with an irrigation system in a garden. Details of the process of nutrition in the ultimate specific tissues ; some are nourished from the blood directly ; in others a series of intermediate stages must precede complete assimilation ; for example, marrow is an intermediate stage between blood and bone.

From the generalisations arrived at in the present work we can deduce the explanation of all kinds of particular phenomena ; an instance is given, showing the co-operation of various factors previously discussed.

GALEN

ON THE NATURAL FACULTIES

BOOK I

ΓΑΛΗΝΟΥ

ΠΕΡΙ ΦΥΣΙΚΩΝ ΔΥΝΑΜΕΩΝ

Α

I

Ἐπειδὴ τὸ μὲν αἰσθάνεσθαί τε καὶ κινεῖσθαι
κατὰ προαίρεσιν ἴδια τῶν ζῴων ἐστί, τὸ δ'
αὐξάνεσθαί τε καὶ τρέφεσθαι κοινὰ καὶ τοῖς
φυτοῖς, εἴη ἂν τὰ μὲν πρότερα τῆς ψυχῆς, τὰ δὲ
δεύτερα τῆς φύσεως ἔργα. εἰ δέ τις καὶ τοῖς
φυτοῖς ψυχῆς μεταδίδωσι καὶ διαιρούμενος αὐτὰς
ὀνομάζει φυτικὴν μὲν ταύτην, αἰσθητικὴν δὲ τὴν
ἑτέραν, λέγει μὲν οὐδ' οὗτος ἄλλα, τῇ λέξει δ'
οὐ πάνυ τῇ συνήθει κέχρηται. ἀλλ' ἡμεῖς γε
μεγίστην λέξεως ἀρετὴν σαφήνειαν εἶναι πε-
2 πεισμένοι καὶ ταύτην εἰδότες ‖ ὑπ' οὐδενὸς οὕτως
ὡς ὑπὸ τῶν ἀσυνήθων ὀνομάτων διαφθειρομένην,
ὡς τοῖς πολλοῖς ἔθος, οὕτως ὀνομάζοντες ὑπὸ μὲν
ψυχῆς θ' ἅμα καὶ φύσεως τὰ ζῷα διοικεῖσθαι
φαμεν, ὑπὸ δὲ φύσεως μόνης τὰ φυτὰ καὶ τό γ'
αὐξάνεσθαί τε καὶ τρέφεσθαι φύσεως ἔργα φαμέν,
οὐ ψυχῆς.

---

[1] That is, "On the Natural Powers," the powers of the
Physis or Nature. By that Galen practically means what
we should call the physiological or biological powers, the
characteristic faculties of the living organism; his Physis
is the subconscious vital principle of the animal or plant.

GALEN

ON THE NATURAL FACULTIES [1]

BOOK I

I

SINCE feeling and voluntary motion are peculiar to animals, whilst growth and nutrition are common to plants as well, we may look on the former as effects [2] of the *soul* [3] and the latter as effects of the *nature*. [4] And if there be anyone who allows a share in soul to plants as well, and separates the two kinds of soul, naming the kind in question *vegetative,* and the other *sensory,* this person is not saying anything else, although his language is somewhat unusual. We, however, for our part, are convinced that the chief merit of language is clearness, and we know that nothing detracts so much from this as do unfamiliar terms ; accordingly we employ those terms which the bulk of people are accustomed to use, and we say that animals are governed at once by their soul and by their nature, and plants by their nature alone, and that growth and nutrition are the effects of nature, not of soul.

Like Aristotle, however, he also ascribes quasi-vital properties to inanimate things, *cf.* Introduction, p. xxvii.
  [2] *Ergon,* here rendered an *effect,* is literally a *work* or *deed* ;
strictly speaking, it is something *done, completed,* as distinguished from *energeia,* which is the actual *doing,* the *activity*
which produces this *ergon. cf.* p. 13, and Introduction, p. xxx.
  [3] Gk. *psyche,* Lat. *anima.*   [4] Gk. *physis,* Lat. *natura.*

II

Καὶ ζητήσομεν κατὰ τόνδε τὸν λόγον, ὑπὸ τίνων γίγνεται δυνάμεων αὐτὰ δὴ ταῦτα καὶ εἰ δή τι ἄλλο φύσεως ἔργον ἐστίν.

Ἀλλὰ πρότερόν γε διελέσθαι τε χρὴ καὶ μηνῦσαι σαφῶς ἕκαστον τῶν ὀνομάτων, οἷς χρησόμεθα κατὰ τόνδε τὸν λόγον, καὶ ἐφ' ὅ τι φέρομεν πρᾶγμα. γενήσεται δὲ τοῦτ' εὐθὺς ἔργων φυσικῶν διδασκαλία σὺν ταῖς τῶν ὀνομάτων ἐξηγήσεσιν.

Ὅταν οὖν τι σῶμα κατὰ μηδὲν ἐξαλλάττηται τῶν προϋπαρχόντων, ἡσυχάζειν αὐτό φαμεν· εἰ δ' ἐξίσταιτό πῃ, κατ' ἐκεῖνο κινεῖσθαι. καὶ τοίνυν ἐπεὶ πολυειδῶς ἐξίσταται, πολυειδῶς καὶ κινηθήσεται. καὶ γὰρ εἰ λευκὸν ὑπάρχον μελαίνοιτο καὶ εἰ μέλαν λευκαίνοιτο, κινεῖται κατὰ χρόαν, 3 καὶ εἰ γλυκὺ τέως ὑπάρχον αὖθις ‖ αὐστηρὸν ἢ ἔμπαλιν ἐξ αὐστηροῦ γλυκὺ γένοιτο, καὶ τοῦτ' ἂν κινεῖσθαι λέγοιτο κατὰ τὸν χυμόν. ἄμφω δὲ ταῦτά τε καὶ τὰ προειρημένα κατὰ τὴν ποιότητα κινεῖσθαι λεχθήσεται καὶ οὐ μόνον γε τὰ κατὰ τὴν χρόαν ἢ τὸν χυμὸν ἐξαλλαττόμενα κινεῖσθαί φαμεν, ἀλλὰ καὶ τὸ θερμότερον ἐκ ψυχροτέρου γενόμενον ἢ ψυχρότερον ἐκ θερμοτέρου κινεῖσθαι καὶ τοῦτο λέγομεν, ὥσπερ γε καὶ εἴ τι ξηρὸν ἐξ

II

THUS we shall enquire, in the course of this treatise, from what *faculties* these effects themselves, as well as any other effects of nature which there may be, take their origin.

First, however, we must distinguish and explain clearly the various terms which we are going to use in this treatise, and to what things we apply them ; and this will prove to be not merely an explanation of terms but at the same time a demonstration of the effects of nature.

When, therefore, such and such a body undergoes no change from its existing state, we say that it is *at rest*; but, if it departs from this in any respect we then say that in this respect it *undergoes motion*.[1] Accordingly, when it departs in various ways from its pre-existing state, it will be said to undergo various kinds of motion. Thus, if that which is white becomes black, or what is black becomes white, it undergoes motion in respect to *colour*; or if what was previously sweet now becomes bitter, or, conversely, from being bitter now becomes sweet, it will be said to undergo motion in respect to *flavour*; to both of these instances, as well as to those previously mentioned, we shall apply the term *qualitative motion*. And further, it is not only things which are altered in regard to colour and flavour which, we say, undergo motion ; when a warm thing becomes cold, and a cold warm, here too we speak of its undergoing motion ; similarly also when any-

---

[1] *Motion* (kinesis) is Aristotle's general term for what we would rather call *change*. It includes various kinds of change, as well as movement proper. *cf.* Introduction, p. xxix.

ὑγροῦ ἢ ὑγρὸν ἐκ ξηροῦ γίγνοιτο. κοινὸν δὲ κατὰ τούτων ἁπάντων ὄνομα φέρομεν τὴν ἀλλοίωσιν.

Ἕν τι τοῦτο γένος κινήσεως. ἕτερον δὲ γένος ἐπὶ τοῖς τὰς χώρας ἀμείβουσι σώμασι καὶ τόπον ἐκ τόπου μεταλλάττειν λεγομένοις, ὄνομα δὲ καὶ τούτῳ φορά.

Αὗται μὲν οὖν αἱ δύο κινήσεις ἁπλαῖ καὶ πρῶται, σύνθετοι δ' ἐξ αὐτῶν αὔξησίς τε καὶ φθίσις, ὅταν ἐξ ἐλάττονός τι μεῖζον ἢ ἐκ μείζονος ἔλαττον γένηται φυλάττον τὸ οἰκεῖον εἶδος. ἕτεραι δὲ δύο κινήσεις γένεσις καὶ φθορά, γένεσις μὲν ἡ εἰς οὐσίαν ἀγωγή, φθορὰ δ' ἡ ἐναντία.

Πάσαις δὲ ταῖς κινήσεσι κοινὸν ἐξάλλαξις 4 τοῦ ‖ προϋπάρχοντος, ὥσπερ οὖν καὶ ταῖς ἡσυχίαις ἡ φυλακὴ τῶν προϋπαρχόντων. ἀλλ' ὅτι μὲν ἐξαλλάττεται καὶ πρὸς τὴν ὄψιν καὶ πρὸς τὴν γεῦσιν καὶ πρὸς τὴν ἀφὴν αἷμα γιγνόμενα τὰ σιτία, συγχωροῦσιν· ὅτι δὲ καὶ κατ' ἀλήθειαν, οὐκέτι τοῦθ' ὁμολογοῦσιν οἱ σοφισταί. οἱ μὲν γάρ τινες αὐτῶν ἅπαντα τὰ τοιαῦτα τῶν ἡμετέρων αἰσθήσεων ἀπάτας τινὰς καὶ παραγωγὰς νομίζουσιν ἄλλοτ' ἄλλως πασχουσῶν, τῆς ὑποκειμένης οὐσίας μηδὲν τούτων, οἷς ἐπονομάζεται, δεχομένης· οἱ δέ τινες εἶναι μὲν ἐν αὐτῇ βούλονται τὰς ποιότητας, ἀμεταβλήτους δὲ καὶ ἀτρέπτους

---

[1] "Conveyance," "transport," "transit"; purely mechanical or passive motion, as distinguished from *alteration* (qualitative change).

[2] "Waxing and waning," the latter literally *phthisis*, a wasting or "decline;" *cf.* Scotch *dwining*, Dutch *verdwijnen*.

[3] Becoming and perishing: Latin, *generatio et corruptio*.

[4] " Ad substantiam productio seu ad formam processus " (Linacre).

thing moist becomes dry, or dry moist. Now, the common term which we apply to all these cases is *alteration*.

This is one kind of motion. But there is another kind which occurs in bodies which change their position, or as we say, pass from one place to another; the name of this is *transference*.[1]

These two kinds of motion, then, are simple and primary, while compounded from them we have *growth* and *decay*,[2] as when a small thing becomes bigger, or a big thing smaller, each retaining at the same time its particular form. And two other kinds of motion are *genesis* and *destruction*,[3] genesis being a coming into existence,[4] and destruction being the opposite.

Now, common to all kinds of motion is *change from the pre-existing state*, while common to all conditions of rest is *retention of the pre-existing state*. The Sophists, however, while allowing that bread in turning into blood becomes changed as regards sight, taste, and touch, will not agree that this change occurs in reality. Thus some of them hold that all such phenomena are tricks and illusions of our senses; the senses, they say, are affected now in one way, now in another, whereas the underlying substance does not admit of any of these changes to which the names are given. Others (such as Anaxagoras)[5] will have it that the qualities do exist in it, but that they

[5] "Preformationist" doctrine of Anaxagoras. To him the apparent alteration in qualities took place when a number of minute pre-existing bodies, all bearing the same quality, came together in sufficient numbers to impress that quality on the senses. The factor which united the minute quality-bearers was Nous. "In the beginning," says Anaxagoras, "all things existed together—then came Nous and brought them into order."

7

ἐξ αἰῶνος εἰς αἰῶνα καὶ τὰς φαινομένας ταύτας
ἀλλοιώσεις τῇ διακρίσει τε καὶ συγκρίσει γίγνεσθαί φασιν ὡς 'Αναξαγόρας.

Εἰ δὴ τούτους ἐκτραπόμενος ἐξελέγχοιμι, μεῖζον
ἄν μοι τὸ πάρεργον τοῦ ἔργου γένοιτο. εἰ μὲν
γὰρ οὐκ ἴσασιν, ὅσα περὶ τῆς καθ' ὅλην τὴν
οὐσίαν ἀλλοιώσεως 'Αριστοτέλει τε καὶ μετ'
αὐτὸν Χρυσίππῳ γέγραπται, παρακαλέσαι χρὴ
τοῖς ἐκείνων αὐτοὺς ὁμιλῆσαι γράμμασιν· εἰ δὲ
γιγνώσκοντες ἔπειθ' ἑκόντες τὰ χείρω πρὸ τῶν
5 βελτιόνων ‖ αἱροῦνται, μάταια δήπου καὶ τὰ
ἡμέτερα νομιοῦσιν. ὅτι δὲ καὶ 'Ιπποκράτης
οὕτως ἐγίγνωσκεν 'Αριστοτέλους ἔτι πρότερος ὤν,
ἐν ἑτέροις ἡμῖν ἀποδέδεικται. πρῶτος γὰρ οὗτος
ἁπάντων ὧν ἴσμεν ἰατρῶν τε καὶ φιλοσόφων ἀποδεικνύειν ἐπεχείρησε τέτταρας εἶναι τὰς πάσας
δραστικὰς εἰς ἀλλήλας ποιότητας, ὑφ' ὧν γίγνεταί
τε καὶ φθείρεται πάνθ', ὅσα γένεσίν τε καὶ φθορὰν
ἐπιδέχεται. καὶ μέντοι καὶ τὸ κεράννυσθαι δι'
ἀλλήλων αὐτὰς ὅλας δι' ὅλων 'Ιπποκράτης ἁπάντων πρῶτος ἔγνω· καὶ τὰς ἀρχάς γε τῶν ἀπο-
δείξεων, ὧν ὕστερον 'Αριστοτέλης μετεχειρίσατο,
παρ' ἐκείνῳ πρώτῳ γεγραμμένας ἔστιν εὑρεῖν.

Εἰ δ' ὥσπερ τὰς ποιότητας οὕτω καὶ τὰς οὐσίας
δι' ὅλων κεράννυσθαι χρὴ νομίζειν, ὡς ὕστερον
ἀπεφήνατο Ζήνων ὁ Κιττιεύς, οὐχ ἡγοῦμαι δεῖν
ἔτι περὶ τούτου κατὰ τόνδε τὸν λόγον ἐπεξιέναι.
μόνην γὰρ εἰς τὰ παρόντα δέομαι γιγνώσκεσθαι

---

[1] "De ea alteratione quae per totam fit substantiam" (Linacre).   [2] The systematizer of Stoicism and successor of Zeno.

[3] Note characteristic impatience with metaphysics. To
Galen, as to Hippocrates and Aristotle, it sufficed to look on

are unchangeable and immutable from eternity to eternity, and that these apparent alterations are brought about by *separation* and *combination*.

Now, if I were to go out of my way to confute these people, my subsidiary task would be greater than my main one. Thus, if they do not know all that has been written, "On Complete Alteration of Substance"[1] by Aristotle, and after him by Chrysippus,[2] I must beg of them to make themselves familiar with these men's writings. If, however, they know these, and yet willingly prefer the worse views to the better, they will doubtless consider my arguments foolish also. I have shown elsewhere that these opinions were shared by Hippocrates, who lived much earlier than Aristotle. In fact, of all those known to us who have been both physicians and philosophers Hippocrates was the first who took in hand to demonstrate that there are, in all, four mutually interacting *qualities*, and that to the operation of these is due the genesis and destruction of all things that come into and pass out of being. Nay, more; Hippocrates was also the first to recognise that all these qualities undergo an intimate mingling with one another; and at least the beginnings of the proofs to which Aristotle later set his hand are to be found first in the writings of Hippocrates.

As to whether we are to suppose that the *substances* as well as their *qualities* undergo this intimate mingling, as Zeno of Citium afterwards declared, I do not think it necessary to go further into this question in the present treatise;[3] for immediate purposes we only

the qualitative differences apprehended by the senses as fundamental. Zeno of Citium was the founder of the Stoic school; on the further analysis by this school of the *qualities* into *bodies cf.* p. 144, note 3.

τὴν δι᾽ ὅλης τῆς οὐσίας ἀλλοίωσιν, ἵνα μή τις
ὀστοῦ καὶ σαρκὸς καὶ νεύρου καὶ τῶν ἄλλων
ἑκάστου μορίων οἱονεὶ μισγάγκειάν τινα τῷ ἄρτῳ
6 νομίσῃ περιέχεσθαι κἄπειτ᾽ ἐν ‖ τῷ σώματι δια-
κρινόμενον ὡς τὸ ὁμόφυλον ἕκαστον ἱέναι. καίτοι
πρό γε τῆς διακρίσεως αἷμα φαίνεται γιγνόμενος
ὁ πᾶς ἄρτος. εἰ γοῦν παμπόλλῳ τις χρόνῳ μηδὲν
ἄλλ᾽ εἴη σιτίον προσφερόμενος, οὐδὲν ἧττον ἐν
ταῖς φλεψὶν αἷμα περιεχόμενον ἕξει. καὶ φανερῶς
τοῦτο τὴν τῶν ἀμετάβλητα τὰ στοιχεῖα τιθεμένων
ἐξελέγχει δόξαν, ὥσπερ οἶμαι καὶ τοὔλαιον εἰς
τὴν τοῦ λύχνου φλόγα καταναλισκόμενον ἅπαν
καὶ τὰ ξύλα πῦρ μικρὸν ὕστερον γιγνόμενα.

Καίτοι τό γ᾽ ἀντιλέγειν αὐτοῖς ἠρνησάμην, ἀλλ᾽
ἐπεὶ τῆς ἰατρικῆς ὕλης ἦν τὸ παράδειγμα καὶ
χρῄζω πρὸς τὸν παρόντα λόγον αὐτοῦ, διὰ τοῦτ᾽
ἐμνημόνευσα. καταλιπόντες οὖν, ὡς ἔφην, τὴν
πρὸς τούτους ἀντιλογίαν, <ἐνὸν> τοῖς βουλομένοις
τὰ τῶν παλαιῶν ἐκμανθάνειν κἀξ ὧν ἡμεῖς ἰδίᾳ
περὶ αὐτῶν ἐπεσκέμμεθα.

Τὸν ἐφεξῆς λόγον ἅπαντα ποιησόμεθα ζητοῦντες
ὑπὲρ ὧν ἐξ ἀρχῆς προὐθέμεθα, πόσαι τε καὶ τίνες
εἰσὶν αἱ τῆς φύσεως δυνάμεις καὶ τί ποιεῖν ἔργον

---

[1] A rallying-ground : lit. a place where two glens meet.

[2] Thus according to Gomperz (*Greek Thinkers*), the hypothesis of Anaxagoras was that "the bread . . . already contained the countless forms of matter as such which the
human body displays. Their minuteness of size would withdraw them from our perception. For the defect or 'weakness' of the senses is the narrowness of their receptive area.

need to recognize the *complete alteration of substance*.
In this way, nobody will suppose that bread repre-
sents a kind of meeting-place [1] for bone, flesh, nerve,
and all the other parts, and that each of these
subsequently becomes separated in the body and
goes to join its own kind; [2] before any separation
takes place, the whole of the bread obviously becomes
blood; (at any rate, if a man takes no other food
for a prolonged period, he will have blood enclosed
in his veins all the same).[3] And clearly this dis-
proves the view of those who consider the elements [4]
unchangeable, as also, for that matter, does the oil
which is entirely used up in the flame of the lamp,
or the faggots which, in a somewhat longer time,
turn into fire.

I said, however, that I was not going to enter into
an argument with these people, and it was only
because the example was drawn from the subject-
matter of medicine, and because I need it for the
present treatise, that I have mentioned it. We shall
then, as I said, renounce our controversy with them,
since those who wish may get a good grasp of the
views of the ancients from our own personal investigations into these matters.

The discussion which follows we shall devote
entirely, as we originally proposed, to an enquiry
into the number and character of the *faculties* of
Nature, and what is the effect which each naturally

These elusive particles are rendered visible and tangible by
the process of *nutrition*, which combines them."

[3] Therefore the blood must have come from the bread. The
food from the alimentary canal was supposed by Galen to be
converted into blood in and by the portal veins. *cf.* p. 17.

[4] By "elements" is meant all homogeneous, amorphous
substances, such as metals, &c., as well as the elementary
tissues.

ἑκάστῃ πέφυκεν. ἔργον δὲ δηλονότι καλῶ τὸ
7 γεγονὸς ἤδη καὶ συμπεπλη‖ρωμένον ὑπὸ τῆς ἐνεργείας αὐτῶν, οἷον τὸ αἷμα, τὴν σάρκα, τὸ νεῦρον·
ἐνέργειαν δὲ τὴν δραστικὴν ὀνομάζω κίνησιν καὶ
τὴν ταύτης αἰτίαν δύναμιν. ἐπεὶ γὰρ ἐν τῷ τὸ
σιτίον αἷμα γίγνεσθαι παθητικὴ μὲν ἡ τοῦ σιτίου,
δραστικὴ δ᾽ ἡ τῆς φλεβὸς γίγνεται κίνησις, ὡσαύτως δὲ κἂν τῷ μεταφέρειν τὰ κῶλα κινεῖ μὲν ὁ
μῦς, κινεῖται δὲ τὰ ὀστᾶ, τὴν μὲν τῆς φλεβὸς καὶ
τῶν μυῶν κίνησιν ἐνέργειαν εἶναί φημι, τὴν δὲ
τῶν σιτίων τε καὶ τῶν ὀστῶν σύμπτωμά τε καὶ
πάθημα· τὰ μὲν γὰρ ἀλλοιοῦται, τὰ δὲ φέρεται.
τὴν μὲν οὖν ἐνέργειαν ἐγχωρεῖ καλεῖν καὶ ἔργον
τῆς φύσεως, οἷον τὴν πέψιν, τὴν ἀνάδοσιν, τὴν
αἱμάτωσιν, οὐ μὴν τό γ᾽ ἔργον ἐξ ἅπαντος ἐνέρ-
γειαν· ἡ γάρ τοι σὰρξ ἔργον μέν ἐστι τῆς φύσεως,
οὐ μὴν ἐνέργειά γε. δῆλον οὖν, ὡς θάτερον μὲν
τῶν ὀνομάτων διχῶς λέγεται, θάτερον δ᾽ οὔ.

III

Ἐμοὶ μὲν οὖν καὶ ἡ φλὲψ καὶ τῶν ἄλλων
ἁπάντων ἕκαστον διὰ τὴν ἐκ τῶν τεττάρων ποιὰν

---

[1] Work or product. Lat. *opus*. *cf.* p. 3, note 2.

[2] Operation, activation, or functioning. Lat. *actio*. *cf. loc. cit.*

[3] *i.e.* a concomitant (secondary) or passive affection. Galen is contrasting active and passive "motion." *cf.* p. 6, note 1.

[4] As already indicated, there is no exact English equivalent for the Greek term *physis*, which is a principle immanent

produces. Now, of course, I mean by an effect [1] that which has already come into existence and has been completed by the *activity* [2] of these faculties—for example, blood, flesh, or nerve. And *activity* is the name I give to the active change or *motion,* and the *cause* of this I call a *faculty.* Thus, when food turns into blood, the motion of the food is passive, and that of the vein active. Similarly, when the limbs have their position altered, it is the muscle which produces, and the bones which undergo the motion. In these cases I call the motion of the vein and of the muscle an *activity,* and that of the food and the bones a *symptom* or *affection,* [3] since the first group undergoes *alteration* and the second group is merely *transported.* One might, therefore, also speak of the *activity* as an *effect* of Nature [4]— for example, digestion, absorption, [5] blood-production; one could not, however, in every case call the effect an activity ; thus flesh is an effect of Nature, but it is, of course, not an activity. It is, therefore, clear that one of these terms is used in two senses, but not the other.

III

It appears to me, then, that the vein, as well as each of the other parts, functions in such and such a way according to the manner in which *the four quali-*

in the animal itself, whereas our term " Nature " suggests something more transcendent ; we are forced often, however, to employ it in default of a better word. *cf.* p. 2, note 1.

[5] In Greek *anadosis.* This process includes two stages : (1) transmission of food from alimentary canal to liver (rather more than our " absorption ") ; (2) further transmission from liver to tissues. *Anadosis* is lit. a yielding-up, a " delivery ;" it may sometimes be rendered " dispersal." " Distribution " (*diadosis*) is a further stage ; *cf.* p. 163, note 4.   13

κρᾶσιν ὡδί πως ἐνεργεῖν δοκεῖ. εἰσὶ δέ γε μὴν οὐκ
8 ὀλίγοι τινὲς ἄνδρες ‖ οὐδ' ἄδοξοι, φιλόσοφοί τε
καὶ ἰατροί, τῷ μὲν θερμῷ καὶ τῷ ψυχρῷ τὸ δρᾶν
ἀναφέροντες, ὑποβάλλοντες δ' αὐτοῖς παθητικὰ
τὸ ξηρόν τε καὶ τὸ ὑγρόν. καὶ πρῶτός γ' Ἀριστοτέλης τὰς τῶν κατὰ μέρος ἁπάντων αἰτίας εἰς
ταύτας ἀνάγειν πειρᾶται τὰς ἀρχάς, ἠκολούθησε
δ' ὕστερον αὐτῷ καὶ ὁ ἀπὸ τῆς στοᾶς χορός. καίτοι τούτοις μέν, ὡς ἂν καὶ αὐτῶν τῶν στοιχείων
τὴν εἰς ἄλληλα μεταβολὴν χύσεσί τέ τισι καὶ
πιλήσεσιν ἀναφέρουσιν, εὔλογον ἦν ἀρχὰς δραστικὰς ποιήσασθαι τὸ θερμὸν καὶ τὸ ψυχρόν,
Ἀριστοτέλει δ' οὐχ οὕτως, ἀλλὰ ταῖς τέτταρσι
ποιότησιν εἰς τὴν τῶν στοιχείων γένεσιν χρωμένῳ
βέλτιον ἦν καὶ τὰς τῶν κατὰ μέρος αἰτίας ἁπάσας
εἰς ταύτας ἀνάγειν. τί δήποτ' οὖν ἐν μὲν τοῖς περὶ
γενέσεως καὶ φθορᾶς ταῖς τέτταρσι χρῆται, ἐν δὲ
τοῖς μετεωρολογικοῖς καὶ τοῖς προβλήμασι καὶ
ἄλλοθι πολλαχόθι ταῖς δύο μόναις; εἰ μὲν γὰρ
ὡς ἐν τοῖς ζῴοις τε καὶ τοῖς φυτοῖς μᾶλλον μὲν
δρᾷ τὸ θερμὸν καὶ τὸ ψυχρόν, ἧττον δὲ τὸ ξηρὸν
καὶ τὸ ὑγρὸν ἀποφαίνοιτό τις, ἴσως ἂν ἔχοι καὶ
τὸν Ἱπποκράτην σύμψηφον· εἰ δ' ὡσαύτως ἐν ‖
9 ἅπασιν, οὐκέτ' οἶμαι συγχωρήσειν τοῦτο μὴ ὅτι
τὸν Ἱπποκράτην ἀλλὰ μηδ' αὐτὸν τὸν Ἀριστοτέλην μεμνῆσθαί γε βουλόμενον ὧν ἐν τοῖς περὶ
γενέσεως καὶ φθορᾶς οὐχ ἁπλῶς ἀλλὰ μετ' ἀποδείξεως αὐτὸς ἡμᾶς ἐδίδαξεν. ἀλλὰ περὶ μὲν
τούτων κἀν τοῖς περὶ κράσεων, εἰς ὅσον ἰατρῷ
χρήσιμον, ἐπεσκεψάμεθα.

---

[1] cf. p. 9.

lies [1] are mixed. There are, however, a considerable number of not undistinguished men—philosophers and physicians—who refer action to the Warm and the Cold, and who subordinate to these, as passive, the Dry and the Moist; Aristotle, in fact, was the first who attempted to bring back the causes of the various special activities to these principles, and he was followed later by the Stoic school. These latter, of course, could logically make active principles of the Warm and Cold, since they refer the change of the elements themselves into one another to certain *diffusions* and *condensations*. [2] This does not hold of Aristotle, however; seeing that he employed the four qualities to explain the genesis of the elements, he ought properly to have also referred the causes of all the special activities to these. How is it that he uses the four qualities in his book " On Genesis and Destruction," whilst in his " Meteorology," his " Problems," and many other works he uses the two only? Of course, if anyone were to maintain that in the case of animals and plants the Warm and Cold are *more* active, the Dry and Moist *less* so, he might perhaps have even Hippocrates on his side ; but if he were to say that this happens in all cases, he would, I imagine, lack support, not merely from Hippocrates, but even from Aristotle himself—if, at least, Aristotle chose to remember what he himself taught us in his work " On Genesis and Destruction," not as a matter of simple statement, but with an accompanying demonstration. I have, however, also investigated these questions, in so far as they are of value to a physician, in my work " On Temperaments."

[2] Since heat and cold tend to cause diffusion and condensation respectively.

IV

. . . ἡ ἐν ταῖς φλεψὶν ἡ αἱματο-
π . . . ευομένη καὶ πᾶσα δ' ἄλλη
δὺ . . . τι νενόηται· πρώτως μὲν
γὰρ . . . ία, ἤδη δὲ καὶ τοῦ ἔργου
κατα . . . λ' εἴπερ ἡ αἰτία πρός
τι, το . . . γενομένου μόνου, τῶν δ'
ἄλλων . . . ν, ὅτι καὶ ἡ δύναμις ἐν τῷ
πρός τι. . . . χρι γ' ἂν ἀγνοῶμεν τὴν οὐσίαν
τῆς ἐνεργούσης αἰτίας, δύναμιν αὐτὴν ὀνομάζομεν,
εἶναί τινα λέγοντες ἐν ταῖς φλεψὶν αἱματοποιητι-
κήν, ὡσαύτως δὲ κἀν τῇ κοιλίᾳ πεπτικὴν κἀν τῇ
καρδίᾳ σφυγμικὴν καὶ καθ' ἕκαστον τῶν ἄλλων
10 ἰδίαν τινὰ τῆς ‖ κατὰ τὸ μόριον ἐνεργείας. εἴπερ
οὖν μεθόδῳ μέλλοιμεν ἐξευρήσειν, ὁπόσαι τε καὶ
ὁποῖαί τινες αἱ δυνάμεις εἰσίν, ἀπὸ τῶν ἔργων
αὐτῶν ἀρκτέον· ἕκαστον γὰρ αὐτῶν ὑπό τινος
ἐνεργείας γίγνεται καὶ τούτων ἑκάστης προηγεῖταί
τις αἰτία.

V

Ἔργα τοίνυν τῆς φύσεως ἔτι μὲν κυουμένου τε
καὶ διαπλαττομένου τοῦ ζῴου τὰ σύμπαντ' ἐστὶ
τοῦ σώματος μόρια, γεννηθέντος δὲ κοινὸν ἐφ'
ἅπασιν ἔργον ἡ εἰς τὸ τέλειον ἑκάστῳ μέγεθος
ἀγωγὴ καὶ μετὰ ταῦθ' ἡ μέχρι τοῦ δυνατοῦ
διαμονή.

Ἐνέργειαι δ' ἐπὶ τρισὶ τοῖς εἰρημένοις ἔργοις
τρεῖς ἐξ ἀνάγκης, ἐφ' ἑκάστῳ μία,· γένεσίς τε καὶ

IV

THE so-called *blood-making*[1] faculty in the veins, then, as well as all the other faculties, fall within the category of relative concepts ; primarily because the faculty is the cause of the activity, but also, accidentally, because it is the cause of the effect. But, if the cause is relative to something—for it is the cause of what results from it, and of nothing else—it is obvious that the faculty also falls into the category of the relative ; and so long as we are ignorant of the true essence of the cause which is operating, we call it a *faculty.* Thus we say that there exists in the veins a blood-making faculty, as also a digestive[2] faculty in the stomach, a pulsatile[3] faculty in the heart, and in each of the other parts a special faculty corresponding to the function or activity of that part. If, therefore, we are to investigate methodically the number and kinds of faculties, we must begin with the effects ; for each of these effects comes from a certain activity, and each of these again is preceded by a cause.

V

THE effects of Nature, then, while the animal is still being formed in the womb, are all the different *parts* of its body ; and after it has been born, an effect in which all parts share is the progress of each to its full size, and thereafter its maintenance of itself as long as possible.

The activities corresponding to the three effects mentioned are necessarily three—one to each—

---

[1] Lit. *haematopoietic. cf.* p. 11, note 3.   [2] Lit. *peptic.*
[3] Lit. *sphygmic.*

αὔξησις καὶ θρέψις. ἀλλ' ἡ μὲν γένεσις οὐχ
ἁπλῆ τις ἐνέργεια τῆς φύσεως, ἀλλ' ἐξ ἀλλοιώσεώς τε καὶ διαπλάσεώς ἐστι σύνθετος. ἵνα μὲν
γὰρ ὀστοῦν γένηται καὶ νεῦρον καὶ φλὲψ καὶ τῶν
ἄλλων ἕκαστον, ἀλλοιοῦσθαι χρὴ τὴν ὑποβεβλη-
μένην οὐσίαν, ἐξ ἧς γίγνεται τὸ ζῷον· ἵνα δὲ καὶ
σχῆμα τὸ δέον καὶ θέσιν καὶ κοιλότητάς τινας
11 καὶ ἀποφύσεις καὶ συμφύσεις καὶ τἄλλα ‖ τὰ
τοιαῦτα κτήσηται, διαπλάττεσθαι χρὴ τὴν ἀλλοιουμένην οὐσίαν, ἣν δὴ καὶ ὕλην τοῦ ζῴου καλῶν,
ὡς τῆς νεὼς τὰ ξύλα καὶ τῆς εἰκόνος τὸν κηρόν,
οὐκ ἂν ἁμάρτοις.

Ἡ δ' αὔξησις ἐπίδοσίς ἐστι καὶ διάστασις κατὰ
μῆκος καὶ πλάτος καὶ βάθος τῶν στερεῶν τοῦ
ζῴου μορίων, ὧνπερ καὶ ἡ διάπλασις ἦν, ἡ δὲ
θρέψις πρόσθεσις τοῖς αὐτοῖς ἄνευ διαστάσεως.

VI

Περὶ πρώτης οὖν τῆς γενέσεως εἴπωμεν, ἣν ἐξ
ἀλλοιώσεώς θ' ἅμα κ‑ διαπλάσεως ἐλέγομεν
γίγνεσθαι.

Καταβληθέντος δὴ τοῦ σπέρματος εἰς τὴν
μήτραν ἢ εἰς τὴν γῆν, οὐδὲν γὰρ διαφέρει, χρόνοις
τισὶν ὡρισμένοις πάμπολλα συνίσταται μόρια
τῆς γεννωμένης οὐσίας ὑγρότητι καὶ ξηρότητι καὶ
ψυχρότητι καὶ θερμότητι καὶ τοῖς ἄλλοις ἅπασιν,

---

[1] *Genesis* corresponds to the intrauterine life, or what we
may call *embryogeny*. *Alteration* here means histogenesis
or tissue-production ; *shaping* or *moulding* (in Greek *diaplasis*)
means the ordering of these tissues into organs (organogenesis).

namely, Genesis, Growth, and Nutrition. Genesis, however, is not a simple activity of Nature, but is compounded of *alteration* and of *shaping*.[1] That is to say, in order that bone, nerve, veins, and all other [tissues] may come into existence, the *underlying substance* from which the animal springs must be *altered*; and in order that the substance so altered may acquire its appropriate shape and position, its cavities, outgrowths, attachments, and so forth, it has to undergo a *shaping* or *formative* process.[2] One would be justified in calling this substance which undergoes alteration the *material* of the animal, just as wood is the material of a ship, and wax of an image.

Growth is an increase and expansion in length, breadth, and thickness of the solid parts of the animal (those which have been subjected to the moulding or shaping process). *Nutrition* is an addition to these, without expansion.

VI

LET us speak then, in the first place, of Genesis, which, as we have said, results from *alteration* together with *shaping*.

The seed having been cast into the womb or into the earth (for there is no difference),[3] then, after a certain definite period, a great number of parts become constituted in the substance which is being generated; these differ as regards moisture, dryness, coldness and warmth,[4] and in all the other qualities

[2] *cf.* p. 25, note 4.
[3] Note inadequate analogy of semen with fertilised seeds of plants (*i.e.* of gamete with zygote). Strictly speaking, of course, semen corresponds to pollen. *cf.* p. 130, note 2.
[4] *i.e.* the four primary qualities; *cf.* chap. iii. *supra.*   19

ὅσα τούτοις ἕπεται, διαφέροντα. τὰ δ' ἑπόμενα
γιγνώσκεις, εἴπερ ὅλως ἐφιλοσόφησάς τι περὶ
γενέσεως καὶ φθορᾶς· αἱ λοιπαὶ γὰρ τῶν ἁπτῶν
ὀνομαζομένων διαφορῶν ταῖς εἰρημέναις ἕπονται
12 πρῶται καὶ μάλιστα, μετὰ δὲ ταύ‖τας αἱ γευσταί
τε καὶ ὀσφρηταὶ καὶ ὁραταί. σκληρότης μὲν οὖν
καὶ μαλακότης καὶ γλισχρότης καὶ κραυρότης καὶ
κουφότης καὶ βαρύτης καὶ πυκνότης καὶ ἀραιότης
καὶ λειότης καὶ τραχύτης καὶ παχύτης καὶ λεπτότης ἁπταὶ ¹διαφοραὶ καὶ εἴρηται περὶ πασῶν
Ἀριστοτέλει καλῶς. οἶσθα δὲ δήπου καὶ τὰς
γευστάς τε καὶ ὀσφρητὰς καὶ ὁρατὰς διαφοράς.
ὥστ', εἰ μὲν τὰς πρώτας τε καὶ στοιχειώδεις
ἀλλοιωτικὰς δυνάμεις ζητοίης, ὑγρότης ἐστὶ καὶ
ξηρότης καὶ ψυχρότης καὶ θερμότης· εἰ δὲ τὰς ἐκ
τῆς τούτων κράσεως γεγομένας, τοσαῦται καθ'
ἕκαστον ἔσονται ζῷον, ὅσαπερ ἂν αὐτοῦ τὰ
αἰσθητὰ στοιχεῖα ὑπάρχῃ· καλεῖται δ' αἰσθητὰ
στοιχεῖα τὰ ὁμοιομερῆ πάντα τοῦ σώματος μόρια·
καὶ ταῦτ' οὐκ ἐκ μεθόδου τινὸς ἀλλ' αὐτόπτην
γενόμενον ἐκμαθεῖν χρὴ διὰ τῶν ἀνατομῶν.

Ὀστοῦν δὴ καὶ χόνδρον καὶ νεῦρον καὶ ὑμένα
καὶ σύνδεσμον καὶ φλέβα καὶ πάνθ' ὅσα τοιαῦτα
κατὰ τὴν πρώτην τοῦ ζῴου γένεσιν ἡ φύσις
ἀπεργάζεται δυνάμει χρωμένη καθόλου μὲν
13 εἰπεῖν τῇ γεννητικῇ τε καὶ ἀλλοιω‖τικῇ, κατὰ
μέρος δὲ θερμαντικῇ τε καὶ ψυκτικῇ καὶ ξηραν-

---

¹ Various secondary or derivative differences in the tissues.
Note pre-eminence of sense of touch.

² De Anima, ii. et seq.

³ Lit. homoeomerous = of similar parts throughout, "the
same all through." He refers to the elementary tissues,
conceived as not being susceptible of further analysis.

which naturally derive therefrom.[1] These derivative qualities, you are acquainted with, if you have given any sort of scientific consideration to the question of genesis and destruction. For, first and foremost after the qualities mentioned come the other so-called *tangible* distinctions, and after them those which appeal to taste, smell, and sight. Now, tangible distinctions are hardness and softness, viscosity, friability, lightness, heaviness, density, rarity, smoothness, roughness, thickness and thinness; all of these have been duly mentioned by Aristotle.[2] And of course you know those which appeal to taste, smell, and sight. Therefore, if you wish to know which alterative faculties are primary and elementary, they are moisture, dryness, coldness, and warmth, and if you wish to know which ones arise from the combination of these, they will be found to be in each animal of a number corresponding to its *sensible elements*. The name *sensible elements* is given to all the *homogeneous*[3] parts of the body, and these are to be detected not by any system, but by personal observation of dissections.[4]

Now Nature constructs bone, cartilage, nerve, membrane, ligament, vein, and so forth, at the first stage of the animal's genesis,[5] employing at this task a faculty which is, in general terms, generative and alterative, and, in more detail, warming, chilling, drying, or moistening; or such as spring from the

[4] That is, by the bodily eye, and not by the mind's eye. The observer is here called an *autoptes* or "eye-witness." Our medical term *autopsy* thus means literally a *personal inspection* of internal parts, ordinarily hidden.

[5] *i.e.* "alteration" is the earlier of the two stages which constitute embryogeny or "genesis." *cf.* p. 18, note 1.

τικῇ καὶ ὑγραντικῇ καὶ ταῖς ἐκ τῆς τούτων
κράσεως γενομέναις, οἷον ὀστοποιητικῇ τε καὶ
νευροποιητικῇ καὶ χονδροποιητικῇ· σαφηνείας
γὰρ ἕνεκα καὶ τούτοις τοῖς ὀνόμασι χρηστέον.

Ἔστι γοῦν καὶ ἡ ἰδία σὰρξ τοῦ ἥπατος ἐκ
τούτου τοῦ γένους καὶ ἡ τοῦ σπληνὸς καὶ ἡ τῶν
νεφρῶν καὶ ἡ τοῦ πνεύμονος καὶ ἡ τῆς καρδίας·
οὕτω δὲ καὶ τοῦ ἐγκεφάλου τὸ ἴδιον σῶμα καὶ
τῆς γαστρὸς καὶ τοῦ στομάχου καὶ τῶν ἐντέρων
καὶ τῶν ὑστερῶν αἰσθητὸν στοιχεῖόν ἐστιν ὁμοιο-
μερές τε καὶ ἁπλοῦν καὶ ἀσύνθετον· ἐὰν γὰρ
ἐξέλῃς ἑκάστου τῶν εἰρημένων τὰς ἀρτηρίας τε
καὶ τὰς φλέβας καὶ τὰ νεῦρα, τὸ ὑπόλοιπον
σῶμα τὸ καθ' ἕκαστον ὄργανον ἁπλοῦν ἐστι καὶ
στοιχειῶδες ὡς πρὸς αἴσθησιν. ὅσα δὲ τῶν
τοιούτων ὀργάνων ἐκ δυοῖν σύγκειται χιτώνων
οὐχ ὁμοίων μὲν ἀλλήλοις, ἁπλοῦ δ' ἑκατέρου,
τούτων οἱ χιτῶνές εἰσι τὰ στοιχεῖα καθάπερ τῆς
τε γαστρὸς καὶ τοῦ στομάχου καὶ τῶν ἐντέρων
καὶ τῶν ἀρτηριῶν, καὶ καθ' ἑκάτερόν γε τῶν
χιτώνων ἴδιος ἡ ἀλλοιωτικὴ δύναμις ἡ ἐκ τοῦ
14 παρὰ τῆς ‖ μητρὸς ἐπιμηνίου γεννήσασα τὸ
μόριον, ὥστε τὰς κατὰ μέρος ἀλλοιωτικὰς δυνάμεις τοσαύτας εἶναι καθ' ἕκαστον ζῷον, ὅσαπερ
ἂν ἔχῃ τὰ στοιχειώδη μόρια. καὶ μέν γε καὶ
τὰς ἐνεργείας ἰδίας ἑκάστῳ τῶν κατὰ μέρος
ἀναγκαῖον ὑπάρχειν ὥσπερ καὶ τὰς χρείας, οἷον
καὶ τῶν ἀπὸ τῶν νεφρῶν εἰς τὴν κύστιν διηκόντων
πόρων, οἳ δὴ καὶ οὐρητῆρες καλοῦνται. οὗτοι

---

[1] The terms Galen actually uses are: ostopoietic, neuropoietic, chondropoietic.

blending of these, for example, the bone-producing,
nerve-producing, and cartilage-producing faculties [1]
(since for the sake of clearness these names must
be used as well).

Now the peculiar [2] flesh of the liver is of this kind
as well, also that of the spleen, that of the kidneys,
that of the lungs, and that of the heart; so also
the proper substance of the brain, stomach, gullet,
intestines, and uterus is *a sensible element*, of similar
parts all through, simple, and uncompounded. That
is to say, if you remove from each of the organs
mentioned its arteries, veins, and nerves,[3] the
substance remaining in each organ is, from the point
of view of the senses, simple and elementary. As
regards those organs consisting of two dissimilar
coats,[4] of which each is simple, of these organs the
coats are the elements—for example, the coats
of the stomach, oesophagus, intestines, and arteries;
each of these two coats has an alterative faculty
peculiar to it, which has engendered it from the
menstrual blood of the mother. Thus the *special*
alterative faculties in each animal are of the same
number as the elementary parts [5]; and further,
the *activities* must necessarily correspond each to one
of the special parts, just as each part has its special
use—for example, those ducts which extend from
the kidneys into the bladder, and which are called
ureters; for these are not arteries, since they do not
pulsate nor do they consist of two coats; and they

[2] As we should say, *parenchyma* (a term used by Erasistratus).

[3] These were all the elementary tissues that Aristotle, for
example, had recognized; other tissues (*e.g.* flesh or muscle)
he believed to be complexes of these.

[4] Or *tunics*.   [5] *i.e.* tissues.

23

γὰρ οὔτ᾿ ἀρτηρίαι εἰσίν, ὅτι μήτε σφύζουσι μήτ᾿
ἐκ δυοῖν χιτώνων συνεστήκασιν, οὔτε φλέβες, ὅτι
μήθ᾿ αἷμα περιέχουσι μήτ᾿ ἔοικεν αὐτῶν ὁ χιτὼν
κατά τι τῷ τῆς φλεβός· ἀλλὰ καὶ νεύρων ἐπὶ
πλέον ἀφεστήκασιν ἢ τῶν εἰρημένων.

Τί ποτ᾿ οὖν εἰσιν; ἐρωτᾷ τις, ὥσπερ ἀναγκαῖον
ὂν ἅπαν μόριον ἢ ἀρτηρίαν ἢ φλέβα ἢ νεῦρον
ὑπάρχειν ἢ ἐκ τούτων πεπλέχθαι καὶ μὴ τοῦτ᾿
αὐτὸ τὸ νῦν λεγόμενον, ὡς ἴδιος ἑκάστῳ τῶν κατὰ
μέρος ὀργάνων ἐστὶν ἡ οὐσία. καὶ γὰρ καὶ αἱ
κύστεις ἑκάτεραι ἥ τε τὸ οὖρον ὑποδεχομένη καὶ
ἡ τὴν ξανθὴν χολὴν οὐ μόνον τῶν ἄλλων ἁπάντων
ἀλλὰ καὶ ἀλλήλων διαφέρουσι καὶ οἱ εἰς τὸ ἧπαρ
15 ἀποφυόμενοι ‖ πόροι, καθάπερ στόμαχοί τινες
ἀπὸ τῆς χοληδόχου κύστεως, οὐδὲν οὔτ᾿ ἀρτηρίαις
οὔτε φλεψὶν οὔτε νεύροις ἐοίκασιν. ἀλλὰ περὶ
μὲν τούτων ἐπὶ πλέον ἐν ἄλλοις τέ τισι κἂν τοῖς
περὶ τῆς Ἱπποκράτους ἀνατομῆς εἴρηται.

Αἱ δὲ κατὰ μέρος ἅπασαι δυνάμεις τῆς φύσεως
αἱ ἀλλοιωτικαὶ αὐτὴν μὲν τὴν οὐσίαν τῶν χιτώνων τῆς κοιλίας καὶ τῶν ἐντέρων καὶ τῶν ὑστερῶν
ἀπετέλεσαν, οἷάπερ ἐστί· τὴν δὲ σύνθεσιν αὐτῶν
καὶ τὴν τῶν ἐμφυομένων πλοκὴν καὶ τὴν εἰς τὸ
ἔντερον ἔκφυσιν καὶ τὴν τῆς ἔνδον κοιλότητος
ἰδέαν καὶ τἆλλ᾿ ὅσα τοιαῦτα δύναμίς τις ἑτέρα
διέπλασεν, ἣν διαπλαστικὴν ὀνομάζομεν, ἣν δὴ
καὶ τεχνικὴν εἶναι λέγομεν, μᾶλλον δ᾿ ἀρίστην
καὶ ἄκραν τέχνην καὶ πάντα τινὸς ἕνεκα ποιοῦσαν,
ὡς μηδὲν ἀργὸν εἶναι μηδὲ περιττὸν μηδ᾿ ὅλως

---

[1] As, for example, Aristotle had held ; cf. p. 23, note 3.
Galen added many new tissues to those described by Aristotle.

are not veins, since they neither contain blood, nor
do their coats in any way resemble those of veins;
from nerves they differ still more than from the
structures mentioned.

"What, then, are they?" someone asks—as
though every part must necessarily be either an
artery, a vein, a nerve, or a complex of these,[1]
and as though the truth were not what I am now
stating, namely, that every one of the various
organs has its own particular substance. For in fact
the two bladders—that which receives the urine,
and that which receives the yellow bile—not only
differ from all other organs, but also from one
another. Further, the ducts which spring out like
kinds of conduits from the gall-bladder and which
pass into the liver have no resemblance either to
arteries, veins or nerves. But these parts have been
treated at a greater length in my work "On the
Anatomy of Hippocrates," as well as elsewhere.

As for the actual substance of the coats of the
stomach, intestine, and uterus, each of these has
been rendered what it is by a special alterative
faculty of Nature; while the bringing of these
together,[2] the combination therewith of the structures
which are inserted into them, the outgrowth into the
intestine,[3] the shape of the inner cavities, and the
like, have all been determined by a faculty which we
call the shaping or formative faculty[4]; this faculty
we also state to be *artistic*—nay, the best and highest
art—doing everything for some purpose, so that

[2] Lit. *synthesis.*

[3] By this is meant the *duodenum*, considered as an outgrowth or prolongation of the stomach towards the intestines.

[4] *cf.* p. 19, note 2.

25

οὕτως ἔχον, ὡς δύνασθαι βέλτιον ἑτέρως ἔχειν.
ἀλλὰ τοῦτο μὲν ἐν τοῖς περὶ χρείας μορίων
ἀποδείξομεν. ‖

VII

16   Ἐπὶ δὲ τὴν αὐξητικὴν ἤδη μεταβάντες δύναμιν
αὐτὸ τοῦθ' ὑπομνήσωμεν πρῶτον, ὡς ὑπάρχει
μὲν καὶ αὐτὴ τοῖς κυουμένοις ὥσπερ καὶ ἡ θρεπ-
τική· ἀλλ' οἷον ὑπηρέτιδές τινές εἰσι τηνικαῦτα
τῶν προειρημένων δυνάμεων, οὐκ ἐν αὐταῖς
ἔχουσαι τὸ πᾶν κῦρος. ἐπειδὰν δὲ τὸ τέλειον
ἀπολάβῃ μέγεθος τὸ ζῷον, ἐν τῷ μετὰ τὴν
ἀποκύησιν χρόνῳ παντὶ μέχρι τῆς ἀκμῆς ἡ μὲν
αὐξητικὴ τηνικαῦτα κρατεῖ· βοηθοὶ δ' αὐτῆς καὶ
οἷον ὑπηρέτιδες ἥ τ' ἀλλοιωτικὴ δύναμίς ἐστι
καὶ ἡ θρεπτική. τί οὖν τὸ ἴδιόν ἐστι τῆς αὐ-
ξητικῆς δυνάμεως; εἰς πᾶν μέρος ἐκτεῖναι τὰ
πεφυκότα. καλεῖται δ' οὕτω τὰ στερεὰ μόρια
τοῦ σώματος, ἀρτηρίαι καὶ φλέβες καὶ νεῦρα καὶ
ὀστᾶ καὶ χόνδροι καὶ ὑμένες καὶ σύνδεσμοι καὶ οἱ
χιτῶνες ἅπαντες, οὓς στοιχειώδεις τε καὶ ὁμοιομερεῖς καὶ ἁπλοῦς ὀλίγον ἔμπροσθεν ἐκαλοῦμεν.
ὅτῳ δὲ τρόπῳ τὴν εἰς πᾶν μέρος ἔκτασιν ἴσχουσιν,
ἐγὼ φράσω παράδειγμά τι πρότερον εἰπὼν ἔνεκα
τοῦ σαφοῦς. ‖

17   Τὰς κύστεις τῶν ὑῶν λαβόντες οἱ παῖδες
πληροῦσί τε πνεύματος καὶ τρίβουσιν ἐπὶ τῆς
τέφρας πλησίον τοῦ πυρός, ὡς ἀλεαίνεσθαι μέν,
βλάπτεσθαι δὲ μηδέν· καὶ πολλή γ' αὕτη ἡ

---

[1] Lit. the auxetic or incremental faculty.

there is nothing ineffective or superfluous, or capable of being better disposed. This, however, I shall demonstrate in my work " On the Use of Parts."

VII

Passing now to the faculty of Growth [1] let us first mention that this, too, is present in the foetus *in utero* as is also the nutritive faculty, but that at that stage these two faculties are, as it were, *handmaids* to those already mentioned,[2] and do not possess in themselves supreme authority. When, however, the animal [3] has attained its complete size, then, during the whole period following its birth and until the acme is reached, the faculty of growth is predominant, while the alterative and nutritive faculties are accessory—in fact, act as its handmaids. What, then, is the property of this faculty of growth ? To extend in every direction that which has already come into existence—that is to say, the solid parts of the body, the arteries, veins, nerves, bones, cartilages, membranes, ligaments, and the various *coats* which we have just called elementary, homogeneous, and simple. And I shall state in what way they gain this extension in every direction, first giving an illustration for the sake of clearness.

Children take the bladders of pigs, fill them with air, and then rub them on ashes near the fire, so as to warm, but not to injure them. This is a common

[2] *i.e.* to the alterative and shaping faculties (histogenetic and organogenetic).
[3] If the reading is correct we can only suppose that Galen meant *the embryo.*

παιδιὰ περί.τε τὴν Ἰωνίαν καὶ ἐν ἄλλοις ἔθνεσιν
οὐκ ὀλίγοις ἐστίν. ἐπιλέγουσι δὲ δὴ καί τιν' ἔπη
τρίβοντες ἐν μέτρῳ τέ τινι καὶ μέλει καὶ ῥυθμῷ
καὶ ἔστι πάντα τὰ ῥήματα ταῦτα παρακέλευσις
τῇ κύστει πρὸς τὴν αὔξησιν. ἐπειδὰν δ' ἱκανῶς
αὐτοῖς διατετάσθαι δοκῇ, πάλιν ἐμφυσῶσί τε καὶ
ἐπιδιατείνουσι καὶ αὖθις τρίβουσι καὶ τοῦτο
πλεονάκις ποιοῦσιν, ἄχρις ἂν αὐτοῖς ἡ κύστις
ἱκανῶς ἔχειν δοκῇ τῆς αὐξήσεως. ἀλλ' ἐν τούτοις
γε τοῖς ἔργοις τῶν παίδων ἐναργῶς, ὅσον εἰς
μέγεθος ἐπιδίδωσιν ἡ ἐντὸς εὐρυχωρία τῆς
κύστεως, τοσοῦτον ἀναγκαῖον εἰς λεπτότητα
καθαιρεῖσθαι τὸ σῶμα καὶ εἴ γε τὴν λεπτότητα
ταύτην ἀνατρέφειν οἷοί τ' ἦσαν οἱ παῖδες, ὁμοίως
ἂν τῇ φύσει τὴν κύστιν ἐκ μικρᾶς μεγάλην
ἀπειργάζοντο. νυνὶ δὲ τοῦτ' αὐτοῖς ἐνδεῖ τὸ
ἔργον οὐδὲ καθ' ἕνα τρόπον εἰς μίμησιν ἐνδεχό-
18 μενον ἀχθῆναι μὴ ὅτι τοῖς ‖ παισὶν ἀλλ' οὐδ' ἄλλῳ
τινί· μόνης γὰρ τῆς φύσεως ἴδιόν ἐστιν.

Ὥστ' ἤδη σοι δῆλον, ὡς ἀναγκαία τοῖς αὐξανομένοις ἡ θρέψις. εἰ γὰρ διατείνοιτο μέν, ἀνατρέ-
φοιτο δὲ μή, φαντασίαν ψευδῆ μᾶλλον, οὐκ
αὔξησιν ἀληθῆ τὰ τοιαῦτα σώματα κτήσεται.
καίτοι καὶ τὸ διατείνεσθαι πάντῃ μόνοις τοῖς ὑπὸ
φύσεως αὐξανομένοις ὑπάρχει. τὰ γὰρ ὑφ' ἡμῶν
διατεινόμενα σώματα κατὰ μίαν τινὰ διάστασιν
τοῦτο πάσχοντα μειοῦται ταῖς λοιπαῖς, οὐδ' ἔστιν
εὑρεῖν οὐδέν, ὃ συνεχὲς ἔτι μένον καὶ ἀδιάσπαστον
εἰς τὰς τρεῖς διαστάσεις ἐπεκτεῖναι δυνάμεθα.
μόνης οὖν τῆς φύσεως τὸ πάντῃ διιστάναι συνεχὲς
ἑαυτῷ μένον ἔτι καὶ τὴν ἀρχαίαν ἅπασαν ἰδέαν
φυλάττον τὸ σῶμα.

game in the district of Ionia, and among not a few other nations. As they rub, they sing songs, to a certain measure, time, and rhythm, and all their words are an exhortation to the bladder to increase in size. When it appears to them fairly well distended, they again blow air into it and expand it further; then they rub it again. This they do several times, until the bladder seems to them to have become large enough. Now, clearly, in these doings of the children, the more the interior cavity of the bladder increases in size, the thinner, necessarily, does its substance become. But, if the children were able to bring nourishment to this thin part, then they would make the bladder big in the same way that Nature does. As it is, however, they cannot do what Nature does, for to imitate this is beyond the power not only of children, but of any one soever; it is a property of Nature alone.

It will now, therefore, be clear to you that *nutrition* is a necessity for growing things. For if such bodies were distended, but not at the same time nourished, they would take on a false appearance of growth, not a true growth. And further, to be distended *in all directions* belongs only to bodies whose growth is directed by Nature; for those which are distended by us undergo this distension in one direction but grow less in the others; it is impossible to find a body which will remain entire and not be torn through whilst we stretch it in the three dimensions. Thus Nature alone has the power to expand a body in all directions so that it remains unruptured and preserves completely its previous form.

Καὶ τοῦτ᾽ ἔστιν ἡ αὔξησις ἄνευ τῆς ἐπιρρεούσης τε καὶ προσπλαττομένης τροφῆς μὴ δυναμένη γενέσθαι.

VIII

Καὶ τοίνυν ὁ λόγος ἥκειν ἔοικεν ὁ περὶ τῆς θρέψεως, ὃς δὴ λοιπός ἐστι καὶ τρίτος ὢν ἐξ ἀρχῆς προὐθέμεθα. τοῦ γὰρ ἐπιρρέοντος ἐν εἴδει 19 τροφῆς παντὶ ‖ μορίῳ τοῦ τρεφομένου σώματος προσπλαττομένου θρέψις μὲν ἡ ἐνέργεια, θρεπτικὴ δὲ δύναμις ἡ αἰτία. ἀλλοίωσις μὲν δὴ κἀνταῦθα τὸ γένος τῆς ἐνεργείας, ἀλλ᾽ οὐχ οἷαπερ ἡ ἐν τῇ γενέσει. ἐκεῖ μὲν γὰρ οὐκ ὂν πρότερον ὕστερον ἐγένετο, κατὰ δὲ τὴν θρέψιν τῷ ἤδη γεγονότι συνεξομοιοῦται τὸ ἐπιρρέον καὶ διὰ τοῦτ᾽ εὐλόγως ἐκείνην μὲν τὴν ἀλλοίωσιν γένεσιν, ταύτην δ᾽ ἐξομοίωσιν ὠνόμασαν.

IX

Ἐπειδὴ δὲ περὶ τῶν τριῶν δυνάμεων τῆς φύσεως αὐτάρκως εἴρηται καὶ φαίνεται μηδεμιᾶς ἄλλης προσδεῖσθαι τὸ ζῷον, ἔχον γε καὶ ὅπως αὐξηθῇ καὶ ὅπως τελειωθῇ καὶ ὅπως ἕως πλείστου διαφυλαχθῇ, δόξειε μὲν ἂν ἴσως ἱκανῶς ἔχειν ὁ λόγος οὗτος ἤδη καὶ πάσας ἐξηγεῖσθαι τὰς τῆς φύσεως δυνάμεις. ἀλλ᾽ εἴ τις πάλιν ἐννοήσειεν, ὡς οὐ-

Such then is *growth*, and it cannot occur without the nutriment which flows to the part and is worked up into it.

VIII

WE have, then, it seems, arrived at the subject of Nutrition, which is the third and remaining consideration which we proposed at the outset. For, when the matter which flows to each part of the body in the form of nutriment is being worked up into it, this activity is *nutrition,* and its cause is the *nutritive faculty.* Of course, the kind of activity here involved is also an *alteration,* but not an alteration like that occurring at the stage of *genesis.*[1] For in the latter case something comes into existence which did not exist previously, while in nutrition the inflowing material becomes assimilated to that which has already come into existence. Therefore, the former kind of alteration has with reason been termed *genesis,* and the latter, *assimilation.*

IX

Now, since the three faculties of Nature have been exhaustively dealt with, and the animal would appear not to need any others (being possessed of the means for growing, for attaining completion, and for maintaining itself as long a time as possible), this treatise might seem to be already complete, and to constitute an exposition of all the faculties of Nature. If, however, one considers that it has not

[1] *i.e.* not the pre-natal development of tissue already described. *cf.* chap. vi.

δενὸς οὐδέπω τῶν τοῦ ζῴου μορίων ἐφήψατο,
κοιλίας λέγω καὶ ἐντέρων καὶ ἥπατος καὶ τῶν
ὁμοίων, οὐδ᾽ ἐξηγήσατο τὰς ἐν αὐτοῖς δυνάμεις,
αὖθις δόξειεν ἂν οἷον προοίμιόν τι μόνον εἰρῆσθαι
20 τῆς χρησίμου διδασκαλίας. ‖ τὸ γὰρ σύμπαν ὧδ᾽
ἔχει. γένεσις καὶ αὔξησις καὶ θρέψις τὰ πρῶτα
καὶ οἷον κεφάλαια τῶν ἔργων ἐστὶ τῆς φύσεως·
ὥστε καὶ αἱ τούτων ἐργαστικαὶ δυνάμεις αἱ
πρῶται τρεῖς εἰσι καὶ κυριώταται· δέονται δ᾽ εἰς
ὑπηρεσίαν, ὡς ἤδη δέδεικται, καὶ ἀλλήλων καὶ
ἄλλων. τίνων μὲν οὖν ἡ γεννητική τε καὶ αὐξητικὴ δέονται, εἴρηται, τίνων δ᾽ ἡ θρεπτική, νῦν
εἰρήσεται.

X

Δοκῶ γάρ μοι δείξειν τὰ περὶ τὴν τῆς τροφῆς
οἰκονομίαν ὄργανά τε καὶ τὰς δυνάμεις αὐτῶν
διὰ ταύτην γεγονότα. ἐπειδὴ γὰρ ἡ ἐνέργεια
ταύτης τῆς δυνάμεως ἐξομοίωσίς ἐστιν, ὁμοιοῦ-
σθαι δὲ καὶ μεταβάλλειν εἰς ἄλληλα πᾶσι τοῖς
οὖσιν ἀδύνατον, εἰ μή τινα ἔχοι κοινωνίαν ἤδη
καὶ συγγένειαν ἐν ταῖς ποιότησι, διὰ τοῦτο
πρῶτον μὲν οὐκ ἐκ πάντων ἐδεσμάτων πᾶν ζῷον
τρέφεσθαι πέφυκεν, ἔπειτα δ᾽ οὐδ᾽ ἐξ ὧν οἷόν τ᾽
ἐστὶν οὐδ᾽ ἐκ τούτων παραχρῆμα, καὶ διὰ ταύτην

---

[1] Administration, lit. "economy."
[2] The *activation* or *functioning* of this faculty, the faculty
in actual operation. cf. p. 3, note 2.

yet touched upon any of *the parts* of the animal (I mean the stomach, intestines, liver, and the like), and that it has not dealt with the faculties resident in these, it will seem as though merely a kind of introduction had been given to the practical parts of our teaching. For the whole matter is as follows: Genesis, growth, and nutrition are the first, and, so to say, the principal effects of Nature; similarly also the faculties which produce these effects—the first faculties—are three in number, and are the most dominating of all. But as has already been shown, these need the service both of each other, and of yet different faculties. Now, these which the faculties of generation and growth require have been stated. I shall now say what ones the nutritive faculty requires.

X

For I believe that I shall prove that the organs which have to do with the disposal[1] of the nutriment, as also their faculties, exist for the sake of this *nutritive faculty*. For since the action of this faculty[2] is *assimilation*, and it is impossible for anything to be assimilated by, and to change into anything else unless they already possess a certain *community and affinity* in their qualities,[3] therefore, in the first place, any animal cannot naturally derive nourishment from any kind of food, and secondly, even in the case of those from which it can do so, it cannot do this at once. Therefore, by reason of

[3] "Un rapport commun et une affinité" (Daremberg). "Societatem aliquam cognationemque in qualitatibus" (Linacre). *cf.* p. 36, note 2.

τὴν ἀνάγκην πλειόνων ὀργάνων ἀλλοιωτικῶν τῆς
21 τροφῆς ἕκαστον ‖ τῶν ζῴων χρῄζει. ἵνα μὲν γὰρ
τὸ ξανθὸν ἐρυθρὸν γένηται καὶ τὸ ἐρυθρὸν ξανθόν,
ἁπλῆς καὶ μιᾶς δεῖται τῆς ἀλλοιώσεως· ἵνα δὲ τὸ
λευκὸν μέλαν καὶ τὸ μέλαν λευκόν, ἁπασῶν τῶν
μεταξύ. καὶ τοίνυν καὶ τὸ μαλακώτατον οὐκ ἂν
ἀθρόως σκληρότατον καὶ τὸ σκληρότατον οὐκ ἂν
ἀθρόως μαλακώτατον γένοιτο, ὥσπερ οὐδὲ τὸ
δυσωδέστατον εὐωδέστατον οὐδ' ἔμπαλιν τὸ εὐωδέστατον δυσωδέστατον ἐξαίφνης γένοιτ' ἄν.

Πῶς οὖν ἐξ αἵματος ὀστοῦν ἄν ποτε γένοιτο μὴ
παχυνθέντος γε πρότερον ἐπὶ πλεῖστον αὐτοῦ καὶ
λευκανθέντος ἢ πῶς ἐξ ἄρτου τὸ αἷμα μὴ κατὰ
βραχὺ μὲν ἀποθεμένου τὴν λευκότητα, κατὰ
βραχὺ δὲ λαμβάνοντος τὴν ἐρυθρότητα; σάρκα
μὲν γὰρ ἐξ αἵματος γενέσθαι ῥᾷστον· εἰ γὰρ εἰς
τοσοῦτον αὐτὸ παχύνειεν ἡ φύσις, ὡς σύστασίν
τινα σχεῖν καὶ μηκέτ' εἶναι ῥυτόν, ἡ πρώτη καὶ
νεοπαγὴς οὕτως ἂν εἴη σάρξ· ὀστοῦν δ' ἵνα γένηται, πολλοῦ μὲν δεῖται χρόνου, πολλῆς δ' ἐργασίας
καὶ μεταβολῆς τῷ αἵματι. ὅτι δὲ καὶ τῷ ἄρτῳ
22 καὶ πολὺ μᾶλλον θρίδα‖κίνῃ καὶ τεύτλῳ καὶ τοῖς
ὁμοίοις παμπόλλης δεῖται τῆς ἀλλοιώσεως εἰς
αἵματος γένεσιν, οὐδὲ τοῦτ' ἄδηλον.

Ἕν μὲν δὴ τοῦτ' αἴτιον τοῦ πολλὰ γενέσθαι τὰ
περὶ τὴν τῆς τροφῆς ἀλλοίωσιν ὄργανα. δεύτερον
δ' ἡ τῶν περιττωμάτων φύσις. ὡς γὰρ ὑπὸ
βοτανῶν οὐδ' ὅλως δυνάμεθα τρέφεσθαι, καίτοι
τῶν βοσκημάτων τρεφομένων, οὕτως ὑπὸ ῥαφανί-

---

[1] Lit. "necessity"; more *restrictive*, however, than our
"law of Nature." *cf.* p. 314, note 1.

[2] His point is that no great change, in colours or in anything else, can take place at one step.

this law,[1] every animal needs several organs for
altering the nutriment. For in order that the yellow
may become red, and the red yellow, one simple
process of alteration is required, but in order that
the white may become black, and the black white,
all the intermediate stages are needed.[2] So also, a
thing which is very soft cannot all at once become
very hard, nor *vice versa*; nor, similarly can anything
which has a very bad smell suddenly become quite
fragrant, nor again, can the converse happen.

How, then, could blood ever turn into bone, without having first become, as far as possible, thickened
and white? And how could bread turn into blood
without having gradually parted with its white-
ness and gradually acquired redness? Thus it is
quite easy for blood to become flesh; for, if Nature
thicken it to such an extent that it acquires a
certain consistency and ceases to be fluid, it thus
becomes original newly-formed flesh; but in order
that blood may turn into bone, much time is needed
and much elaboration and transformation of the
blood. Further, it is quite clear that bread, and,
more particularly lettuce, beet, and the like, require a great deal of alteration in order to become
blood.

This, then, is one reason why there are so many
organs concerned in the alteration of food. A
second reason is the nature of the *superfluities*.[3] For,
as we are unable to draw any nourishment from
grass, although this is possible for cattle, similarly
we can derive nourishment from radishes, albeit not

[3] Not quite our "waste *products*," since these are considered as being partly synthetic, whereas the Greek *peritto-
mata* were simply superfluous substances which could not be
used and were thrown aside.

δος τρεφόμεθα μέν, ἀλλ᾽ οὐχ ὡς ὑπὸ τῶν κρεῶν.
τούτων μὲν γὰρ ὀλίγου δεῖν ὅλων ἡ φύσις ἡμῶν
κρατεῖ καὶ μεταβάλλει καὶ ἀλλοιοῖ καὶ χρηστὸν
ἐξ αὐτῶν αἷμα συνίστησιν· ἐν δὲ τῇ ῥαφανίδι τὸ
μὲν οἰκεῖόν τε καὶ μεταβληθῆναι δυνάμενον, μόγις
καὶ τοῦτο καὶ σὺν πολλῇ τῇ κατεργασίᾳ, παντάπασιν ἐλάχιστον· ὅλη δ᾽ ὀλίγου δεῖν ἐστι περιτ-
τωματικὴ καὶ διεξέρχεται τὰ τῆς πέψεως ὄργανα,
βραχέος ἐξ αὐτῆς εἰς τὰς φλέβας ἀναληφθέντος
αἵματος καὶ οὐδὲ τούτου τελέως χρηστοῦ. δευτέρας οὖν αὖθις ἐδέησε διακρίσεως τῇ φύσει τῶν ἐν
ταῖς φλεψὶ περιττωμάτων. καὶ χρεία καὶ τού-
23 τοις ὁδῶν τέ τινων ἑτέρων ἐπὶ τὰς ἐκ‖κρίσεις αὐτὰ
παραγουσῶν, ὡς μὴ λυμαίνοιτο τοῖς χρηστοῖς,
ὑποδοχῶν τέ τινων οἷον δεξαμενῶν, ἐν αἷς ὅταν
εἰς ἱκανὸν πλῆθος ἀφίκηται, τηνικαῦτ᾽ ἐκκριθή-
σεται.

Δεύτερον δή σοι καὶ τοῦτο τὸ γένος τῶν ἐν τῷ
σώματι μορίων ἐξεύρηται τοῖς περιττώμασι τῆς
τροφῆς ἀνακείμενον. ἄλλο δὲ τρίτον ὑπὲρ τοῦ
πάντη φέρεσθαι, καθάπερ τινὲς ὁδοὶ πολλαὶ διὰ
τοῦ σώματος ὅλου κατατετμημέναι.

Μία μὲν γὰρ εἴσοδος ἡ διὰ τοῦ στόματος ἅπασι
τοῖς σιτίοις, οὐχ ἓν δὲ τὸ τρεφόμενον ἀλλὰ
πάμπολλά τε καὶ πάμπολυ διεστῶτα. μὴ τοίνυν
θαύμαζε τὸ πλῆθος τῶν ὀργάνων, ὅσα θρέψεως
ἕνεκεν ἡ φύσις ἐδημιούργησε. τὰ μὲν γὰρ ἀλλοι-

---

[1] Note "our natures," cf. p. 12, note 4 ; p. 47, note 1.
[2] The term οἰκεῖος, here rendered *appropriate*, is explained
on p. 33. cf. also footnote on same page. Linacre often
translated it *conveniens*, and it may usually be rendered
proper, peculiar, own special, or *own particular* in English.
Sometimes it is almost equal to *akin, cognate, related*: cf.

to the same extent as from meat; for almost the whole of the latter is mastered by our natures [1]; it is transformed and altered and constituted useful blood; but, in the radish, what is appropriate [2] and capable of being altered (and that only with difficulty, and with much labour) is the very smallest part; almost the whole of it is surplus matter, and passes through the digestive organs, only a very little being taken up into the veins as blood—nor is this itself entirely utilisable blood. Nature, therefore, had need of a second process of separation for the superfluities in the veins. Moreover, these superfluities need, on the one hand, certain fresh routes to conduct them to the outlets, so that they may not spoil the useful substances, and they also need certain *reservoirs*, as it were, in which they are collected till they reach a sufficient quantity, and are then discharged.

Thus, then, you have discovered bodily parts of a second kind, consecrated in this case to the [removal of the] superfluities of the food. There is, however, also a third kind, for carrying the pabulum in every direction; these are like a number of roads intersecting the whole body.

Thus there is one entrance—that through the mouth—for all the various articles of food. What receives nourishment, however, is not one single part, but a great many parts, and these widely separated; do not be surprised, therefore, at the abundance of organs which Nature has created for the purpose of nutrition. For those of them which have to do with

p. 319, note 2. With Galen's οἰκεῖος and ἀλλότριος we may compare the German terms *eigen* and *fremd* used by Aberhalden in connection with his theory of defensive ferments in the blood-serum.

οὖντα προπαρασκευάζει τὴν ἐπιτήδειον ἑκάστῳ
μορίῳ τροφήν, τὰ δὲ διακρίνει τὰ περιττώματα,
τὰ δὲ παραπέμπει, τὰ δ' ὑποδέχεται, τὰ δ'
ἐκκρίνει, τὰ δ' ὁδοὶ τῆς πάντη φορᾶς εἰσι τῶν
χρηστῶν χυμῶν, ὥστ', εἴπερ βούλει τὰς δυνάμεις
τῆς φύσεως ἁπάσας ἐκμαθεῖν, ὑπὲρ ἑκάστου
τούτων ἂν εἴη σοι τῶν ὀργάνων ἐπισκεπτέον.

24   Ἀρχὴ δ' αὐτῶν τῆς διδασκαλίας, ὅσα ‖ τοῦ
τέλους ἐγγὺς ἔργα τε τῆς φύσεώς ἐστι καὶ μόρια
καὶ δυνάμεις αὐτῶν.

XI

Αὐτοῦ δὲ δὴ πάλιν ἀναμνηστέον ἡμῖν τοῦ
τέλους, οὗπερ ἕνεκα τοσαῦτά τε καὶ τοιαῦτα τῇ
φύσει δεδημιούργηται μόρια. τὸ μὲν οὖν ὄνομα
τοῦ πράγματος, ὥσπερ καὶ πρότερον εἴρηται,
θρέψις· ὁ δὲ κατὰ τοὔνομα λόγος ὁμοίωσις τοῦ
τρέφοντος τῷ τρεφομένῳ. ἵνα δ' αὕτη γένηται,
προηγήσασθαι χρὴ πρόσφυσιν, ἵνα δ' ἐκείνη,
πρόσθεσιν. ἐπειδὰν γὰρ ἐκπέσῃ τῶν ἀγγείων
ὁ μέλλων θρέψειν ὁτιοῦν τῶν τοῦ ζῴου μορίων
χυμός, εἰς ἅπαν αὐτὸ διασπείρεται πρῶτον,
ἔπειτα προστίθεται κἄπειτα προσφύεται καὶ
τελέως ὁμοιοῦται.

---

[1] Transit, cf. p. 6, note 1.
[2] i.e. of the living organism, cf. p. 2, note 1.
[3] i.e. with nutrition.
[4] We might perhaps say, more shortly, "assimilation of
food to feeder," or, "of food to fed"; Linacre renders,
"nutrimenti cum nutrito assimilatio."

alteration prepare the nutriment suitable for each part; others separate out the superfluities; some pass these along, others store them up, others excrete them; some, again, are paths for the transit [1] in all directions of the *utilisable* juices. So, if you wish to gain a thorough acquaintance with all the faculties of Nature,[2] you will have to consider each one of these organs.

Now in giving an account of these we must begin with those effects of Nature, together with their corresponding parts and faculties, which are closely connected with the purpose to be achieved.[3]

XI

LET us once more, then, recall the actual purpose for which Nature has constructed all these parts. Its name, as previously stated, is *nutrition*, and the definition corresponding to the name is: *an assimilation of that which nourishes to that which receives nourishment.*[4] And in order that this may come about, we must assume a preliminary process of *adhesion*,[5] and for that, again, one of *presentation.*[6] For whenever the juice which is destined to nourish any of the parts of the animal is emitted from the vessels, it is in the first place dispersed all through this part, next it is presented, and next it adheres, and becomes completely assimilated.

[5] Lit. *prosphysis, i.e.* attachment, implantation.

[6] Lit. *prosthesis,* "apposition." One is almost tempted to retain the terms *prosthesis* and *prosphysis* in translation, as they obviously correspond much more closely to Galen's physiological conceptions than any English or semi-English words can.

Δηλοῦσι δ' αἱ καλούμεναι λεῦκαι τὴν διαφορὰν
ὁμοιώσεώς τε καὶ προσφύσεως, ὥσπερ τὸ γένος
ἐκεῖνο τῶν ὑδέρων, ὅ τινες ὀνομάζουσιν ἀνὰ
σάρκα, διορίζει σαφῶς πρόσθεσιν προσφύσεως.
οὐ γὰρ ἐνδείᾳ δήπου τῆς ἐπιρρεούσης ὑγρότητος,
ὡς ἔνιαι τῶν ἀτροφιῶν τε καὶ φθίσεων, ἡ τοῦ
25 τοιούτου γένεσις ὑδέρου ‖ συντελεῖται. φαίνεται
γὰρ ἱκανῶς ἥ τε σὰρξ ὑγρὰ καὶ διάβροχος
ἕκαστόν τε τῶν στερεῶν τοῦ σώματος μορίων
ὡσαύτως διακείμενον. ἀλλὰ πρόσθεσις μέν τις
γίγνεται τῆς ἐπιφερομένης τροφῆς, ἅτε δ' ὑδατω-
δεστέρας οὔσης ἔτι καὶ μὴ πάνυ τι κεχυμωμένης
μηδὲ τὸ γλίσχρον ἐκεῖνο καὶ κολλῶδες, ὃ δὴ
τῆς ἐμφύτου θερμασίας οἰκονομίᾳ προσγίγνεται,
κεκτημένης ἡ πρόσφυσις ἀδύνατός ἐστιν ἐπιτελεῖσθαι πλήθει λεπτῆς ὑγρότητος ἀπέπτου
διαρρεούσης τε καὶ ῥᾳδίως ὀλισθαινούσης ἀπὸ
τῶν στερεῶν τοῦ σώματος μορίων τῆς τροφῆς.
ἐν δὲ ταῖς λεύκαις πρόσφυσις μέν τις γίγνεται
τῆς τροφῆς, οὐ μὴν ἐξομοίωσίς γε. καὶ δῆλον ἐν
τῷδε τὸ μικρῷ πρόσθεν ῥηθὲν ὡς ὀρθῶς ἐλέγετο
τὸ δεῖν πρόσθεσιν μὲν πρῶτον, ἐφεξῆς δὲ πρόσφυσιν, ἔπειτ' ἐξομοίωσιν γενέσθαι τῷ μέλλοντι
τρέφεσθαι.

Κυρίως μὲν οὖν τὸ τρέφον ἤδη τροφή, τὸ δ' οἷον
μὲν τροφή, οὔπω δὲ τρέφον, ὁποῖόν ἐστι τὸ
προσφυόμενον ἢ προστιθέμενον, τροφὴ μὲν οὐ

---

[1] Lit. *phthisis. cf.* p. 6, note 2.   Now means *tuberculosis*
only.

[2] More literally, "chymified." In *anasarca* the sub-
cutaneous tissue is soft, and pits on pressure. In the "white"
disease referred to here (by which is probably meant *nodular
leprosy*) the same tissues are indurated and "brawny." The

The so-called white [leprosy] shows the difference between assimilation and adhesion, in the same way that the kind of dropsy which some people call *anasarca* clearly distinguishes presentation from adhesion. For, of course, the genesis of such a dropsy does not come about as do some of the conditions of atrophy and wasting,[1] from an insufficient supply of moisture; the flesh is obviously moist enough,—in fact it is thoroughly saturated,—and each of the solid parts of the body is in a similar condition. While, however, the nutriment conveyed to the part does undergo presentation, it is still too watery, and is not properly transformed into a *juice*,[2] nor has it acquired that viscous and agglutinative quality which results from the operation of *innate heat*;[3] therefore, adhesion cannot come about, since, owing to this abundance of thin, crude liquid, the pabulum runs off and easily slips away from the solid parts of the body. In white [leprosy], again, there is adhesion of the nutriment but no real assimilation. From this it is clear that what I have just said is correct, namely, that in that part which is to be nourished there must first occur presentation, next adhesion, and finally assimilation proper.

Strictly speaking, then, *nutriment* is that which is actually nourishing, while the *quasi-nutriment* which is not yet nourishing (*e.g.* matter which is undergoing adhesion or presentation) is not, strictly speaking, nutriment, but is so called only by an equivocation.

principle of certain diseases being best explained as cases of *arrest* at various stages of the metabolic path is recognized in modern pathology, although of course the instances given by Galen are too crude to stand.

[3] The effects of *oxidation* attributed to the heat which accompanies it? *cf.* p. 141, note 1 ; p. 254, note 1.

κυρίως, ὁμωνύμως δὲ τροφή· τὸ δ' ἐν ταῖς φλεψὶν
26 ἔτι περιεχόμενον ‖ καὶ τούτου μᾶλλον ἔτι τὸ κατὰ
τὴν γαστέρα τῷ μέλλειν ποτὲ θρέψειν, εἰ καλῶς
κατεργασθείη, κέκληται τροφή. κατὰ ταὐτὰ δὲ
καὶ τῶν ἐδεσμάτων ἕκαστον τροφὴν ὀνομάζομεν
οὔτε τῷ τρέφειν ἤδη τὸ ζῷον οὔτε τῷ τοιοῦτον
ὑπάρχειν οἷον τὸ τρέφον, ἀλλὰ τῷ δύνασθαί τε
καὶ μέλλειν τρέφειν, εἰ καλῶς κατεργασθείη.

Τοῦτο γὰρ ἦν καὶ τὸ πρὸς Ἱπποκράτους
λεγόμενον· " Τροφὴ δὲ τὸ τρέφον, τροφὴ καὶ τὸ
οἷον τροφὴ καὶ τὸ μέλλον." τὸ μὲν γὰρ ὁμοιού-
ενον ἤδη τροφὴν ὠνόμασε, τὸ δ' οἷον μὲν ἐκεῖνο
προστιθέμενον ἢ προσφυόμενον οἷον τροφήν· τὸ
δ' ἄλλο πᾶν, ὅσον ἐν τῇ γαστρὶ καὶ ταῖς φλεψὶ
περιέχεται, μέλλον.

XII

Ὅτι μὲν οὖν ἀναγκαῖον ὁμοίωσίν τιν' εἶναι τοῦ
τρέφοντος τῷ τρεφομένῳ τὴν θρέψιν, ἄντικρυς
δῆλον. οὐ μὴν ὑπάρχουσάν γε ταύτην τὴν ὁμοί-
ωσιν, ἀλλὰ φαινομένην μόνον εἶναί φασιν οἱ μήτε
τεχνικὴν οἰόμενοι τὴν φύσιν εἶναι μήτε προνοητικὴν τοῦ ζῴου μήθ' ὅλως τινὰς οἰκείας ἔχειν
δυνάμεις, αἷς χρωμένη τὰ μὲν ἀλλοιοῖ, τὰ δ'
27 ἕλκει, ‖ τὰ δ' ἐκκρίνει.

Καὶ αὗται δύο γεγόνασιν αἱρέσεις κατὰ γένος
ἐν ἰατρικῇ τε καὶ φιλοσοφίᾳ τῶν ἀποφηναμένων

Also, that which is still contained in the veins, and still more, that which is in the stomach, from the fact that it is destined to nourish if properly elaborated, has been called "nutriment." Similarly we call the various kinds of food "nutriment," not because they are already nourishing the animal, nor because they exist in the same state as the material which actually is nourishing it, but because they are able and destined to nourish it if they be properly elaborated.

This was also what Hippocrates said, viz., "Nutriment is what is engaged in nourishing, as also is quasi-nutriment, and what is destined to be nutriment." For to that which is already being assimilated he gave the name of *nutriment*; to the similar material which is being presented or becoming adherent, the name of *quasi-nutriment*; and to everything else—that is, contained in the stomach and veins—the name of *destined nutriment*.

XII

It is quite clear, therefore, that nutrition must necessarily be a process of assimilation of that which is nourishing to that which is being nourished. Some, however, say that this assimilation does not occur in reality, but is merely apparent; these are the people who think that Nature is not artistic, that she does not show forethought for the animal's welfare, and that she has absolutely no native powers whereby she alters some substances, attracts others, and discharges others.

Now, speaking generally, there have arisen the following two sects in medicine and philosophy

τι περὶ φύσεως ἀνδρῶν, ὅσοι γ' αὐτῶν γιγνώσκουσιν, ὅ τι λέγουσι, καὶ τὴν ἀκολουθίαν ὧν
ὑπέθεντο θεωροῦσι θ' ἅμα καὶ·διαφυλάττουσιν.
ὅσοι δὲ μηδ' αὐτὸ τοῦτο συνιᾶσιν, ἀλλ' ἁπλῶς,
ὅ τι ἂν ἐπὶ γλῶτταν ἔλθῃ, ληροῦσιν, ἐν οὐδετέρᾳ
τῶν αἱρέσεων ἀκριβῶς καταμένοντες, οὐδὲ μεμ-
νῆσθαι τῶν τοιούτων προσήκει.

Τίνες οὖν αἱ δύο αἱρέσεις αὗται καὶ τίς ἡ τῶν
ἐν αὐταῖς ὑποθέσεων ἀκολουθία; τὴν ὑποβεβλη-
μένην οὐσίαν γενέσει καὶ φθορᾷ πᾶσαν ἡνωμένην
θ' ἅμα καὶ ἀλλοιοῦσθαι δυναμένην ὑπέθετο
θάτερον γένος τῆς αἱρέσεως, ἀμετάβλητον δὲ καὶ
ἀναλλοίωτον καὶ κατατετμημένην εἰς λεπτὰ καὶ
κεναῖς ταῖς μεταξὺ χώραις διειλημμένην ἡ λοιπή.

Καὶ τοίνυν ὅσοι γε τῆς ἀκολουθίας τῶν ὑποθέσεων αἰσθάνονται, κατὰ μὲν τὴν δευτέραν
αἵρεσιν οὔτε φύσεως οὔτε ψυχῆς ἰδίαν τινὰ νομί-
ζουσιν οὐσίαν ἢ δύναμιν ὑπάρχειν, || ἀλλ' ἐν τῇ
ποιᾷ συνόδῳ τῶν πρώτων ἐκείνων σωμάτων τῶν
ἀπαθῶν ἀποτελεῖσθαι. κατὰ δὲ τὴν προτέραν
εἰρημένην αἵρεσιν οὐχ ὑστέρα τῶν σωμάτων ἡ
φύσις, ἀλλὰ πολὺ προτέρα τε καὶ πρεσβυτέρα.
καὶ τοίνυν κατὰ μὲν τούτους αὕτη τὰ σώματα
τῶν τε φυτῶν καὶ τῶν ζῴων συνίστησι δυνάμεις
τινὰς ἔχουσα τὰς μὲν ἑλκτικάς θ' ἅμα καὶ
ὁμοιωτικὰς τῶν οἰκείων, τὰς δ' ἀποκριτικὰς τῶν

---

[1] Here follows a contrast between the Vitalists and the
Epicurean Atomists. *cf.* p. 153 *et seq.*
[2] A unity or *continuum*, an *individuum*.

among those who have made any definite pronouncement regarding Nature. I speak, of course, of such
of them as know what they are talking about, and
who realize the logical sequence of their hypotheses,
and stand by them ; as for those who cannot understand even this, but who simply talk any nonsense
that comes to their tongues, and who do not remain
definitely attached either to one sect or the other—
such people are not even worth mentioning.

What, then, are these sects, and what are the
logical consequences of their hypotheses?[1] The one
class supposes that all substance which is subject to
genesis and destruction is at once *continuous*[2] and
susceptible of *alteration*. The other school assumes
substance to be unchangeable, unalterable, and subdivided into fine particles, which are separated from
one another by empty spaces.

All people, therefore, who can appreciate the logical
sequence of an hypothesis hold that, according to
the second teaching, there does not exist any substance or faculty peculiar either to Nature or to
Soul,[3] but that these result from the way in which
the primary corpuscles,[4] which are unaffected by
change, come together. According to the first-
mentioned teaching, on the other hand, Nature is not
posterior to the corpuscles, but is a long way prior to
them and older than they ; and therefore in their
view it is Nature which puts together the bodies
both of plants and animals ; and this she does by
virtue of certain faculties which she possesses—these
being, on the one hand, attractive and assimilative of
what is appropriate, and, on the other, expulsive of

---

[3] Lit. to the *physis* or the *psyche* ; that is, a denial of the
autonomy of physiology and psychology.   [4] Lit. *somata*.

ἀλλοτρίων, καὶ τεχνικῶς ἅπαντα διαπλάττει τε γεννῶσα καὶ προνοεῖται τῶν γεννωμένων ἑτέραις αὖθίς τισι δυνάμεσι, στερκτικῇ μέν τινι καὶ προνοητικῇ τῶν ἐγγόνων, κοινωνικῇ δὲ καὶ φιλικῇ τῶν ὁμογενῶν. κατὰ δ' αὖ τοὺς ἑτέρους οὔτε τούτων οὐδὲν ὑπάρχει ταῖς φύσεσιν οὔτ' ἔννοιά τίς ἐστι τῇ ψυχῇ σύμφυτος ἐξ ἀρχῆς οὐκ ἀκολουθίας οὐ μάχης, οὐ διαιρέσεως οὐ συνθέσεως, οὐ δικαίων οὐκ ἀδίκων, οὐ καλῶν οὐκ αἰσχρῶν, ἀλλ' ἐξ αἰσθήσεώς τε καὶ δι' αἰσθήσεως ἅπαντα τὰ τοιαῦθ' ἡμῖν ἐγγίγνεσθαί φασι καὶ φαντασίαις τισὶ καὶ μνήμαις οἰακίζεσθαι τὰ ζῷα.

29   Ἔνιοι ‖ δ' αὐτῶν καὶ ῥητῶς ἀπεφήναντο μηδεμίαν εἶναι τῆς ψυχῆς δύναμιν, ᾗ λογιζόμεθα, ἀλλ' ὑπὸ τῶν αἰσθητῶν ἄγεσθαι παθῶν ἡμᾶς καθάπερ βοσκήματα πρὸς μηδὲν ἀνανεῦσαι μηδ' ἀντειπεῖν δυναμένους. καθ' οὓς δηλονότι καὶ ἀνδρεία καὶ φρόνησις καὶ σωφροσύνη καὶ ἐγκράτεια λῆρός ἐστι μακρὸς καὶ φιλοῦμεν οὔτ' ἀλλήλους οὔτε τὰ ἔγγονα καὶ τοῖς θεοῖς οὐδὲν ἡμῶν μέλει. καταφρονοῦσι δὲ καὶ τῶν ὀνειράτων καὶ τῶν οἰωνῶν καὶ τῶν συμβόλων καὶ πάσης ἀστρολογίας, ὑπὲρ ὧν ἡμεῖς μὲν ἰδίᾳ δι' ἑτέρων γραμμάτων ἐπὶ πλέον ἐσκεψάμεθα περὶ τῶν Ἀσκληπιάδου τοῦ ἰατροῦ σκοπούμενοι δογμάτων. ἔνεστι δὲ τοῖς βουλομένοις κἀκείνοις μὲν ὁμιλῆσαι τοῖς λόγοις καὶ νῦν δ' ἤδη σκοπεῖν, ὥσπερ τινῶν δυοῖν ὁδῶν ἡμῖν προκειμένων, ὁποτέραν βέλτιόν ἐστι τρέπεσθαι. Ἱπποκράτης μὲν γὰρ τὴν προτέραν ῥηθεῖσαν ἐτράπετο, καθ' ἣν ἥνωται μὲν ἡ οὐσία καὶ ἀλλοιοῦται καὶ σύμπνουν ὅλον ἐστὶ καὶ σύρρουν τὸ

what is foreign. Further, she skilfully moulds everything during the stage of genesis; and she also provides for the creatures after birth, employing here other faculties again, namely, one of affection and forethought for offspring, and one of sociability and friendship for kindred. According to the other school, none of these things exist in the natures [1] [of living things], nor is there in the soul any original innate idea, whether of agreement or difference, ot separation or synthesis, of justice or injustice, of the beautiful or ugly; all such things, they say, arise in us *from sensation and through sensation*, and animals are steered by certain images and memories.

Some of these people have even expressly declared that the soul possesses no reasoning faculty, but that we are led like cattle by the impression of our senses, and are unable to refuse or dissent from anything. In their view, obviously, courage, wisdom, temperance, and self-control are all mere nonsense, we do not love either each other or our offspring, nor do the gods care anything for us. This school also despises dreams, birds, omens, and the whole of astrology, subjects with which we have dealt at greater length in another work,[2] in which we discuss the views of Asclepiades the physician.[3] Those, who wish to do so may familiarize themselves with these arguments, and they may also consider at this point which of the two roads lying before us is the better one to take. Hippocrates took the first-mentioned. According to this teaching, substance is one and is subject to *alteration*; there is a consensus in the move-

---

[1] For "natures" in the plural, involving the idea of a separate nature immanent in each individual, *cf.* p. 36, note 1.
[2] A lost work.
[3] For Asclepiades *v.* p. 49, note 5.

σῶμα καὶ ἡ φύσις ἅπαντα τεχνικῶς καὶ δικαίως
πράττει δυνάμεις ἔχουσα, καθ᾽ ἃς ἕκαστον τῶν
30 μορίων ἕλκει μὲν ‖ ἐφ᾽ ἑαυτὸ τὸν οἰκεῖον ἑαυτῷ
χυμόν, ἕλξαν δὲ προσφύει τε παντὶ μέρει τῶν ἐν
αὐτῷ καὶ τελέως ἐξομοιοῖ, τὸ δὲ μὴ κρατηθὲν ἐν
τούτῳ μηδὲ τὴν παντελῆ δυνηθὲν ἀλλοίωσίν τε
καὶ ὁμοιότητα τοῦ τρεφομένου καταδέξασθαι δι᾽
ἑτέρας αὖ τινος ἐκκριτικῆς δυνάμεως ἀποτρίβεται.

XIII

Μαθεῖν δ᾽ ἔνεστιν οὐ μόνον ἐξ ὧν οἱ τἀναντία
τιθέμενοι διαφέρονται τοῖς ἐναργῶς φαινομένοις,
εἰς ὅσον ὀρθότητός τε καὶ ἀληθείας ἥκει τὰ Ἱπποκράτους δόγματα, ἀλλὰ κἀξ αὐτῶν τῶν κατὰ
μέρος ἐν τῇ φυσικῇ θεωρίᾳ ζητουμένων τῶν τ᾽
ἄλλων ἁπάντων καὶ τῶν ἐν τοῖς ζῴοις ἐνεργειῶν.
ὅσοι γὰρ οὐδεμίαν οὐδενὶ μορίῳ νομίζουσιν ὑπάρχειν ἑλκτικὴν τῆς οἰκείας ποιότητος δύναμιν,
ἀναγκάζονται πολλάκις ἐναντία λέγειν τοῖς ἐναργῶς φαινομένοις, ὥσπερ καὶ Ἀσκληπιάδης ὁ
ἰατρὸς ἐπὶ τῶν νεφρῶν ἐποίησεν, οὓς οὐ μόνον
Ἱπποκράτης ἢ Διοκλῆς ἢ Ἐρασίστρατος ἢ

---

[1] "Le corps tout entier a unité de souffle (*perspiration et
expiration*) et unité de flux (*courants, circulation des liquides*)"
(Daremberg). "Conspirabile et confluxile corpus esse" (Linacre). Apparently Galen refers to the pneuma and the various
humours. *cf.* p. 293, note 2.

[2] *i.e.* "appropriated"; very nearly "assimilated."

ments of air and fluid throughout the whole body;[1] Nature acts throughout in an artistic and equitable manner, having certain faculties, by virtue of which each part of the body draws to itself the juice which is proper to it, and, having done so, attaches it to every portion of itself, and completely assimilates it; while such part of the juice as has not been mastered,[2] and is not capable of undergoing complete alteration and being assimilated to the part which is being nourished, is got rid of by yet another (an expulsive) faculty.

XIII

Now the extent of exactitude and truth in the doctrines of Hippocrates may be gauged, not merely from the way in which his opponents are at variance with obvious facts, but also from the various subjects of natural research themselves—the functions of animals, and the rest. For those people who do not believe that there exists in any part of the animal a faculty for attracting *its own special quality*[3] are compelled repeatedly to deny obvious facts.[4] For instance, Asclepiades, the physician,[5] did this in the case of the kidneys. That these are organs for secreting [separating out] the urine, was the belief not only of Hippocrates, Diocles,

---

[3] "Attractricem convenientis qualitatis vim" (Linacre). *cf.* p. 36, note 2.   [4] Lit. "obvious phenomena."
[5] Asclepiades of Bithynia, who flourished in the first half of the first century B.C., was an adherent of the atomistic philosophy of Democritus, and is the typical representative of the Mechanistic school in Graeco-Roman medicine; he disbelieved in any principle of individuality ("nature") in the organism, and his methods of treatment, in accordance with his pathology, were mechano-therapeutical. *cf.* p. 64, note 3.

Πραξαγόρας ἤ τις ἄλλος ἰατρὸς ἄριστος ὄργανα διακριτικὰ τῶν οὔρων πεπιστεύκασιν ὑπάρχειν, 31 ἀλλὰ καὶ οἱ ‖ μάγειροι σχεδὸν ἅπαντες ἴσασιν, ὁσημέραι θεώμενοι τήν τε θέσιν αὐτῶν καὶ τὸν ἀφ' ἑκατέρου πόρον εἰς τὴν κύστιν ἐμβάλλοντα, τὸν οὐρητῆρα καλούμενον, ἐξ αὐτῆς τῆς κατασκευῆς ἀναλογιζόμενοι τήν τε χρείαν αὐτῶν καὶ τὴν δύναμιν. καὶ πρό γε τῶν μαγείρων ἅπαντες ἄνθρωποι καὶ δυσουροῦντες πολλάκις καὶ παντάπασιν ἰσχουροῦντες, ὅταν ἀλγῶσι μὲν τὰ κατὰ τὰς ψόας, ψαμμώδη δ' ἐξουρῶσιν, νεφριτικοὺς ὀνομάζουσι σφᾶς αὐτούς.

Ἀσκληπιάδην δ' οἶμαι μηδὲ λίθον οὐρηθέντα ποτὲ θεάσασθαι πρὸς τῶν οὕτω πασχόντων μηδ' ὡς προηγήσατο κατὰ τὴν μεταξὺ τῶν νεφρῶν καὶ τῆς κύστεως χώραν ὀδύνη τις ὀξεῖα διερχομένου τοῦ λίθου τὸν οὐρητῆρα μηδ' ὡς οὐρηθέντος αὐτοῦ τά τε τῆς ὀδύνης καὶ τὰ τῆς ἰσχουρίας ἐπαύσατο παραχρῆμα. πῶς οὖν εἰς τὴν κύστιν τῷ λόγῳ παράγει τὸ οὖρον, ἄξιον ἀκοῦσαι καὶ θαυμάσαι τἀνδρὸς τὴν σοφίαν, ὃς καταλιπὼν οὕτως εὐρείας ὁδοὺς ἐναργῶς φαινομένας ἀφανεῖς 32 καὶ στενὰς καὶ παντάπασιν ἀναισθήτους ‖ ὑπέθετο. βούλεται γὰρ εἰς ἀτμοὺς ἀναλυόμενον τὸ πινόμενον ὑγρὸν εἰς τὴν κύστιν διαδίδοσθαι κἄπειτ' ἐξ ἐκείνων αὖθις ἀλλήλοις συνιόντων οὕτως ἀπολαμβάνειν αὐτὸ τὴν ἀρχαίαν ἰδέαν καὶ γίγνεσθαι πάλιν ὑγρὸν ἐξ ἀτμῶν ἀτεχνῶς ὡς περὶ σπογγιᾶς τινος ἢ ἐρίου τῆς κύστεως διανοούμενος, ἀλλ' οὐ σώματος ἀκριβῶς πυκνοῦ καὶ στεγανοῦ δύο χιτῶνας ἰσχυροτάτους κεκτημένου,

Erasistratus, Praxagoras,[1] and all other physicians of eminence, but practically every butcher is aware of this, from the fact that he daily observes both the position of the kidneys and the duct (termed the ureter) which runs from each kidney into the bladder, and from this arrangement he infers their characteristic use and faculty. But, even leaving the butchers aside, all people who suffer either from frequent dysuria or from retention of urine call themselves "nephritics,"[2] when they feel pain in the loins and pass sandy matter in their water.

I do not suppose that Asclepiades ever saw a stone which had been passed by one of these sufferers, or observed that this was preceded by a sharp pain in the region between kidneys and bladder as the stone traversed the ureter, or that, when the stone was passed, both the pain and the retention at once ceased. It is worth while, then, learning how his theory accounts for the presence of urine in the bladder, and one is forced to marvel at the ingenuity of a man who puts aside these broad, clearly visible routes,[3] and postulates others which are narrow, invisible—indeed, entirely imperceptible. His view, in fact, is that the fluid which we drink passes into the bladder by being resolved into vapours, and that, when these have been again condensed, it thus regains its previous form, and turns from vapour into fluid. He simply looks upon the bladder as a sponge or a piece of wool, and not as the perfectly compact and impervious body that it is, with two very

[1] Diocles of Carystus was the chief representative of the Dogmatic or Hippocratic school in the first half of the fourth century B.C. Praxagoras was his disciple, and followed him in the leadership of the school. For Erasistratus, cf. p. 95 et seq.

[2] Sufferers from kidney-trouble.   [3] The ureters.

δι' ὧν εἴπερ διέρχεσθαι φήσομεν τοὺς ἀτμούς, τί
δήποτ' οὐχὶ διὰ τοῦ περιτοναίου καὶ τῶν φρενῶν
διελθόντες ἐνέπλησαν ὕδατος τό τ' ἐπιγάστριον
ἅπαν καὶ τὸν θώρακα; ἀλλὰ παχύτερος, φησίν,
ἐστὶ δηλαδὴ καὶ στεγανώτερος ὁ περιτόναιος
χιτὼν τῆς κύστεως καὶ διὰ τοῦτ' ἐκεῖνος μὲν
ἀποστέγει τοὺς ἀτμούς, ἡ δὲ κύστις παραδέχεται.
ἀλλ' εἴπερ ἀνατετμήκει ποτέ, τάχ' ἂν ἠπίστατο
τὸν μὲν ἔξωθεν χιτῶνα τῆς κύστεως ἀπὸ τοῦ
περιτοναίου πεφυκότα τὴν αὐτὴν ἐκείνῳ φύσιν
ἔχειν, τὸν δ' ἔνδοθεν τὸν αὐτῆς τῆς κύστεως ἴδιον
πλέον ἢ διπλάσιον ἐκείνου τὸ πάχος ὑπάρχειν.

33   Ἀλλ' ἴσως οὔτε τὸ ‖ πάχος οὔθ' ἡ λεπτότης
τῶν χιτώνων, ἀλλ' ἡ θέσις τῆς κύστεως αἰτία τοῦ
φέρεσθαι τοὺς ἀτμοὺς εἰς αὐτήν. καὶ μὴν εἰ καὶ
διὰ τἆλλα πάντα πιθανὸν ἦν αὐτοὺς ἐνταυθοῖ
συναθροίζεσθαι, τό γε τῆς θέσεως μόνης αὔταρκες
κωλῦσαι. κάτω μὲν γὰρ ἡ κύστις κεῖται, τοῖς δ'
ἀτμοῖς σύμφυτος ἡ πρὸς τὸ μετέωρον φορά, ὥστε
πολὺ πρότερον ἂν ἔπλησαν ἅπαντα τὰ κατὰ τὸν
θώρακά τε καὶ τὸν πνεύμονα, πρὶν ἐπὶ τὴν κύστιν
ἀφικέσθαι.

Καίτοι τί θέσεως κύστεως καὶ περιτοναίου καὶ
θώρακος μνημονεύω; διεκπεσόντες γὰρ δήπου
τούς τε τῆς κοιλίας καὶ τῶν ἐντέρων χιτῶνας οἱ
ἀτμοὶ κατὰ τὴν μεταξὺ χώραν αὐτῶν τε τούτων
καὶ τοῦ περιτοναίου συναθροισθήσονται καὶ ὑγρὸν
ἐνταυθοῖ γενήσονται, ὥσπερ καὶ τοῖς ὑδερικοῖς ἐν
τούτῳ τῷ χωρίῳ τὸ πλεῖστον ἀθροίζεται τοῦ

---

[1] Unless otherwise stated, "peritoneum" stands for parietal peritoneum alone.

strong coats. For if we say that the vapours pass
through these coats, why should they not pass through
the peritoneum [1] and the diaphragm, thus filling
the whole abdominal cavity and thorax with water?
" But," says he, "of course the peritoneal coat is
more impervious than the bladder, and this is why
it keeps out the vapours, while the bladder admits
them." Yet if he had ever practised anatomy, he
might have known that the outer coat of the bladder
springs from the peritoneum and is essentially the
same as it, and that the inner coat, which is peculiar
to the bladder, is more than twice as thick as the
former.

Perhaps, however, it is not the thickness or thinness of the coats, but the *situation* of the bladder,
which is the reason for the vapours being carried
into it? On the contrary, even if it were probable
for every other reason that the vapours accumulate
there, yet the situation of the bladder would be
enough in itself to prevent this. For the bladder is
situated below, whereas vapours have a natural
tendency to rise upwards; thus they would fill all
the region of the thorax and lungs long before they
came to the bladder.

But why do I mention the situation of the bladder,
peritoneum, and thorax? For surely, when the vapours
have passed through the coats of the stomach and
intestines, it is in the space between these and
the peritoneum [2] that they will collect and become
liquefied (just as in dropsical subjects it is in this
region that most of the water gathers).[3] Otherwise
the vapours must necessarily pass straight forward

2 In the peritoneal cavity.
3 Contrast, however, *anasarca*, p. 41.

ὕδατος, ἢ πάντως αὐτοὺς χρὴ φέρεσθαι πρόσω
διὰ πάντων τῶν ὁπωσοῦν ὁμιλούντων καὶ μηδέ-
ποθ' ἵστασθαι. ἀλλ' εἰ καὶ τοῦτό τις ὑπόθοιτο,
διεκπεσόντες ἂν οὕτως οὐ τὸ περιτόναιον μόνον
ἀλλὰ καὶ τὸ ἐπιγάστριον, εἰς τὸ περιέχον σκε-
34 δασθεῖεν ἢ πάντως ἂν ὑπὸ τῷ δέρματι ‖ συν-
αθροισθεῖεν.

Ἀλλὰ καὶ πρὸς ταῦτ' ἀντιλέγειν οἱ νῦν
Ἀσκληπιάδειοι πειρῶνται, καίτοι πρὸς ἁπάντων
ἀεὶ τῶν παρατυγχανόντων αὐτοῖς, ὅταν περὶ
τούτων ἐρίζωσι, καταγελώμενοι. οὕτως ἄρα
δυσαπότριπτόν τι κακόν ἐστιν ἡ περὶ τὰς αἱρέσεις
φιλοτιμία καὶ δυσέκνιπτον ἐν τοῖς μάλιστα καὶ
ψώρας ἁπάσης δυσιατότερον.

Τῶν γοῦν καθ' ἡμᾶς τις σοφιστῶν τά τ' ἄλλα
καὶ περὶ τοὺς ἐριστικοὺς λόγους ἱκανῶς συγκε-
κροτημένος καὶ δεινὸς εἰπεῖν, εἴπερ τις ἄλλος,
ἀφικόμενος ἐμοί ποθ' ὑπὲρ τούτων εἰς λόγους,
τοσοῦτον ἀπέδει τοῦ δυσωπεῖσθαι πρός τινος
τῶν εἰρημένων, ὥστε καὶ θαυμάζειν ἔφασκεν
ἐμοῦ τὰ σαφῶς φαινόμενα λόγοις ληρώδεσιν
ἀνατρέπειν ἐπιχειροῦντος. ἐναργῶς γὰρ ὁσημέραι
θεωρεῖσθαι τὰς κύστεις ἁπάσας, εἴ τις αὐτὰς
ἐμπλήσειεν ὕδατος ἢ ἀέρος, εἶτα δήσας τὸν
τράχηλον πιέζοι πανταχόθεν, οὐδαμόθεν μεθιεί-
σας οὐδέν, ἀλλ' ἀκριβῶς ἅπαν ἐντὸς ἑαυτῶν
στεγούσας. καίτοι γ' εἴπερ ἦσάν τινες ἐκ τῶν
νεφρῶν εἰς αὐτὰς ἥκοντες αἰσθητοὶ καὶ μεγάλοι
πόροι, πάντως ἄν, ἔφη, δι' ἐκείνων, ὥσπερ εἰσῄει
35 τὸ ‖ ὑγρὸν εἰς αὐτάς, οὕτω καὶ θλιβόντων
ἐξεκρίνετο. ταῦτα καὶ τὰ τοιαῦτ' εἰπὼν ἐξαίφνης

54

through everything which in any way comes in contact with them, and will never come to a standstill. But, if this be assumed, then they will traverse not merely the peritoneum but also the epigastrium, and will become dispersed into the surrounding air; otherwise they will certainly collect under the skin.

Even these considerations, however, our present-day Asclepiadeans attempt to answer, despite the fact that they always get soundly laughed at by all who happen to be present at their disputations on these subjects—so difficult an evil to get rid of is this sectarian partizanship, so excessively resistant to all cleansing processes, harder to heal than any itch!

Thus, one of our Sophists who is a thoroughly hardened disputer and as skilful a master of language as there ever was, once got into a discussion with me on this subject; so far from being put out of countenance by any of the above-mentioned considerations, he even expressed his surprise that I should try to overturn obvious facts by ridiculous arguments! "For," said he, "one may clearly observe any day in the case of any bladder, that, if one fills it with water or air and then ties up its neck and squeezes it all round, it does not let anything out at any point, but accurately retains all its contents. And surely," said he, "if there were any large and perceptible channels coming into it from the kidneys the liquid would run out through these when the bladder was squeezed, in the same way that it entered?" [1]   Having abruptly made these and

---

[1] Regurgitation, however, is prevented by the fact that the ureter runs for nearly one inch obliquely through the bladder wall before opening into its cavity, and thus an efficient *valve* is produced.

GALEN

ἀπταίστῳ καὶ σαφεῖ τῷ στόματι τελευτῶν
ἀναπηδήσας ἀπήει καταλιπὼν ἡμᾶς ὡς οὐδὲ
πιθανῆς τινος ἀντιλογίας εὐπορῆσαι δυναμένους.

Οὕτως οὐ μόνον ὑγιὲς οὐδὲν ἴσασιν οἱ ταῖς
αἱρέσεσι δουλεύοντες, ἀλλ' οὐδὲ μαθεῖν ὑπο-
μένουσι. δέον γὰρ ἀκοῦσαι τὴν αἰτίαν, δι' ἣν
εἰσιέναι μὲν δύναται διὰ τῶν οὐρητήρων εἰς τὴν
κύστιν τὸ ὑγρόν, ἐξιέναι δ' αὖθις ὀπίσω τὴν
αὐτὴν ὁδὸν οὐκέθ' οἷόν τε, καὶ θαυμάσαι τὴν
τέχνην¹ τῆς φύσεως, οὔτε μαθεῖν ἐθέλουσι καὶ
λοιδοροῦνται προσέτι μάτην ὑπ' αὐτῆς ἄλλα τε
πολλὰ καὶ τοὺς νεφροὺς γεγονέναι φάσκοντες.²
εἰσὶ δ' οἳ καὶ δειχθῆναι παρόντων αὐτῶν τοὺς
ἀπὸ τῶν νεφρῶν εἰς τὴν κύστιν ἐμφυομένους
οὐρητῆρας ὑπομείναντες ἐτόλμησαν εἰπεῖν οἱ μέν,
ὅτι μάτην καὶ οὗτοι γεγόνασιν, οἱ δ', ὅτι σπερ-
ματικοί τινές εἰσι πόροι καὶ διὰ τοῦτο κατὰ τὸν
τράχηλον αὐτῆς, οὐκ εἰς τὸ κύτος ἐμφύονται.
δείξαντες οὖν ἡμεῖς αὐτοῖς τοὺς ὡς ἀληθῶς
σπερματικοὺς πόρους κατωτέρω τῶν οὐρητήρων ‖
36 ἐμβάλλοντας εἰς τὸν τράχηλον, νῦν γοῦν, εἰ καὶ
μὴ πρότερον, ᾠήθημεν ἀπάξειν τε τῶν ψευδῶς
ὑπειλημμένων ἐπί τε τἀναντία μεταστήσειν
αὐτίκα. οἱ δὲ καὶ πρὸς τοῦτ' ἀντιλέγειν ἐτόλμων
οὐδὲν εἶναι θαυμαστὸν εἰπόντες, ἐν ἐκείνοις μὲν
ὡς ἂν στεγανωτέροις οὖσιν ἐπὶ πλέον ὑπομένειν
τὸ σπέρμα, κατὰ δὲ τοὺς ἀπὸ τῶν νεφρῶν ὡς ἂν
ἱκανῶς ἀνευρυσμένους ἐκρεῖν διὰ ταχέων. ἡμεῖς

---

¹ On the τέχνη (artistic or creative skill) shown by the
living organism (φύσις) v. pp. 25, 45, 47; Introduction, p. xxix.
² Direct denial of Aristotle's dictum that "Nature does
nothing in vain." We are reminded of the view of certain

similar remarks in precise and clear tones, he con-
cluded by jumping up and departing—leaving me
as though I were quite incapable of finding any
plausible answer !

The fact is that those who are enslaved to their
sects are not merely devoid of all sound knowledge, but they will not even stop to learn ! Instead of listening, as they ought, to the reason
why liquid can enter the bladder through the
ureters, but is unable to go back again the same way,
—instead of admiring Nature's artistic skill [1]—they
refuse to learn ; they even go so far as to scoff, and
maintain that the kidneys, as well as many other
things, have been made by Nature *for no purpose !* [2]
And some of them who had allowed themselves to
be shown the ureters coming from the kidneys and
becoming implanted in the bladder, even had the
audacity to say that these also existed for no purpose ;
and others said that they were spermatic ducts, and
that this was why they were inserted into the neck
of the bladder and not into its cavity. When, therefore, we had demonstrated to them the real spermatic ducts [3] entering the neck of the bladder lower
down than the ureters, we supposed that, if we had
not done so before, we would now at least draw
them away from their false assumptions, and convert
them forthwith to the opposite view. But even this
they presumed to dispute, and said that it was not to
be wondered at that the semen should remain longer
in these latter ducts, these being more constricted,
and that it should flow quickly down the ducts
which came from the kidneys, seeing that these were

modern laboratory physicians and surgeons that the *colon* is
a "useless" organ. *cf.* Erasistratus, p. 143.

[3] The *vasa deferentia*.

οὖν ἠναγκάσθημεν αὐτοῖς τοῦ λοιποῦ δεικνύειν εἰσρέον τῇ κύστει διὰ τῶν οὐρητήρων τὸ οὖρον ἐναργῶς ἐπὶ ζῶντος ἔτι τοῦ ζῴου, μόγις ἂν οὕτω ποτὲ τὴν φλυαρίαν αὐτῶν ἐπισχήσειν ἐλπίζοντες.

Ὁ δὲ τρόπος τῆς δείξεώς ἐστι τοιόσδε. διελεῖν χρὴ τὸ πρὸ τῶν οὐρητήρων περιτόναιον, εἶτα βρόχοις αὐτοὺς ἐκλαβεῖν κἄπειτ' ἐπιδήσαντας ἐᾶσαι τὸ ζῷον· οὐ γὰρ ἂν οὐρήσειεν ἔτι. μετὰ δὲ ταῦτα λύειν μὲν τοὺς ἔξωθεν δεσμούς, δεικνύναι δὲ κενὴν μὲν τὴν κύστιν, μεστοὺς δ' ἱκανῶς καὶ διατεταμένους τοὺς οὐρητῆρας καὶ κινδυνεύοντας ῥαγῆναι κἄπειτα τοὺς βρόχους αὐτῶν ἀφελόντας ἐναργῶς ὁρᾶν ἤδη πληρουμένην οὔρου τὴν κύστιν.

57 Ἐπὶ δὲ τούτῳ ‖ φανέντι, πρὶν οὐρῆσαι τὸ ζῷον, βρόχον αὐτοῦ περιβαλεῖν χρὴ τῷ αἰδοίῳ κἄπειτα θλίβειν πανταχόθεν τὴν κύστιν. οὐδὲ γὰρ ἂν οὐδὲν ἔτι διὰ τῶν οὐρητήρων ἐπανέλθοι πρὸς τοὺς νεφρούς. κἂν τούτῳ δῆλον γίγνεται τὸ μὴ μόνον ἐπὶ τεθνεῶτος ἀλλὰ καὶ περιόντος ἔτι τοῦ ζῴου κωλύεσθαι μεταλαμβάνειν αὖθις ἐκ τῆς κύστεως τοὺς οὐρητῆρας τὸ οὖρον. ἐπὶ τούτοις ὀφθεῖσιν ἐπιτρέπειν ἤδη τὸ ζῷον οὐρεῖν λύοντας αὐτοῦ τὸν ἐπὶ τῷ αἰδοίῳ βρόχον, εἶτ' αὖθις ἐπιβαλεῖν μὲν θατέρῳ τῶν οὐρητήρων, ἐᾶσαι δὲ τὸν ἕτερον εἰς τὴν κύστιν συρρεῖν καί τινα διαλιπόντας χρόνον ἐπιδεικνύειν ἤδη, πῶς ὁ μὲν ἕτερος αὐτῶν ὁ δεδεμένος μεστὸς καὶ διατεταμένος κατὰ τὰ πρὸς τῶν νεφρῶν μέρη φαίνεται, ὁ δ' ἕτερος ὁ λελυμένος αὐτὸς μὲν χαλαρός ἐστι, πεπλήρωκε δ' οὔρου τὴν κύστιν. εἶτ' αὖθις διατεμεῖν πρῶτον μὲν τὸν πλήρη καὶ δεῖξαι, πῶς ἐξακοντίζεται τὸ

58

well dilated. We were, therefore, further compelled to show them in a still living animal, the urine plainly running out through the ureters into the bladder; even thus we hardly hoped to check their nonsensical talk.

Now the method of demonstration is as follows. One has to divide the peritoneum in front of the ureters, then secure these with ligatures, and next, having bandaged up the animal, let him go (for he will not continue to urinate). After this one loosens the external bandages and shows the bladder empty and the ureters quite full and distended—in fact almost on the point of rupturing; on removing the ligature from them, one then plainly sees the bladder becoming filled with urine.

When this has been made quite clear, then, before the animal urinates, one has to tie a ligature round his penis and then to squeeze the bladder all over; still nothing goes back through the ureters to the kidneys. Here, then, it becomes obvious that not only in a dead animal, but in one which is still living, the ureters are prevented from receiving back the urine from the bladder. These observations having been made, one now loosens the ligature from the animal's penis and allows him to urinate, then again ligatures one of the ureters and leaves the other to discharge into the bladder. Allowing, then, some time to elapse, one now demonstrates that the ureter which was ligatured is obviously full and distended on the side next to the kidneys, while the other one—that from which the ligature had been taken—is itself flaccid, but has filled the bladder with urine. Then, again, one must divide the full ureter, and demonstrate how

οὖρον ἐξ αὐτοῦ, καθάπερ ἐν ταῖς φλεβοτομίαις
τὸ αἷμα, μετὰ ταῦτα δὲ καὶ τὸν ἕτερον αὖθις
διατεμεῖν κἄπειτ' ἐπιδῆσαι τὸ ζῷον ἔξωθεν, ἀμ-
38 φοτέρων διῃρημένων, || εἶθ' ὅταν ἱκανῶς ἔχειν
δοκῇ, λῦσαι τὸν δεσμόν. εὑρεθήσεται γὰρ ἡ μὲν
κύστις κενή, πλῆρες δ' οὔρου τὸ μεταξὺ τῶν
ἐντέρων τε καὶ τοῦ περιτοναίου χωρίον ἅπαν,
ὡς ἂν εἰ καὶ ὑδερικὸν ἦν τὸ ζῷον. ταῦτ' οὖν εἴ
τις αὐτὸς καθ' ἑαυτὸν βουληθείη βασανίζειν ἐπὶ
ζῴου, μεγάλως μοι δοκεῖ καταγνώσεσθαι τῆς
Ἀσκληπιάδου προπετείας. εἰ δὲ δὴ καὶ τὴν
αἰτίαν μάθοι, δι' ἣν οὐδὲν ἐκ τῆς κύστεως εἰς
τοὺς οὐρητῆρας ἀντεκρεῖ, πεισθῆναι ἄν μοι δοκεῖ
καὶ διὰ τοῦδε τὴν εἰς τὰ ζῷα πρόνοιάν τε καὶ
τέχνην τῆς φύσεως.

Ἱπποκράτης μὲν οὖν ὧν ἴσμεν ἰατρῶν τε καὶ
φιλοσόφων πρῶτος ἁπάντων, ὡς ἂν καὶ πρῶτος
ἐπιγνοὺς τὰ τῆς φύσεως ἔργα, θαυμάζει τε καὶ
διὰ παντὸς αὐτὴν ὑμνεῖ δικαίαν ὀνομάζων καὶ
μόνην ἐξαρκεῖν εἰς ἅπαντα τοῖς ζῴοις φησίν,
αὐτὴν ἐξ αὑτῆς ἀδιδάκτως πράττουσαν ἅπαντα
τὰ δέοντα· τοιαύτην δ' οὖσαν αὐτὴν εὐθέως
καὶ δυνάμεις ὑπέλαβεν ἔχειν ἑλκτικὴν μὲν τῶν
οἰκείων, ἀποκριτικὴν δὲ τῶν ἀλλοτρίων καὶ
39 τρέφειν τε καὶ αὔξειν αὐ||τὴν τὰ ζῷα καὶ κρίνειν
τὰ νοσήματα· καὶ διὰ τοῦτ' ἐν τοῖς σώμασιν
ἡμῶν σύμπνοιάν τε μίαν εἶναί φησι καὶ σύρροιαν
καὶ πάντα συμπαθέα. κατὰ δὲ τὸν Ἀσκληπιάδην

---

[1] "De l'habileté et de la prévoyance de la nature à l'égard
des animaux" (Daremberg). cf. p. 56, note 1.
[2] cf. p. 36, note 2.

the urine spurts out of it, like blood in the operation of venesection ; and after this one cuts through the other also, and both being thus divided, one bandages up the animal externally. Then when enough time seems to have elapsed, one takes off the bandages ; the bladder will now be found empty, and the whole region between the intestines and the peritoneum full of urine, as if the animal were suffering from dropsy. Now, if anyone will but test this for himself on an animal, I think he will strongly condemn the rashness of Asclepiades, and if he also learns the reason why nothing regurgitates from the bladder into the ureters, I think he will be persuaded by this also of the forethought and art shown by Nature in relation to animals.[1]

Now Hippocrates, who was the first known to us of all those who have been both physicians and philosophers inasmuch as he was the first to recognize what Nature effects, expresses his admiration of her, and is constantly singing her praises and calling her "just." Alone, he says, she suffices for the animal in every respect, performing of her own accord and without any teaching all that is required. Being such, she has, as he supposes, certain *faculties*, one attractive of what is appropriate,[2] and another eliminative of what is foreign, and she nourishes the animal, makes it grow, and expels its diseases by crisis.[3] Therefore he says that there is in our bodies a concordance in the movements of air and fluid, and that everything is in sympathy. According to Asclepiades, however, nothing is

[3] The morbid material passed successively through the stages of "crudity," "coction" (*pepsis*), and "elimination" (*crisis*). For "critical days" *cf.* p. 74, note 1.

οὐδὲν οὐδενὶ συμπαθές ἐστι φύσει, διῃρημένης τε
καὶ κατατεθραυσμένης εἰς ἄναρμα στοιχεῖα καὶ
ληρώδεις ὄγκους ἁπάσης τῆς οὐσίας. ἐξ ἀνάγκης
οὖν ἄλλα τε μυρία τοῖς ἐναργῶς φαινομένοις
ἐναντίως ἀπεφήνατο καὶ τῆς φύσεως ἠγνόησε
τήν τε τῶν οἰκείων ἐπισπαστικὴν δύναμιν καὶ
τὴν τῶν ἀλλοτρίων ἀποκριτικήν. ἐπὶ μὲν οὖν
τῆς ἐξαιματώσεώς τε καὶ ἀναδόσεως ἐξεῦρέ τινα
ψυχρὰν ἀδολεσχίαν· εἰς δὲ τὴν τῶν περιττωμάτων
κάθαρσιν οὐδὲν ὅλως εὑρὼν εἰπεῖν οὐκ ὤκνησεν
ὁμόσε χωρῆσαι τοῖς φαινομένοις, ἐπὶ μὲν τῆς
τῶν οὔρων διακρίσεως ἀποστερήσας μὲν τῶν τε
νεφρῶν καὶ τῶν οὐρητήρων τὴν ἐνέργειαν, ἀδήλους
δέ τινας πόρους εἰς τὴν κύστιν ὑποθέμενος· τοῦτο
γὰρ ἦν δηλαδὴ μέγα καὶ σεμνὸν ἀπιστήσαντα
τοῖς φαινομένοις πιστεῦσαι τοῖς ἀδήλοις.

40   Ἐπὶ ‖ δὲ τῆς ξανθῆς χολῆς ἔτι μεῖζον αὐτῷ
καὶ νεανικώτερόν ἐστι τὸ τόλμημα· γεννᾶσθαι
γὰρ αὐτὴν ἐν τοῖς χοληδόχοις ἀγγείοις, οὐ διακρίνεσθαι λέγει.

Πῶς οὖν τοῖς ἰκτερικοῖς ἅμ' ἄμφω συμπίπτει,
τὰ μὲν διαχωρήματα μηδὲν ὅλως ἐν αὐτοῖς
ἔχοντα χολῆς, ἀνάπλεων δ' αὐτοῖς γιγνόμενον
ὅλον τὸ σῶμα; ληρεῖν πάλιν ἐνταῦθ' ἀναγκάζεται
τοῖς ἐπὶ τῶν οὔρων εἰρημένοις παραπλησίως.
ληρεῖ δ' οὐδὲν ἧττον καὶ περὶ τῆς μελαίνης χολῆς
καὶ τοῦ σπληνὸς οὔτε τί ποθ' ὑφ' Ἱπποκράτους
εἴρηται συνιεὶς ἀντιλέγειν τ' ἐπιχειρῶν οἷς οὐκ
οἶδεν ἐμπλήκτῳ τινὶ καὶ μανικῷ στόματι.

---

[1] This was the process by which nutriment was taken up
from the alimentary canal; "absorption," "dispersal;" cf.

naturally in sympathy with anything else, all
substance being divided and broken up into inharmonious elements and absurd "molecules."
Necessarily, then, besides making countless other
statements in opposition to plain fact, he was ignorant
of Nature's faculties, both that attracting what is
appropriate, and that expelling what is foreign. Thus
he invented some wretched nonsense to explain
blood-production and *anadosis*,[1] and, being utterly
unable to find anything to say regarding the
clearing-out[2] of superfluities, he did not hesitate
to join issue with obvious facts, and, in this matter of
urinary secretion, to deprive both the kidneys and
the ureters of their activity, by assuming that there
were certain invisible channels opening into the
bladder. It was, of course, a grand and impressive
thing to do, to mistrust the obvious, and to pin one's
faith in things which could not be seen!

Also, in the matter of the yellow bile, he makes
an even grander and more spirited venture; for he
says this is actually generated in the bile-ducts, not
merely separated out.

How comes it, then, that in cases of jaundice two
things happen at the same time—that the dejections
contain absolutely no bile, and that the whole body
becomes full of it? He is forced here again to talk
nonsense, just as he did in regard to the urine. He
also talks no less nonsense about the black bile and
the spleen, not understanding what was said by
Hippocrates; and he attempts in stupid—I might
say insane—language, to contradict what he knows
nothing about.

[1] p. 13, note 5. The subject is dealt with more fully in
chap. xvi.
[2] Lit. *catharsis*.

GALEN

Τί δὴ τὸ κέρδος ἐκ τῶν τοιούτων δογμάτων εἰς τὰς θεραπείας ἐκτήσατο; μήτε νεφριτικόν τι νόσημα δύνασθαι θεραπεῦσαι μήτ᾽ ἰκτερικὸν μήτε μελαγχολικόν, ἀλλὰ καὶ περὶ τοῦ πᾶσιν ἀνθρώποις οὐχ Ἱπποκράτει μόνον ὁμολογουμένου τοῦ καθαίρειν τῶν φαρμάκων ἔνια μὲν τὴν ξανθὴν χολήν, ἔνια δὲ τὴν μέλαιναν, ἄλλα δέ τινα φλέγμα καί τινα τὸ λεπτὸν καὶ ὑδατῶδες περίττωμα, μηδὲ περὶ τούτων συγχωρεῖν, ἀλλ᾽ ὑπ᾽ αὐτῶν τῶν φαρμάκων γίγνεσθαι λέγειν τοιοῦτον ἕκαστον
41 τῶν κενουμένων, ὥσπερ ὑπὸ τῶν χολη‖δόχων πόρων τὴν χολήν· καὶ μηδὲν διαφέρειν κατὰ τὸν θαυμαστὸν Ἀσκληπιάδην ἢ ὑδραγωγὸν διδόναι τοῖς ὑδεριῶσιν ἢ χολαγωγὸν φάρμακον· ἅπαντα γὰρ ὁμοίως κενοῦν καὶ συντήκειν τὸ σῶμα καὶ τὸ σύντηγμα τοιόνδε τι φαίνεσθαι ποιεῖν, μὴ πρότερον ὑπάρχον τοιοῦτον.

Ἆρ᾽ οὖν οὐ μαίνεσθαι νομιστέον αὐτὸν ἢ παντάπασιν ἄπειρον εἶναι τῶν ἔργων τῆς τέχνης; τίς γὰρ οὐκ οἶδεν, ὡς, εἰ μὲν φλέγματος ἀγωγὸν δοθείη φάρμακον τοῖς ἰκτεριῶσιν, οὐκ ἂν οὐδὲ τέτταρας κυάθους καθαρθεῖεν· οὕτω δ᾽ οὐδ᾽ εἰ τῶν ὑδραγωγῶν τι· χολαγωγῷ δὲ φαρμάκῳ πλεῖστον μὲν ἐκκενοῦται χολῆς, αὐτίκα δὲ καθαρὸς τοῖς οὕτω καθαρθεῖσιν ὁ χρὼς γίγνεται. πολλοὺς γοῦν ἡμεῖς μετὰ τὸ θεραπεῦσαι τὴν ἐν τῷ ἥπατι διάθεσιν ἅπαξ καθήραντες ἀπηλλάξαμεν τοῦ παθήματος. οὐ μὴν οὐδ᾽ εἰ φλέγματος ἀγωγῷ καθαίροις φαρμάκῳ, πλέον ἄν τι διαπράξαιο.

---

[1] i.e. urine. [2] On use of κενόω v. p. 67, note 9.
[3] i.e. bile and phlegm had no existence as such before the

64

And what profit did he derive from these opinions from the point of view of treatment? He neither was able to cure a kidney ailment, nor jaundice, nor a disease of black bile, nor would he agree with the view held not merely by Hippocrates but by all men regarding drugs—that some of them purge away yellow bile, and others black, some again phlegm, and others the thin and watery superfluity [1]; he held that all the substances evacuated [2] were *produced by the drugs themselves*, just as yellow bile is produced by the biliary passages! It matters nothing, according to this extraordinary man, whether we give a hydragogue or a cholagogue in a case of dropsy, for these all equally purge [2] and dissolve the body, and produce a solution having such and such an appearance, which did not exist as such before! [3]

Must we not, therefore, suppose he was either mad, or entirely unacquainted with practical medicine? For who does not know that if a drug for attracting phlegm be given in a case of jaundice it will not even evacuate four *cyathi* [4] of phlegm? Similarly also if one of the hydragogues be given. A cholagogue, on the other hand, clears away a great quantity of bile, and the skin of patients so treated at once becomes clear. I myself have, in many cases, after treating the liver condition, then removed the disease by means of a single purgation; whereas, if one had employed a drug for removing phlegm one would have done no good.

drugs were given; they are the products of dissolved tissue. Asclepiades did not believe that diseases were due to a *materia peccans*, but to disturbances in the movements of the molecules (ὄγκοι) which constitute the body; thus, in opposition to the humoralists such as Galen, he had no use for drugs. *cf.* p. 49, note 5. [4] About 4 oz., or one-third of a pint.

Καὶ ταῦτ' οὐχ Ἱπποκράτης μὲν οὕτως οἶδε
γιγνόμενα, τοῖς δ' ἀπὸ τῆς ἐμπειρίας μόνης ὁρμω-
μένοις ἑτέρως ἔγνωσται, ἀλλὰ κἀκεί‖νοις ὡσαύτως
καὶ πᾶσιν ἰατροῖς, οἷς μέλει τῶν ἔργων τῆς τέχνης, οὕτω δοκεῖ πλὴν Ἀσκληπιάδου. προδοσίαν
γὰρ εἶναι νενόμικε τῶν στοιχείων ὧν ὑπέθετο τὴν
ἀληθῆ περὶ τῶν τοιούτων ὁμολογίαν. εἰ γὰρ
ὅλως εὑρεθείη τι φάρμακον ἑλκτικὸν τοῦδέ τινος
τοῦ χυμοῦ μόνου, κίνδυνος κρατεῖν δηλαδὴ τῷ
λόγῳ τὸ ἐν ἑκάστῳ τῶν σωμάτων εἶναί τινα
δύναμιν ἐπισπαστικὴν τῆς οἰκείας ποιότητος. διὰ
τοῦτο κνῆκον μὲν καὶ κόκκον τὸν κνίδιον καὶ
ἱπποφαὲς οὐχ ἕλκειν ἐκ τοῦ σώματος ἀλλὰ ποιεῖν
τὸ φλέγμα φησίν· ἄνθος δὲ χαλκοῦ καὶ λεπίδα
καὶ αὐτὸν τὸν κεκαυμένον χαλκὸν καὶ χαμαίδρυν
καὶ χαμαιλέοντα εἰς ὕδωρ ἀναλύειν τὸ σῶμα καὶ
τοὺς ὑδερικοὺς ὑπὸ τούτων οὐ καθαιρομένους ὀνί-
νασθαι ἀλλὰ κενουμένους συναυξόντων δηλαδὴ τὸ
πάθος. εἰ γὰρ οὐ κενοῖ τὸ περιεχόμενον ἐν τοῖς
σώμασιν ὑδατῶδες ὑγρὸν ἀλλ' αὐτὸ γεννᾷ, τῷ
νοσήματι προστιμωρεῖται. καὶ μέν γε καὶ ἡ
σκαμμωνία πρὸς τῷ μὴ κενοῦν ἐκ τοῦ σώματος
τῶν ἰκτερικῶν τὴν χολὴν ἔτι καὶ τὸ χρηστὸν αἷμα
χολὴν ἐργαζομένη ‖ καὶ συντήκουσα τὸ σῶμα καὶ
τηλικαῦτα κακὰ δρῶσα καὶ τὸ πάθος ἐπαύξουσα
κατά γε τὸν Ἀσκληπιάδου λόγον.

Ὅμως ἐναργῶς ὁρᾶται πολλοὺς ὠφελοῦσα.
ναί, φησίν, ὀνίνανται μέν, ἀλλ' αὐτῷ μόνῳ τῷ

---

[1] The Empiricists. *cf.* Introduction, p. xiii.
[2] His ὄγκοι or molecules.
[3] He does not say "organized" or "living" body; inanimate things were also thought to possess "natures"; *cf.* p. 2, note 1.

Nor is Hippocrates the only one who knows this to be so, whilst those who take experience alone as their starting-point [1] know otherwise; they, as well as all physicians who are engaged in the practice of medicine, are of this opinion. Asclepiades, however, is an exception; he would hold it a betrayal of his assumed " elements " [2] to confess the truth about such matters. For if a single drug were to be discovered which attracted such and such a humour only, there would obviously be danger of the opinion gaining ground that there is in every body [3] a faculty which attracts its own particular quality. He therefore says that safflower,[4] the Cnidian berry,[5] and *Hippophaes*,[6] do not draw phlegm from the body, but actually make it. Moreover, he holds that the flower and scales of bronze, and burnt bronze itself, and germander,[7] and wild mastich [8] dissolve the body into water, and that dropsical patients derive benefit from these substances, not because they are purged by them, but because they are rid of substances which actually help to increase the disease; for, if the medicine does not evacuate [9] the dropsical fluid contained in the body, but generates it, it aggravates the condition further. Moreover, scammony, according to the Asclepiadean argument, not only fails to evacuate [9] the bile from the bodies of jaundiced subjects, but actually turns the useful blood into bile, and dissolves the body; in fact it does all manner of evil and increases the disease.

And yet this drug may be clearly seen to do good to numbers of people! "Yes," says he, "they derive

---

[4] Carthamus tinctorius. [5] Daphne Gnidium.
[6] Euphorbia acanthothamnos. [7] Teucrium chamaedrys.
[8] Atractylis gummifera. [9] On use of κενόω *cf.* p. 98, note 1.

λόγῳ τῆς κενώσεως. καὶ μὴν εἰ φλέγματος ἀγωγὸν αὐτοῖς δοίης φάρμακον, οὐκ ὀνήσονται. καὶ
τοῦθ' οὕτως ἐναργές ἐστιν, ὥστε καὶ οἱ ἀπὸ μόνης
τῆς ἐμπειρίας ὁρμώμενοι γιγνώσκουσιν αὐτό.
καίτοι τούτοις γε τοῖς ἀνδράσιν αὐτὸ δὴ τοῦτ'
ἔστι φιλοσόφημα, τὸ μηδενὶ λόγῳ πιστεύειν ἀλλὰ
μόνοις τοῖς ἐναργῶς φαινομένοις. ἐκεῖνοι μὲν οὖν
σωφρονοῦσιν· Ἀσκληπιάδης δὲ παραπαίει ταῖς
αἰσθήσεσιν ἡμᾶς ἀπιστεῖν κελεύων, ἔνθα τὸ φαινόμενον ἀνατρέπει σαφῶς αὐτοῦ τὰς ὑποθέσεις.
καίτοι μακρῷ γ' ἦν ἄμεινον οὐχ ὁμόσε χωρεῖν
τοῖς φαινομένοις ἀλλ' ἐκείνοις ἀναθέσθαι τὸ πᾶν.

Ἀρ' οὖν ταῦτα μόνον ἐναργῶς μάχεται τοῖς
Ἀσκληπιάδου δόγμασιν ἢ καὶ τὸ θέρους μὲν
πλείονα κενοῦσθαι τὴν ξανθὴν χολὴν ὑπὸ τῶν
αὐτῶν φαρμάκων, χειμῶνος δὲ τὸ φλέγμα, καὶ
νεανίσκῳ μὲν πλείονα τὴν χολήν, πρεσβύτῃ δὲ τὸ
44 φλέγμα; φαίνεται ‖ γὰρ ἕκαστον ἕλκειν τὴν
οὖσαν, οὐκ αὐτὸ γεννᾶν τὴν οὐκ οὖσαν. εἰ γοῦν
ἐθελήσαις νεανίσκῳ τινὶ τῶν ἰσχνῶν καὶ θερμῶν
ὥρᾳ θέρους μήτ' ἀργῶς βεβιωκότι μήτ' ἐν πλησ-
μονῇ φλέγματος ἀγωγὸν δοῦναι φάρμακον, ὀλί-
γιστον μὲν καὶ μετὰ βίας πολλῆς ἐκκενώσεις τοῦ
χυμοῦ, βλάψεις δ' ἐσχάτως τὸν ἄνθρωπον· ἔμπαλιν δ' εἰ χολαγωγὸν δοίης, καὶ πάμπολυ κενώσεις
καὶ βλάψεις οὐδέν.

Ἀρ' ἀπιστοῦμεν ἔτι τῷ μὴ οὐχ ἕκαστον τῶν
φαρμάκων ἐπάγεσθαι τὸν οἰκεῖον ἑαυτῷ χυμόν;

---

[1] Empiricist physicians.

benefit certainly, but merely in proportion to the evacuation." . . . But if you give these cases a drug which draws off phlegm they will not be benefited. This is so obvious that even those who make experience alone their starting-point [1] are aware of it; and these people make it a cardinal point of their teaching to trust to no arguments, but only to what can be clearly seen. In this, then, they show good sense; whereas Asclepiades goes far astray in bidding us distrust our senses where obvious facts plainly overturn his hypotheses. Much better would it have been for him not to assail obvious facts, but rather to devote himself entirely to these.

Is it, then, these facts only which are plainly irreconcilable with the views of Asclepiades? Is not also the fact that in summer yellow bile is evacuated in greater quantity by the same drugs, and in winter phlegm, and that in a young man more bile is evacuated, and in an old man more phlegm? Obviously each drug attracts something which already exists, and does not generate something previously non-existent. Thus if you give in the summer season a drug which attracts phlegm to a young man of a lean and warm habit, who has lived neither idly nor too luxuriously, you will with great difficulty evacuate a very small quantity of this humour, and you will do the man the utmost harm. On the other hand, if you give him a cholagogue, you will produce an abundant evacuation and not injure him at all.

Do we still, then, disbelieve that each drug attracts *that humour which is proper to it?* [2] Possibly the

[2] Note that drugs also have "natures"; *cf.* p. 66, note 3, and pp. 83–84.

ἴσως φήσουσιν οἱ ἀπ' Ἀσκληπιάδου, μᾶλλον δ'
οὐκ ἴσως, ἀλλὰ πάντως ἀπιστεῖν ἐροῦσιν, ἵνα μὴ
προδῶσι τὰ φίλτατα.

XIV

Πάλιν οὖν καὶ ἡμεῖς ἐφ' ἑτέραν μεταβῶμεν
ἀδολεσχίαν· οὐ γὰρ ἐπιτρέπουσιν οἱ σοφισταὶ
τῶν ἀξίων τι ζητημάτων προχειρίζεσθαι καίτοι
παμπόλλων ὑπαρχόντων, ἀλλὰ κατατρίβειν ἀναγ-
κάζουσι τὸν χρόνον εἰς τὴν τῶν σοφισμάτων, ὧν
προβάλλουσι, λύσιν.

Τίς οὖν ἡ ἀδολεσχία; ἡ ἔνδοξος αὕτη καὶ
45 πολυθρύλητος λίθος ἡ τὸν σίδηρον ‖ ἐπισπωμένη.
τάχα γὰρ ἂν αὕτη ποτὲ τὴν ψυχὴν αὐτῶν ἐπι-
σπάσαιτο πιστεύειν εἶναί τινας ἐν ἑκάστῳ τῶν σωμάτων ἑλκτικὰς τῶν οἰκείων ποιοτήτων δυνάμεις.

Ἐπίκουρος μὲν οὖν καίτοι παραπλησίοις Ἀσ-
κληπιάδῃ στοιχείοις πρὸς τὴν φυσιολογίαν χρώ-
μενος ὅμως ὁμολογεῖ, πρὸς μὲν τῆς ἡρακλείας
λίθου τὸν σίδηρον ἕλκεσθαι, πρὸς δὲ τῶν ἠλέκ-
τρων τὰ κυρήβια καὶ πειρᾶταί γε καὶ τὴν αἰτίαν
ἀποδιδόναι τοῦ φαινομένου. τὰς γὰρ ἀπορρεούσας
ἀτόμους ἀπὸ τῆς λίθου ταῖς ἀπορρεούσαις ἀπὸ
τοῦ σιδήρου τοῖς σχήμασιν οἰκείας εἶναί φησιν,
ὥστε περιπλέκεσθαι ῥᾳδίως. προσκρουούσας οὖν
αὐτὰς τοῖς συγκρίμασιν ἑκατέροις τῆς τε λίθου
καὶ τοῦ σιδήρου κἄπειτ' εἰς τὸ μέσον ἀποπαλ-
λομένας οὕτως ἀλλήλαις τε περιπλέκεσθαι καὶ

---

[1] Pun here.   [2] Lit. *physiology, i.e. nature-lore*, almost
our "Natural Philosophy"; *cf.* Introduction, p. xxvi.

adherents of Asclepiades will assent to this—or rather, they will—not possibly, but certainly—declare that they disbelieve it, lest they should betray their darling prejudices.

XIV

LET us pass on, then, again to another piece of nonsense ; for the sophists do not allow one to engage in enquiries that are of any worth, albeit there are many such ; they compel one to spend one's time in dissipating the fallacious arguments which they bring forward.

What, then, is this piece of nonsense ? It has to do with the famous and far-renowned stone which draws iron [the lodestone]. It might be thought that this would draw[1] their minds to a belief that there are in all bodies certain *faculties* by which they attract their own proper qualities.

Now Epicurus, despite the fact that he employs in his *Physics*[2] elements similar to those of Asclepiades,[3] yet allows that iron is attracted by the lodestone,[4] and chaff by amber. He even tries to give the cause of the phenomenon. His view is that the atoms which flow from the stone are related in shape to those flowing from the iron, and so they become easily interlocked with one another ; thus it is that, after colliding with each of the two compact masses (the stone and the iron) they then rebound into the middle and so become entangled with each other,

---

[3] The ultimate particle of Epicurus was the ἄτομος or atom (lit. "non-divisible"), of Asclepiades, the ὄγκος or molecule. Asclepiades took his atomic theory from Epicurus, and he again from Democritus ; *cf.* p. 49, note 5.

[4] Lit. *Herculean stone.*

συνεπισπᾶσθαι τὸν σίδηρον. τὸ μὲν οὖν τῶν
ὑποθέσεων εἰς τὴν αἰτιολογίαν ἀπίθανον ἄντικρυς
δῆλον, ὅμως δ᾽ οὖν ὁμολογεῖ τὴν ὁλκήν. καὶ οὕτω
γε καὶ κατὰ τὰ σώματα τῶν ζῴων φησὶ γίγνεσθαι
τάς τ᾽ ἀναδόσεις καὶ τὰς διακρίσεις τῶν περιττωμάτων καὶ τὰς τῶν καθαιρόντων φαρμάκων
ἐνεργείας.

Ἀσκληπιάδης δὴ τό τε τῆς εἰρημένης αἰτίας
46 ἀπίθανον ‖ ὑπιδόμενος καὶ μηδεμίαν ἄλλην ἐφ᾽
οἷς ὑπέθετο στοιχείοις ἐξευρίσκων πιθανὴν ἐπὶ τὸ
μηδ᾽ ὅλως ἕλκεσθαι λέγειν ὑπὸ μηδενὸς μηδὲν
ἀναισχυντήσας ἐτράπετο, δέον, εἰ μήθ᾽ οἷς Ἐπίκουρος εἶπεν ἠρέσκετο μήτ᾽ ἄλλα βελτίω λέγειν
εἶχεν, ἀποστῆναι τῶν ὑποθέσεων καὶ τήν τε φύσιν
εἰπεῖν τεχνικὴν καὶ τὴν οὐσίαν τῶν ὄντων ἑνου-
μένην τε πρὸς ἑαυτὴν ἀεὶ καὶ ἀλλοιουμένην ὑπὸ
τῶν ἑαυτῆς μορίων εἰς ἄλληλα δρώντων τε καὶ
πασχόντων. εἰ γὰρ ταῦθ᾽ ὑπέθετο, χαλεπὸν οὐδὲν
ἦν τὴν τεχνικὴν ἐκείνην φύσιν ὁμολογῆσαι δύνα-
μεις ἔχειν ἐπισπαστικὴν μὲν τῶν οἰκείων, ἀποκριτικὴν δὲ τῶν ἀλλοτρίων. οὐ γὰρ δι᾽ ἄλλο τί
γ᾽ ἦν αὐτῇ τὸ τεχνικῇ τ᾽ εἶναι καὶ τοῦ ζῴου
διασωστικῇ καὶ τῶν νοσημάτων κριτικῇ παρὰ τὸ
προσίεσθαι μὲν καὶ φυλάττειν τὸ οἰκεῖον, ἀποκρίνειν δὲ τὸ ἀλλότριον.

Ἀλλ᾽ Ἀσκληπιάδης κἀνταῦθα τὸ μὲν ἀκόλουθον ταῖς ἀρχαῖς αἷς ὑπέθετο συνεῖδεν, οὐ μὴν τήν
γε πρὸς τὸ φαινόμενον ἐναργῶς ἠδέσθη μάχην,
47 ἀλλ᾽ ὁμόσε ‖ χωρεῖ καὶ περὶ τούτου πᾶσιν οὐκ
ἰατροῖς μόνον ἀλλ᾽ ἤδη καὶ τοῖς ἄλλοις ἀνθρώποις

---

[1] Lit. *aetiology*.   [2] *Anadosis* ; *cf.* p. 62, note 1.

and draw the iron after them. So far, then, as his
hypotheses regarding causation[1] go, he is perfectly
unconvincing; nevertheless, he does grant that
there is an attraction. Further, he says that it is
on similar principles that there occur in the bodies of
animals the dispersal of nutriment[2] and the discharge
of waste matters, as also the actions of cathartic
drugs.

Asclepiades, however, who viewed with suspicion
the incredible character of the cause mentioned,
and who saw no other credible cause on the basis
of his supposed elements, shamelessly had recourse
to the statement that nothing is in any way attracted
by anything else. Now, if he was dissatisfied with
what Epicurus said, and had nothing better to say
himself, he ought to have refrained from making
hypotheses, and should have said that Nature
is a constructive artist and that the substance of
things is always tending towards unity and also
towards alteration because its own parts act upon
and are acted upon by one another.[3] For, if he had
assumed this, it would not have been difficult to allow
that this constructive nature has powers which
attract appropriate and expel alien matter. For in
no other way could she be constructive, preservative
of the animal, and eliminative of its diseases,[4] unless
it be allowed that she conserves what is appropriate
and discharges what is foreign.

But in this matter, too, Asclepiades realized the
logical sequence of the principles he had assumed;
he showed no scruples, however, in opposing plain
fact; he joins issue in this matter also, not merely
with all physicians, but with everyone else, and

[3] cf. p. 45.   [4] The *vis conservatrix et medicatrix Naturae.*

οὔτε κρίσιν εἶναί τινα λέγων οὔθ᾽ ἡμέραν κρίσιμον
οὔθ᾽ ὅλως οὐδὲν ἐπὶ σωτηρίᾳ τοῦ ζῴου πραγμα-
τεύσασθαι τὴν φύσιν. ἀεὶ γὰρ τὸ μὲν ἀκόλουθον
φυλάττειν βούλεται, τὸ δ᾽ ἐναργῶς φαινόμενον
ἀνατρέπειν ἔμπαλιν Ἐπικούρῳ. τιθεὶς γὰρ ἐκεῖνος
ἀεὶ τὸ φαινόμενον αἰτίαν αὐτοῦ ψυχρὰν ἀποδίδωσι.
τὰ γὰρ ἀποπαλλόμενα σμικρὰ σώματα τῆς ἡρακλείας λίθου τοιούτοις ἑτέροις περιπλέκεσθαι μορίοις τοῦ σιδήρου κἄπειτα διὰ τῆς περιπλοκῆς
ταύτης μηδαμοῦ φαινομένης ἐπισπᾶσθαι βαρεῖαν
οὕτως οὐσίαν οὐκ οἶδ᾽ ὅπως ἄν τις πεισθείη. καὶ
γὰρ εἰ τοῦτο συγχωρήσομεν, τό γε τῷ σιδήρῳ
πάλιν ἕτερον προστεθέν τι συνάπτεσθαι τὴν
αὐτὴν αἰτίαν οὐκέτι προσίεται.

Τί γὰρ ἐροῦμεν; ἢ δηλαδὴ τῶν ἀπορρεόντων
τῆς λίθου μορίων ἔνια μὲν προσκρούσαντα τῷ
σιδήρῳ πάλιν ἀποπάλλεσθαι καὶ ταῦτα μὲν εἶναι,
δι᾽ ὧν κρεμάννυσθαι συμβαίνει τὸν σίδηρον, τὰ δ᾽
48 εἰς αὐτὸν εἰσδυόμενα διὰ τῶν ‖ κενῶν πόρων διεξέρχεσθαι τάχιστα κἄπειτα τῷ παρακειμένῳ
σιδήρῳ προσκρούοντα μήτ᾽ ἐκεῖνον διαδῦναι δύνασθαι, καίτοι τόν γε πρῶτον διαδύντα, παλινδρο-
μοῦντα δ᾽ αὖθις ἐπὶ τὸν πρότερον ἑτέρας αὖθις
ἐργάζεσθαι ταῖς προτέραις ὁμοίας περιπλοκάς;

Ἐναργῶς γὰρ ἐνταῦθα τὸ ληρῶδες τῆς αἰτίας
ἐλέγχεται. γραφεῖα γοῦν οἶδά ποτε σιδηρᾶ πέντε
κατὰ τὸ συνεχὲς ἀλλήλοις συναφθέντα, τοῦ πρώτου μὲν μόνου τῆς λίθου ψαύσαντος, ἐξ ἐκείνου

---

[1] cf. p. 61, note 3. The *crisis* or resolution in fevers was
observed to take place with a certain regularity; hence
arose the doctrine of " critical days."

maintains that there is no such thing as a crisis, or critical day,[1] and that Nature does absolutely nothing for the preservation of the animal. For his constant aim is to follow out logical consequences and to upset obvious fact, in this respect being opposed to Epicurus; for the latter always stated the observed fact, although he gives an ineffective explanation of it. For, that these small corpuscles belonging to the lodestone rebound, and become entangled with other similar particles of the iron, and that then, by means of this entanglement (which cannot be seen anywhere) such a heavy substance as iron is attracted—I fail to understand how anybody could believe this. Even if we admit this, the same principle will not explain the fact that, when the iron has another piece brought in contact with it, this becomes attached to it.

For what are we to say? That, forsooth, some of the particles that flow from the lodestone collide with the iron and then rebound back, and that it is by these that the iron becomes suspended? that others penetrate into it, and rapidly pass through it by way of its empty channels?[2] that these then collide with the second piece of iron and are not able to penetrate it although they penetrated the first piece? and that they then course back to the first piece, and produce entanglements like the former ones?

The hypothesis here becomes clearly refuted by its absurdity. As a matter of fact, I have seen five writing-stylets of iron attached to one another in a line, only the first one being in contact with the

[2] These were hypothetical spaces or channels between the atoms; cf. Introduction, p. xiv.

δ' εἰς τἆλλα τῆς δυνάμεως διαδοθείσης· καὶ οὐκ ἔστιν εἰπεῖν, ὡς, εἰ μὲν τῷ κάτω τοῦ γραφείου πέρατι προσάγοις ἕτερον, ἔχεταί τε καὶ συνάπτεται καὶ κρέμαται τὸ προσενεχθέν· εἰ δ' ἄλλῳ τινὶ μέρει τῶν πλαγίων προσθείης, οὐ συνάπτεται. πάντη γὰρ ὁμοίως ἡ τῆς λίθου διαδίδοται δύναμις, εἰ μόνον ἅψαιτο κατά τι τοῦ πρώτου γραφείου. καὶ μέντοι κἀκ τούτου πάλιν εἰς τὸ δεύτερον ὅλον ἡ δύναμις ἅμα νοήματι διαρρεῖ κἀξ ἐκείνου πάλιν εἰς τὸ τρίτον ὅλον. εἰ δὴ νοήσαις σμικράν τινα λίθον ἡρακλείαν ἐν οἴκῳ τινὶ κρεμαμένην, εἶτ' ἐν κύκλῳ ψαύοντα πάμπολλα σιδήρια κἀκείνων πάλιν ἕτερα κἀκείνων ἄλλα καὶ τοῦτ' ἄχρι πλεί-
49 ονος, ἅπαντα ‖ δήπου πίμπλασθαι δεῖ τὰ σιδήρια τῶν ἀπορρεόντων τῆς λίθου σωμάτων. καὶ κινδυνεύει διαφορηθῆναι τὸ σμικρὸν ἐκεῖνο λιθίδιον εἰς τὰς ἀπορροὰς διαλυθέν. καίτοι, κἂν εἰ μηδὲν παρακέοιτ' αὐτῷ σιδήριον, εἰς τὸν ἀέρα σκεδάννυται, μάλιστ' εἰ καὶ θερμὸς ὑπάρχοι.

Ναί, φησί, σμικρὰ γὰρ αὐτὰ χρὴ πάνυ νοεῖν, ὥστε τῶν ἐμφερομένων τῷ ἀέρι ψηγμάτων τούτων δὴ τῶν σμικροτάτων ἐκείνων ἔνια μυριοστὸν εἶναι μέρος. εἶτ' ἐξ οὕτω σμικρῶν τολμᾶτε λέγειν κρεμάννυσθαι βάρη τηλικαῦτα σιδήρου; εἰ γὰρ ἕκαστον αὐτῶν μυριοστόν ἐστι μέρος τῶν ἐν τῷ ἀέρι φερομένων ψηγμάτων, πηλίκον χρὴ νοῆσαι τὸ πέρας αὐτῶν τὸ ἀγκιστροειδές, ᾧ περιπλέκεται πρὸς ἄλληλα; πάντως γὰρ δήπου τοῦτο σμικρότατόν ἐστιν ὅλου τοῦ ψήγματος.

---

[1] He means the specific drawing power or faculty of the lodestone.   [2] cf. our modern "radium-emanations."

lodestone, and the power[1] being transmitted through
it to the others. Moreover, it cannot be said that if
you bring a second stylet into contact with the lower
end of the first, it becomes held, attached, and suspended, whereas, if you apply it to any other part of
the side it does not become attached. For the power
of the lodestone is distributed in all directions; it
merely needs to be in contact with the first stylet at
any point; from this stylet again the power flows, as
quick as a thought, all through the second, and from
that again to the third. Now, if you imagine a small
lodestone hanging in a house, and in contact with it
all round a large number of pieces of iron, from them
again others, from these others, and so on,—all these
pieces of iron must surely become filled with the
corpuscles which emanate from the stone; therefore,
this first little stone is likely to become dissipated by
disintegrating into these emanations.[2] Further, even
if there be no iron in contact with it, it still disperses
into the air, particularly if this be also warm.

"Yes," says Epicurus, "but these corpuscles must
be looked on as exceedingly small, so that some
of them are a ten-thousandth part of the size of
the very smallest particles carried in the air." Then
do you venture to say that so great a weight of iron
can be suspended by such small bodies? If each of
them is a ten-thousandth part as large as the dust
particles which are borne in the atmosphere, how big
must we suppose the hook-like extremities by which
they interlock with each other[3] to be? For of
course this is quite the smallest portion of the whole
particle.

[3] cf. Ehrlich's hypothesis of "receptors" in explanation of
the "affinities" of animal cells.

Εἶτα μικρὸν μικρῷ, κινούμενον κινουμένῳ περι-
πλακὲν οὐκ εὐθὺς ἀποπάλλεται. καὶ γὰρ δὴ καὶ
ἀλλ' ἄττα πάντως αὐτοῖς, τὰ μὲν ἄνωθεν, τὰ δὲ
κάτωθεν, καὶ τὰ μὲν ἔμπροσθεν, τὰ δ' ὄπισθεν,
τὰ δ' ἐκ τῶν δεξιῶν, τὰ δ' ἐκ τῶν ἀριστερῶν ‖
50 ἐκρηγνύμενα σείει τε καὶ βράττει καὶ μένειν οὐκ
ἐᾷ. καὶ μέντοι καὶ πολλὰ χρὴ νοεῖν ἐξ ἀνάγκης
ἕκαστον ἐκείνων τῶν σμικρῶν σωμάτων ἔχειν
ἀγκιστρώδη πέρατα. δι' ἑνὸς μὲν γὰρ ἀλλήλοις
συνάπτεται, δι' ἑτέρου δ' ἑνὸς τοῦ μὲν ὑπερκειμένου τῇ λίθῳ, τοῦ δ' ὑποκειμένου τῷ σιδήρῳ.
εἰ γὰρ ἄνω μὲν ἐξαφθείη τῆς λίθου, κάτω δὲ τῷ
σιδήρῳ μὴ συμπλακείη, πλέον οὐδέν. ὥστε τοῦ
μὲν ὑπερκειμένου τὸ ἄνω μέρος ἐκκρέμασθαι χρὴ
τῆς λίθου, τοῦ δ' ὑποκειμένου τῷ κάτω πέρατι
συνῆφθαι τὸν σίδηρον. ἐπεὶ δὲ κἀκ τῶν πλαγίων
ἀλλήλοις περιπλέκεται, πάντως που κἀνταῦθα
ἔχει τὰ ἄγκιστρα. καὶ μέμνησό μοι πρὸ πάντων,
ὅπως ὄντα σμικρὰ τὰς τοιαύτας καὶ τοσαύτας
ἀποφύσεις ἔχει. καὶ τούτου μᾶλλον ἔτι, πῶς, ἵνα
τὸ δεύτερον σιδήριον συναφθῇ τῷ πρώτῳ καὶ τῷ
δευτέρῳ τὸ τρίτον κἀκείνῳ τὸ τέταρτον, ἅμα μὲν
διεξέρχεσθαι χρὴ τοὺς πόρους ταυτὶ τὰ σμικρὰ
καὶ ληρώδη ψήγματα, ἅμα δ' ἀποπάλλεσθαι τοῦ
51 μετ' αὐτὸ ‖ τεταγμένου, καίτοι κατὰ πᾶν ὁμοίου
τὴν φύσιν ὑπάρχοντος.

Οὐδὲ γὰρ ἡ τοιαύτη πάλιν ὑπόθεσις ἄτολμος,
ἀλλ', εἰ χρὴ τἀληθὲς εἰπεῖν, μακρῷ τῶν ἔμπροσθεν ἀναισχυντοτέρα, πέντε σιδηρίων ὁμοίων ἀλλή-

Then, again, when a small body becomes entangled with another small body, or when a body
in motion becomes entangled with another also in
motion, they do not rebound at once. For, further,
there will of course be others which break in upon
them from above, from below, from front and rear,
from right and left, and which shake and agitate
them and never let them rest. Moreover, we must
perforce suppose that each of these small bodies has
a large number of these hook-like extremities. For
by one it attaches itself to its neighbours, by another
—the topmost one—to the lodestone, and by the
bottom one to the iron. For if it were attached to
the stone above and not interlocked with the iron
below, this would be of no use.[1] Thus, the upper part
of the superior extremity must hang from the lodestone, and the iron must be attached to the lower
end of the inferior extremity ; and, since they interlock with each other by their sides as well, they
must, of course, have hooks there too. Keep in
mind also, above everything, what small bodies these
are which possess all these different kinds of outgrowths. Still more, remember how, in order that
the second piece of iron may become attached to the
first, the third to the second, and to that the fourth,
these absurd little particles must both penetrate the
passages in the first piece of iron and at the same
time rebound from the piece coming next in the
series, although this second piece is naturally in
every way similar to the first.

Such an hypothesis, once again, is certainly not
lacking in audacity ; in fact, to tell the truth, it is far
more shameless than the previous ones ; according

---

[1] *i.e.* from the point of view of the theory.

λοις ἐφεξῆς τεταγμένων διὰ τοῦ πρώτου διαδυό-
μενα ῥαδίως τῆς λίθου τὰ μόρια κατὰ τὸ δεύτερον
ἀποπάλλεσθαι καὶ μὴ διὰ τούτου κατὰ τὸν αὐτὸν
τρόπον ἑτοίμως διεξέρχεσθαι. καὶ μὴν ἑκατέρως
ἄτοπον. εἰ μὲν γὰρ ἀποπάλλεται, πῶς εἰς τὸ
τρίτον ὠκέως διεξέρχεται; εἰ δ᾽ οὐκ ἀποπάλλεται,
πῶς κρεμάννυται τὸ δεύτερον ἐκ τοῦ πρώτου; τὴν
γὰρ ἀπόπαλσιν αὐτὸς ὑπέθετο δημιουργὸν τῆς
ὁλκῆς.

Ἀλλ᾽, ὅπερ ἔφην, εἰς ἀδολεσχίαν ἀναγκαῖον
ἐμπίπτειν, ἐπειδάν τις τοιούτοις ἀνδράσι διαλέγη-
ται. σύντομον οὖν τινα καὶ κεφαλαιώδη λόγον
εἰπὼν ἀπαλλάττεσθαι βούλομαι. τοῖς Ἀσκληπιάδου γράμμασιν εἴ τις ἐπιμελῶς ὁμιλήσειε, τήν
τε πρὸς τὰς ἀρχὰς ἀκολουθίαν τῶν τοιούτων
δογμάτων ἀκριβῶς ἂν ἐκμάθοι καὶ τὴν πρὸς τὰ
φαινόμενα μάχην. ὁ μὲν οὖν Ἐπίκουρος τὰ
52 φαινόμενα φυλάττειν βουλόμενος ἀσχημονεῖ ‖ φι-
λοτιμούμενος ἐπιδεικνύειν αὐτὰ ταῖς ἀρχαῖς ὁμο-
λογοῦντα· ὁ δ᾽ Ἀσκληπιάδης τὸ μὲν ἀκόλουθον
ταῖς ἀρχαῖς φυλάττει, τοῦ φαινομένου δ᾽ οὐδὲν
αὐτῷ μέλει. ὅστις οὖν βούλεται τὴν ἀτοπίαν
ἐξελέγχειν τῶν ὑποθέσεων, εἰ μὲν πρὸς Ἀσκληπιάδην ὁ λόγος αὐτῷ γίγνοιτο, τῆς πρὸς τὸ
φαινόμενον ὑπομιμνησκέτω μάχης· εἰ δὲ πρὸς
Ἐπίκουρον, τῆς πρὸς τὰς ἀρχὰς διαφωνίας. αἱ
δ᾽ ἄλλαι σχεδὸν αἱρέσεις αἱ τῶν ὁμοίων ἀρχῶν
ἐχόμεναι τελέως ἀπέσβησαν, αὗται δ᾽ ἔτι μόναι
διαρκοῦσιν οὐκ ἀγεννῶς. καίτοι τὰ μὲν Ἀσκληπιάδου Μηνόδοτος ὁ ἐμπειρικὸς ἀφύκτως
ἐξελέγχει, τήν τε πρὸς τὰ φαινόμενα μάχην ὑπο-
μιμνήσκων αὐτὸν καὶ τὴν πρὸς ἄλληλα· τὰ δ᾽

to it, when five similar pieces of iron are arranged in a line, the particles of the lodestone which easily traverse the first piece of iron rebound from the second, and do not pass readily through it in the same way. Indeed, it is nonsense, whichever alternative is adopted. For, if they do rebound, how then do they pass through into the third piece ? And if they do not rebound, how does the second piece become suspended to the first ? For Epicurus himself looked on the rebound as the active agent in attraction.

But, as I have said, one is driven to talk nonsense whenever one gets into discussion with such men. Having, therefore, given a concise and summary statement of the matter, I wish to be done with it. For if one diligently familiarizes oneself with the writings of Asclepiades, one will see clearly their logical dependence on his first principles, but also their disagreement with observed facts. Thus, Epicurus, in his desire to adhere to the facts, cuts an awkward figure by aspiring to show that these agree with his principles, whereas Asclepiades safeguards the sequence of principles, but pays no attention to the obvious fact. Whoever, therefore, wishes to expose the absurdity of their hypotheses, must, if the argument be in answer to Asclepiades, keep in mind his disagreement with observed fact ; or if in answer to Epicurus, his discordance with his principles. Almost all the other sects depending on similar principles are now entirely extinct, while these alone maintain a respectable existence still. Yet the tenets of Asclepiades have been unanswerably confuted by Menodotus the Empiricist, who draws his attention to their opposition to phenomena and to each other ;

Ἐπικούρου πάλιν ὁ Ἀσκληπιάδης ἐχόμενος ἀεὶ
τῆς ἀκολουθίας, ἧς ἐκεῖνος οὐ πάνυ τι φαίνεται
φροντίζων.

Ἀλλ᾽ οἱ νῦν ἄνθρωποι, πρὶν καὶ ταύτας
ἐκμαθεῖν τὰς αἱρέσεις καὶ τὰς ἄλλας τὰς
βελτίους κἄπειτα χρόνῳ πολλῷ κρῖναί τε καὶ
βασανίσαι τὸ καθ᾽ ἑκάστην αὐτῶν ἀληθές τε καὶ
ψεῦδος, οἱ μὲν ἰατροὺς ἑαυτούς, οἱ δὲ φιλοσόφους
53 ὀνομάζουσι μηδὲν εἰδότες. ‖ οὐδὲν οὖν θαυμαστὸν
ἐπίσης τοῖς ἀληθέσι τὰ ψευδῆ τετιμῆσθαι. ὅτῳ
γὰρ ἂν ἕκαστος πρώτῳ περιτύχῃ διδασκάλῳ,
τοιοῦτος ἐγένετο, μὴ περιμείνας μηδὲν ἔτι παρ᾽
ἄλλου μαθεῖν. ἔνιοι δ᾽ αὐτῶν, εἰ καὶ πλείοσιν
ἐντύχοιεν, ἀλλ᾽ οὕτω γ᾽ εἰσὶν ἀσύνετοί τε καὶ
βραδεῖς τὴν διάνοιαν, ὥστε καὶ γεγηρακότες οὔπω
συνιᾶσιν ἀκολουθίαν λόγου. πάλαι δὲ τοὺς τοιούτους ἐπὶ τὰς βαναύσους ἀπέλυον τέχνας. ἀλλὰ
ταῦτα μὲν ἐς ὅ τι τελευτήσει θεὸς οἶδεν.

Ἡμεῖς δ᾽ ἐπειδή, καίτοι φεύγοντες ἀντιλέγειν
τοῖς ἐν αὐταῖς ταῖς ἀρχαῖς εὐθὺς ἐσφαλμένοις,
ὅμως ἠναγκάσθημεν ὑπ᾽ αὐτῆς τῶν πραγμάτων
τῆς ἀκολουθίας εἰπεῖν τινα καὶ διαλεχθῆναι πρὸς
αὐτούς, ἔτι καὶ τοῦτο προσθήσομεν τοῖς εἰρημένοις, ὡς οὐ μόνον τὰ καθαίροντα φάρμακα
πέφυκεν ἐπισπᾶσθαι τὰς οἰκείας ποιότητας ἀλλὰ
καὶ τὰ τοὺς σκόλοπας ἀνάγοντα καὶ τὰς τῶν
βελῶν ἀκίδας εἰς πολὺ βάθος σαρκὸς ἐμπεπαρ-
μένας ἐνίοτε. καὶ μέντοι καὶ ὅσα τοὺς ἰοὺς τῶν
θηρίων ἢ τοὺς ἐμπεφαρμαγμένους τοῖς βέλεσιν
ἀνέλκει, καὶ ταῦτα τὴν αὐτὴν ταῖς ἡρακλείαις
54 λίθοις ἐπι‖δείκνυται δύναμιν. ἔγωγ᾽ οὖν οἶδά ποτε
καταπεπαρμένον ἐν ποδὶ νεανίσκου σκόλοπα τοῖς

and, again, those of Epicurus have been confuted by Asclepiades, who adhered always to logical sequence, about which Epicurus evidently cares little.

Now people of the present day do not begin by getting a clear comprehension of these sects, as well as of the better ones, thereafter devoting a long time to judging and testing the true and false in each of them; despite their ignorance, they style themselves, some "physicians" and others "philosophers." No wonder, then, that they honour the false equally with the true. For everyone becomes like the first teacher that he comes across, without waiting to learn anything from anybody else. And there are some of them, who, even if they meet with more than one teacher, are yet so unintelligent and slow-witted that even by the time they have reached old age they are still incapable of understanding the steps of an argument. . . . In the old days such people used to be set to menial tasks. . . . What will be the end of it God knows!

Now, we usually refrain from arguing with people whose principles are wrong from the outset. Still, having been compelled by the natural course of events to enter into some kind of a discussion with them, we must add this further to what was said— that it is not only cathartic drugs which naturally attract their special qualities,[1] but also those which remove thorns and the points of arrows such as sometimes become deeply embedded in the flesh. Those drugs also which draw out animal poisons or poisons applied to arrows all show the same faculty as does the lodestone. Thus, I myself have seen a thorn which was embedded in a young man's foot fail to

[1] cf. p. 69, note 2.

μὲν δακτύλοις ἕλκουσιν ἡμῖν βιαίως οὐκ ἀκολου-
θήσαντα, φαρμάκου δ' ἐπιτεθέντος ἀλύπως τε καὶ
διὰ ταχέων ἀνελθόντα. καίτοι καὶ πρὸς τοῦτό
τινες ἀντιλέγουσι φάσκοντες, ὅταν ἡ φλεγμονὴ
λυθῇ τοῦ μέρους, αὐτόματον ἐξιέναι τὸν σκόλοπα
πρὸς οὐδενὸς ἀνελκόμενον. ἀλλ' οὗτοί γε πρῶτον μὲν ἀγνοεῖν ἐοίκασιν, ὡς ἄλλα μέν ἐστι
φλεγμονῆς, ἄλλα δὲ τῶν οὕτω καταπεπαρμένων
ἑλκτικὰ φάρμακα· καίτοι γ' εἴπερ ἀφλεγμάντων
γενομένων ἐξεκρίνετο τὰ παρὰ φύσιν, ὅσα φλεγμονῆς ἐστι λυτικά, ταῦτ' εὐθὺς ἂν ἦν κἀκείνων
ἑλκτικά.

Δεύτερον δ', ὃ καὶ μᾶλλον ἄν τις θαυμάσειεν,
ὡς οὐ μόνον ἄλλα μὲν τοὺς σκόλοπας, ἄλλα δὲ
τοὺς ἰοὺς ἐξάγει φάρμακα, ἀλλὰ καὶ αὐτῶν τῶν
τοὺς ἰοὺς ἑλκόντων τὰ μὲν τὸν τῆς ἐχίδνης, τὰ δὲ
τὸν τῆς τρυγόνος, τὰ δ' ἄλλου τινὸς ἐπισπᾶται
καὶ σαφῶς ἔστιν ἰδεῖν τοῖς φαρμάκοις ἐπικει-
μένους αὐτούς. ἐνταῦθ' οὖν Ἐπίκουρον μὲν
55 ἐπαινεῖν χρὴ τῆς πρὸς ‖ τὸ φαινόμενον αἰδοῦς,
μέμφεσθαι δὲ τὸν λόγον τῆς αἰτίας. ὃν γὰρ
ἡμεῖς ἕλκοντες τοῖς δακτύλοις οὐκ ἀνηγάγομεν
σκόλοπα, τοῦτον ὑπὸ τῶν σμικρῶν ἐκείνων ἀνέλ-
κεσθαι ψηγμάτων, πῶς οὐ παντάπασιν ἄτοπον
εἶναι χρὴ νομίζειν;

Ἆρ' οὖν ἤδη πεπείσμεθα τῶν ὄντων ἑκάστῳ
δύναμίν τιν' ὑπάρχειν, ᾗ τὴν οἰκείαν ἕλκει
ποιότητα, τὸ μὲν μᾶλλον, τὸ δ' ἧττον;

Ἢ καὶ τὸ τῶν πυρῶν ἔτι παράδειγμα προ-

---

[1] That is to say, the two properties should go together in
all cases—which they do not.   [2] *Trygon pastinaca.*

come out when we exerted forcible traction with
our fingers, and yet come away painlessly and rapidly
on the application of a medicament. Yet even to
this some people will object, asserting that when the
inflammation is dispersed from the part the thorn
comes away of itself, without being pulled out by
anything. But these people seem, in the first place,
to be unaware that there are certain drugs for
drawing out inflammation and different ones for
drawing out embedded substances; and surely if it
was on the cessation of an inflammation that the
abnormal matters were expelled, then all drugs which
disperse inflammations ought, *ipso facto*, to possess the
power of extracting these substances as well.[1]

And secondly, these people seem to be unaware of
a still more surprising fact, namely, that not merely
do certain medicaments draw out thorns and others
poisons, but that of the latter there are some which
attract the poison of the viper, others that of the
sting-ray,[2] and others that of some other animal; we
can, in fact, plainly observe these poisons deposited
on the medicaments. Here, then, we must praise
Epicurus for the respect he shows towards obvious
facts, but find fault with his views as to causation.
For how can it be otherwise than extremely foolish to
suppose that a thorn which we failed to remove by
digital traction could be drawn out by these minute
particles?

Have we now, therefore, convinced ourselves that
everything which exists[3] possesses a faculty by which
it attracts its proper quality, and that some things do
this more, and some less?

Or shall we also furnish our argument with the

[3] *cf.* p. 66, note 3.

χειρισόμεθα τῷ λόγῳ; φανήσονται γὰρ οἶμαι
καὶ τῶν γεωργῶν αὐτῶν ἀμαθέστεροι περὶ τὴν
φύσιν οἱ μηδὲν ὅλως ὑπὸ μηδενὸς ἕλκεσθαι
συγχωροῦντες· ὡς ἔγωγε πρῶτον, μὲν ἀκούσας
τὸ γιγνόμενον ἐθαύμασα καὶ αὐτὸς ἠβουλήθην
αὐτόπτης αὐτοῦ καταστῆναι. μετὰ ταῦτα δέ,
ὡς καὶ τὰ τῆς πείρας ὡμολόγει, τὴν αἰτίαν
σκοπούμενος ἐν παμπόλλῳ χρόνῳ κατὰ πάσας
τὰς αἱρέσεις οὐδεμίαν ἄλλην εὑρεῖν οἷός τ' ἦν
οὐδ' ἄχρι τοῦ πιθανοῦ προϊοῦσαν ἀλλὰ κατα-
γελάστους τε καὶ σαφῶς ἐξελεγχομένας τὰς ἄλλας
ἁπάσας πλὴν τῆς τὴν ὁλκὴν πρεσβευούσης.

Ἔστι δὲ τὸ γιγνόμενον τοιόνδε. κατακομί-
56 ζοντες οἱ παρ' ἡμῖν γεωργοὶ τοὺς ‖ ἐκ τῶν ἀγρῶν
πυροὺς εἰς τὴν πόλιν ἐν ἀμάξαις τισίν, ὅταν
ὑφελέσθαι βουληθῶσιν, ὥστε μὴ φωραθῆναι,
κεράμι' ἄττα πληρώσαντες ὕδατος μέσοις αὐτοῖς
ἐνιστᾶσιν. ἕλκοντες οὖν ἐκεῖνοι διὰ τοῦ κεραμίου
τὸ ὑγρὸν εἰς αὑτοὺς ὄγκον μὲν καὶ βάρος
προσκτῶνται, κατάδηλοι δ' οὐ πάνυ γίγνονται τοῖς
ὁρῶσιν, εἰ μή τις προπεπυσμένος ἤδη περιεργό-
τερον ἐπισκοποῖτο. καίτοι γ' εἰ βουληθείης ἐν
ἡλίῳ καταθεῖναι πάνυ θερμῷ ταὐτὸν ἀγγεῖον,
ἐλάχιστον παντελῶς εὑρήσεις τὸ δαπανώμενον
ἐφ' ἑκάστης ἡμέρας. οὕτως ἄρα καὶ τῆς ἡλιακῆς
θερμασίας τῆς σφοδρᾶς ἰσχυροτέραν οἱ πυροὶ
δύναμιν ἔχουσιν ἕλκειν εἰς ἑαυτοὺς τὴν πλησιάζουσαν ὑγρότητα. λῆρος οὖν ἐνταῦθα μακρὸς
ἡ πρὸς τὸ λεπτομερὲς φορὰ τοῦ περιέχοντος
ἡμᾶς ἀέρος καὶ μάλισθ' ὅταν ἱκανῶς ᾖ θερμός,

---

[1] The way that corn can attract moisture.

illustration afforded by *corn* ?[1] For those who refuse
to admit that anything is attracted by anything else,
will, I imagine, be here proved more ignorant regarding Nature than the very peasants. When, for
my own part, I first learned of what happens, I was
surprised, and felt anxious to see it with my own
eyes. Afterwards, when experience also had con-
firmed its truth, I sought long among the various
sects for an explanation, and, with the exception
of that which gave the first place to *attraction,* I
could find none which even approached plausibility,
all the others being ridiculous and obviously quite
untenable.

What happens, then, is the following. When our
peasants are bringing corn from the country into the
city in wagons, and wish to filch some away without
being detected, they fill earthen jars with water and
stand them among the corn ; the corn then draws
the moisture into itself through the jar and acquires
additional bulk and weight, but the fact is never
detected by the onlookers unless someone who knew
about the trick before makes a more careful inspection. Yet, if you care to set down the same vessel in
the very hot sun, you will find the daily loss to be
very little indeed. Thus corn has a greater power
than extreme solar heat of drawing to itself the
moisture in its neighbourhood.[2] Thus the theory
that the water is carried towards the rarefied part of
the air surrounding us [3] (particularly when that is
distinctly warm) is utter nonsense ; for although it is

[2] Specific attraction of the "proper" quality ; *cf.* p. 85,
note 3.

[3] Theory of evaporation insufficient to account for it. *cf.*
p 104, note 1.

πολὺ μὲν ὑπάρχοντος ἢ κατὰ τοὺς πυροὺς λεπτο-
μερεστέρου, δεχομένου δ' οὐδὲ τὸ δέκατον μέρος
τῆς εἰς ἐκείνους μεταλαμβανομένης ὑγρότητος.

XV

Ἐπεὶ δ' ἱκανῶς ἠδολεσχήσαμεν οὐχ ἑκόντες,
ἀλλ', ὡς ἡ παροιμία φησί, μαινομένοις ἀναγ-
57 κασθέντες συμ‖μανῆναι, πάλιν ἐπὶ τὴν τῶν οὔρων
ἐπανέλθωμεν διάκρισιν, ἐν ᾗ τῶν μὲν Ἀσκληπιάδου λήρων ἐπιλαθώμεθα, μετὰ δὲ τῶν πε-
πεισμένων διηθεῖσθαι τὰ οὖρα διὰ τῶν νεφρῶν,
τίς ὁ τρόπος τῆς ἐνεργείας ἐστίν, ἐπισκεψώμεθα·
πάντως γὰρ ἢ ἐξ αὐτῶν ἐπὶ τοὺς νεφροὺς φέρεται
τὰ οὖρα τοῦτο βέλτιον εἶναι νομίζοντα, καθάπερ
ἡμεῖς, ὁπόταν εἰς τὴν ἀγορὰν ἀπίωμεν· ἤ, εἰ τοῦτ'
ἀδύνατον, ἕτερόν τι χρὴ τῆς φορᾶς αὐτῶν ἐξευρεῖν
αἴτιον. τί δὴ τοῦτ' ἔστιν; εἰ γὰρ μὴ τοῖς νεφροῖς
δώσομέν τινα δύναμιν ἑλκτικὴν τῆς τοιαύτης
ποιότητος, ὡς Ἱπποκράτης ἐνόμιζεν, οὐδὲν ἕτερον
ἐξευρήσομεν. ὅτι μὲν γὰρ ἤτοι τούτους ἕλκειν
αὐτὸ προσῆκεν ἢ τὰς φλέβας πέμπειν, εἴπερ γε μὴ
ἐξ ἑαυτοῦ φέρεται, παντί που δῆλον. ἀλλ' εἰ μὲν
αἱ φλέβες περιστελλόμεναι προωθοῖεν, οὐκ ἐκεῖνο
μόνον, ἀλλὰ σὺν αὐτῷ καὶ τὸ πᾶν αἷμα τὸ
περιεχόμενον ἐν ἑαυταῖς εἰς τοὺς νεφροὺς ἐκ-
θλίψουσιν· εἰ δὲ τοῦτ' ἀδύνατον, ὡς δείξομεν,
λείπεται τοὺς νεφροὺς ἕλκειν.

much more rarefied there than it is amongst the corn, yet it does not take up a tenth part of the moisture which the corn does.

XV

Since, then, we have talked sufficient nonsense—not willingly, but because we were forced, as the proverb says, " to behave madly among madmen "—let us return again to the subject of urinary secretion. Here let us forget the absurdities of Asclepiades, and, in company with those who are persuaded that the urine does pass through the kidneys, let us consider what is the character of this function. For, most assuredly, either the urine is conveyed by its own motion to the kidneys, considering this the better course (as do we when we go off to market![1]), or, if this be impossible, then some other reason for its conveyance must be found. What, then, is this? If we are not going to grant the kidneys a faculty for attracting this particular quality,[2] as Hippocrates held, we shall discover no other reason. For, surely everyone sees that either the kidneys must attract the urine, or the veins must propel it—if, that is, it does not move of itself. But if the veins did exert a propulsive action when they contract, they would squeeze out into the kidneys not merely the urine, but along with it the whole of the blood which they contain.[3] And if this is impossible, as we shall show, the remaining explanation is that the kidneys do exert traction.

[1] Playful suggestion of free-will in the urine.
[2] Specific attraction. *cf.* p. 87, note 2.
[3] *i.e.* there would be no selective action.

Πῶς οὖν ἀδύνατον τοῦτο; τῶν νεφρῶν ἡ θέσις ἀντιβαίνει. οὐ γὰρ δὴ οὕτω γ᾽ ὑπόκεινται τῇ
58 κοίλῃ φλεβὶ ‖ καθάπερ τοῖς ἐξ ἐγκεφάλου περιττώμασιν ἔν τε τῇ ῥινὶ καὶ κατὰ τὴν ὑπερῴαν οἱ τοῖς ἠθμοῖς ὅμοιοι πόροι, ἀλλ᾽ ἑκατέρωθεν αὐτῇ παράκεινται. καὶ μήν, εἴπερ ὁμοίως τοῖς ἠθμοῖς ὅσον ἂν ᾖ λεπτότερον καὶ τελέως ὀρρῶδες, τοῦτο μὲν ἑτοίμως διαπέμπουσι, τὸ δὲ παχύτερον ἀποστέγουσιν, ἅπαν ἐπ᾽ αὐτοὺς ἰέναι χρὴ τὸ αἷμα τὸ περιεχόμενον ἐν τῇ κοίλῃ φλεβί, καθάπερ εἰς τοὺς τρυγητοὺς ὁ πᾶς οἶνος ἐμβάλλεται. καὶ μέν γε καὶ τὸ τοῦ γάλακτος τοῦ τυρουμένου παράδειγμα σαφῶς ἄν, ὃ βούλομαι λέγειν, ἐνδείξαιτο. καὶ γὰρ καὶ τοῦτο πᾶν ἐμβληθὲν εἰς τοὺς ταλάρους οὐ πᾶν διηθεῖται, ἀλλ᾽ ὅσον μὲν ἂν ᾖ λεπτότερον τῆς εὐρύτητος τῶν πλοκάμων, εἰς τὸ κάταντες φέρεται καὶ τοῦτο μὲν ὀρρὸς ἐπονομάζεται· τὸ λοιπὸν δὲ τὸ παχὺ τὸ μέλλον ἔσεσθαι τυρός, ὡς ἂν οὐ παραδεχομένων αὐτὸ τῶν ἐν τοῖς ταλάροις πόρων, οὐ διεκπίπτει κάτω. καὶ τοίνυν, εἴπερ οὕτω μέλλει διηθεῖσθαι τῶν νεφρῶν ὁ τοῦ αἵματος ὀρρός, ἅπαν ἐπ᾽ αὐτοὺς ἥκειν χρὴ τὸ αἷμα καὶ μὴ τὸ μὲν ναί, τὸ δ᾽ οὔ. ‖

59 Πῶς οὖν ἔχει τὸ φαινόμενον ἐκ τῆς ἀνατομῆς;

Τὸ μὲν ἕτερον μέρος τῆς κοίλης ἄνω πρὸς τὴν καρδίαν ἀναφέρεται, τὸ λοιπὸν δ᾽ ἐπιβαίνει τῇ ῥάχει καθ᾽ ὅλης αὐτῆς ἐκτεινόμενον ἄχρι τῶν σκελῶν, ὥστε τὸ μὲν ἕτερον οὐδ᾽ ἐγγὺς ἀφικνεῖται

---

[1] Nasal mucus was supposed to be the non-utilizable part of the nutriment conveyed to the brain. *cf.* p. 214, note 3.

And how is propulsion by the veins impossible?
The situation of the kidneys is against it. They
do not occupy a position beneath the hollow vein
[vena cava] as does the sieve-like [ethmoid] passage
in the nose and palate in relation to the surplus
matter from the brain;[1] they are situated on both
sides of it. Besides, if the kidneys are like sieves,
and readily let the thinner serous [whey-like] portion
through, and keep out the thicker portion, then
the whole of the blood contained in the vena cava
must go to them, just as the whole of the wine
is thrown into the filters. Further, the example
of milk being made into cheese will show clearly
what I mean. For this, too, although it is all
thrown into the wicker strainers, does not all
percolate through; such part of it as is too fine in
proportion to the width of the meshes passes downwards, and this is called *whey* [serum]; the remaining
thick portion which is destined to become cheese
cannot get down, since the pores of the strainers will
not admit it. Thus it is that, if the blood-serum has
similarly to percolate through the kidneys, the whole
of the blood must come to them, and not merely one
part of it.

What, then, is the appearance as found on dissection?

One division of the vena cava is carried upwards[2]
to the heart, and the other mounts upon the spine
and extends along its whole length as far as the legs;
thus one division does not even come near the

[2] He means from its origin in the liver (*i.e.* in the three
hepatic veins). His idea was that the upper division took
nutriment to heart, lungs, head, etc., and the lower division
to lower part of body. On the relation of right auricle to
vena cava and right ventricle, *cf.* p. 321, notes 4 and 5.

τῶν νεφρῶν, τὸ λοιπὸν δὲ πλησιάζει μέν, οὐ μὴν
εἰς αὐτούς γε καταφύεται. ἐχρῆν δ', εἴπερ
ἔμελλεν ὡς δι' ἠθμῶν αὐτῶν καθαρθήσεσθαι τὸ
αἷμα, πᾶν ἐμπίπτειν εἰς αὐτοὺς κἄπειτα κάτω
μὲν φέρεσθαι τὸ λεπτόν, ἴσχεσθαι δ' ἄνω τὸ
παχύ. νυνὶ δ' οὐχ οὕτως ἔχει· πλάγιοι γὰρ
ἑκατέρωθεν τῆς κοίλης φλεβὸς οἱ νεφροὶ κεῖνται.
οὔκουν ὡς ἠθμοὶ διηθοῦσι, πεμπούσης μὲν ἐκείνης,
αὐτοὶ δ' οὐδεμίαν εἰσφερόμενοι δύναμιν, ἀλλ'
ἕλκουσι δηλονότι· τοῦτο γὰρ ἔτι λείπεται.

Πῶς οὖν ἕλκουσιν; εἰ μέν, ὡς Ἐπίκουρος οἴεται
τὰς ὁλκὰς ἁπάσας γίγνεσθαι κατὰ τὰς τῶν
ἀτόμων ἀποπάλσεις τε καὶ περιπλοκάς, ἄμεινον
ἦν ὄντως εἰπεῖν αὐτοὺς μηδ' ἕλκειν ὅλως· πολὺ
γὰρ ἂν οὕτω γε τῶν ἐπὶ τῆς ἡρακλείας λίθου
60 μικρῷ πρόσθεν εἰρη‖μένων ὁ λόγος ἐξεταζόμενος
εὑρεθείη γελοιότερος· ἀλλ' ὡς Ἱπποκράτης ἠβού-
λετο. λεχθήσεται δὲ σαφέστερον ἐπὶ προήκοντι
τῷ λόγῳ. νυνὶ γὰρ οὐ τοῦτο πρόκειται διδάσκειν,
ἀλλ' ὡς οὔτ' ἄλλο τι δυνατὸν εἰπεῖν αἴτιον εἶναι
τῆς τῶν οὔρων διακρίσεως πλὴν τῆς ὁλκῆς τῶν
νεφρῶν οὔθ' οὕτω γίγνεσθαι τὴν ὁλκήν, ὡς οἱ
μηδεμίαν οἰκείαν διδόντες τῇ φύσει δύναμιν
οἴονται γίγνεσθαι.

Τούτου γὰρ ὁμολογηθέντος, ὡς ἔστιν ὅλως τις
ἐν τοῖς ὑπὸ φύσεως διοικουμένοις δύναμις ἑλκτική,
ληρώδης νομίζοιτ' ἂν ὁ περὶ ἀναδόσεως τροφῆς
ἄλλο τι λέγειν ἐπιχειρῶν.

---

[1] We arrive at our belief by excluding other possibilities.
[2] *i.e.* the mechanistic physicists. *cf.* pp. 45–47.
[3] *cf.* p. 85, note 3.

kidneys, while the other approaches them but is certainly not inserted into them. Now, if the blood were destined to be purified by them as if they were sieves, the whole of it would have to fall into them, the thin part being thereafter conveyed downwards, and the thick part retained above. But, as a matter of fact, this is not so. For the kidneys lie on either side of the vena cava. They therefore do not act like sieves, filtering fluid sent to them by the vena cava, and themselves contributing no force. They obviously exert traction ; for this is the only remaining alternative.

How, then, do they exert this traction ? If, as Epicurus thinks, all attraction takes place by virtue of the *rebounds* and *entanglements* of atoms, it would be certainly better to maintain that the kidneys have no attractive action at all; for his theory, when examined, would be found as it stands to be much more ridiculous even than the theory of the lodestone, mentioned a little while ago. Attraction occurs in the way that Hippocrates laid down ; this will be stated more clearly as the discussion proceeds ; for the present our task is not to demonstrate this, but to point out that no other cause of the secretion of urine can be given except that of attraction by the kidneys,[1] and that this attraction does not take place in the way imagined by people who do not allow Nature a faculty of her own.[2]

For if it be granted that there is any attractive faculty at all in those things which are governed by Nature,[3] a person who attempted to say anything else about the absorption of nutriment[4] would be considered a fool.

[4] The subject of *anadosis* is taken up in the next chapter. *cf.* also p. 62, note 1.

XVI

Ἐρασίστρατος δ' οὐκ οἶδ' ὅπως ἑτέραις μέν τισι
δόξαις εὐήθεσιν ἀντεῖπε διὰ μακρῶν, ὑπερέβη δὲ
τελέως τὴν Ἱπποκράτους, οὐδ' ἄχρι τοῦ μνημονεῦ-
σαι μόνον αὐτῆς, ὡς ἐν τοῖς περὶ καταπόσεως
ἐποίησεν, ἀξιώσας. ἐν ἐκείνοις μὲν γὰρ ἄχρι
τοσούτου φαίνεται μνημονεύων, ὡς τοὔνομ'
εἰπεῖν τῆς ὁλκῆς μόνον ὧδέ πως γράφων·

" Ὁλκὴ μὲν οὖν τῆς κοιλίας οὐδεμία φαίνεται
61 εἶναι"· περὶ δὲ τῆς ‖ ἀναδόσεως τὸν λόγον ποι-
ούμενος οὐδ' ἄχρι συλλαβῆς μιᾶς ἐμνημόνευσε
τῆς Ἱπποκρατείου δόξης. καίτοι γ' ἐπήρκεσεν
ἂν ἡμῖν, εἰ καὶ τοῦτ' ἔγραψε μόνον, ὡς Ἱπποκράτης εἰπὼν "Σάρκες ὁλκοὶ καὶ ἐκ κοιλίης καὶ
ἔξωθεν" ψεύδεται· οὔτε γὰρ ἐκ τῆς κοιλίας οὔτ'
ἔξωθεν ἕλκειν δύνανται. εἰ δὲ καὶ ὅτι μήτρας
αἰτιώμενος ἄρρωστον αὐχένα κακῶς εἶπεν "Οὐ
γὰρ δύναται αὐτέης ὁ στόμαχος εἰρύσαι τὴν γονήν,"
ἢ εἰ καί τι τοιοῦτον ἄλλο γράφειν ὁ Ἐρασίστρατος ἠξίωσε, τότ' ἂν καὶ ἡμεῖς πρὸς αὐτὸν
ἀπολογούμενοι εἴπομεν·

Ὦ γενναῖε, μὴ ῥητορικῶς ἡμῶν κατάτρεχε
χωρὶς ἀποδείξεως, ἀλλ' εἰπέ τινα κατηγορίαν
τοῦ δόγματος, ἵν' ἢ πεισθῶμέν σοι ὡς καλῶς
ἐξέλεγχοντι τὸν παλαιὸν λόγον ἢ μεταπείσωμεν

---

[1] On Erasistratus v. Introd. p. xii. His view that the
stomach exerts no *holké*, or attraction, is dealt with more
fully in Book III., chap. viii.

XVI

Now, while Erasistratus [1] for some reason replied at great length to certain other foolish doctrines, he entirely passed over the view held by Hippocrates, not even thinking it worth while to mention it, as he did in his work " On Deglutition "; in that work, as may be seen, he did go so far as at least to make mention of the word *attraction*, writing somewhat as follows :

" Now, the stomach does not appear to exercise any attraction." [1] But when he is dealing with *anadosis* he does not mention the Hippocratic view even to the extent of a single syllable. Yet we should have been satisfied if he had even merely written this : " Hippocrates lies in saying ' The flesh [2] attracts both from the stomach and from without,' for it cannot attract either from the stomach or from without." Or if he had thought it worth while to state that Hippocrates was wrong in criticizing the weakness of the neck of the uterus, " seeing that the orifice of the uterus has no power of attracting semen," [3] or if he [Erasistratus] had thought proper to write any other similar opinion, then we in our turn would have defended ourselves in the following terms :

" My good sir, do not run us down in this rhetorical fashion without some proof ; state some definite objection to our view, in order that either you may convince us by a brilliant refutation of the ancient doctrine, or that, on the other hand, we may convert you from your ignorance." Yet why do I

2 *i.e.* the tissues.   3 *cf.* p. 291.

ὡς ἀγνοοῦντα. καίτοι τί λέγω ῥητορικῶς; μὴ
γάρ, ἐπειδή τινες τῶν ῥητόρων, ἃ μάλιστ᾽ ἀδυνα-
τοῦσι διαλύεσθαι, ταῦτα διαγελάσαντες οὐδ᾽
ἐπιχειροῦσιν ἀντιλέγειν, ἤδη που τοῦτο καὶ ἡμεῖς
ἡγώμεθ᾽ εἶναι τὸ ῥητορικῶς· τὸ γὰρ διὰ λόγου
62 πιθανοῦ ἐστι τὸ ‖ ῥητορικῶς, τὸ δ᾽ ἄνευ λόγου
βωμολοχικόν, οὐ ῥητορικόν. οὔκουν οὔτε ῥητορικῶς οὔτε διαλεκτικῶς ἀντεῖπεν ὁ Ἐρασίστρατος
ἐν τῷ περὶ τῆς καταπόσεως λόγῳ. τί γάρ φησιν;
"Ὁλκὴ μὲν οὖν τῆς κοιλίας οὐδεμία φαίνεται
εἶναι." πάλιν οὖν αὐτῷ παρ᾽ ἡμῶν ἀντιμαρτυρῶν
ὁ αὐτὸς λόγος ἀντιπαραβαλλέσθω· περιστολὴ
μὲν οὖν τοῦ στομάχου οὐδεμία φαίνεται εἶναι.
καὶ πῶς οὐ φαίνεται; τάχ᾽ ἂν ἴσως εἴποι τις τῶν
ἀπ᾽ αὐτοῦ· τὸ γὰρ ἀεὶ τῶν ἄνωθεν αὐτοῦ μερῶν
συστελλομένων διαστέλλεσθαι τὰ κάτω πῶς οὐκ
ἔστι τῆς περιστολῆς ἐνδεικτικόν; αὖθις οὖν ἡμεῖς,
καὶ πῶς οὐ φαίνεται, φήσομεν, ἡ τῆς κοιλίας
ὁλκή; τὸ γὰρ ἀεὶ τῶν κάτωθεν μερῶν τοῦ
στομάχου διαστελλομένων συστέλλεσθαι τὰ ἄνω
πῶς οὐκ ἔστι τῆς ὁλκῆς ἐνδεικτικόν; εἰ δὲ
σωφρονήσειέ ποτε καὶ γνοίη τὸ φαινόμενον τοῦτο
μηδὲν μᾶλλον τῆς ἑτέρας τῶν δοξῶν ὑπάρχειν
ἐνδεικτικὸν ἀλλ᾽ ἀμφοτέρων εἶναι κοινόν, οὕτως
ἂν ἤδη δείξαιμεν αὐτῷ τὴν ὀρθὴν ὁδὸν τῆς τοῦ
ἀληθοῦς εὑρέσεως.

Ἀλλὰ περὶ μὲν τῆς κοιλίας αὖθις. ἡ δὲ τῆς
63 τροφῆς ἀνάδοσις οὐδὲν δεῖται ‖ τῆς πρὸς τὸ κενούμενον ἀκολουθίας ἅπαξ γε τῆς ἑλκτικῆς δυνάμεως

say "rhetorical"? For we too are not to suppose that when certain rhetoricians pour ridicule upon that which they are quite incapable of refuting, without any attempt at argument, their words are really thereby constituted rhetoric. For rhetoric proceeds by persuasive reasoning; words without reasoning are buffoonery rather than rhetoric. Therefore, the reply of Erasistratus in his treatise " On Deglutition " was neither rhetoric nor logic. For what is it that he says? " Now, the stomach does not appear to exercise any traction." Let us testify against him in return, and set our argument beside his in the same form. *Now, there appears to be no peristalsis[1] of the gullet.* " And how does this appear?" one of his adherents may perchance ask. " For is it not indicative of *peristalsis* that always when the upper parts of the gullet contract the lower parts dilate?" Again, then, we say, "And in what way does the attraction of the stomach not appear? For is it not indicative of *attraction* that always when the lower parts of the gullet dilate the upper parts contract?" Now, if he would but be sensible and recognize that this phenomenon is not more indicative of the one than of the other view, but that it applies equally to both,[2] we should then show him without further delay the proper way to the discovery of truth.

We will, however, speak about the stomach again. And the dispersal of nutriment [anadosis] need not make us have recourse to the theory regarding the

[1] *Peristalsis* may be used here to translate Gk. *peristolé,* meaning the contraction and dilation of muscle-fibres *circularly* round a lumen. *cf.* p. 263, note 2.

[2] For a demonstration that this phenomenon is a conclusive proof neither of *peristolé* nor of real vital *attraction,* but is found even in dead bodies *v.* p. 267.

ἐπὶ τῶν νεφρῶν ὡμολογημένης, ἢν καίτοι πάνυ
σαφῶς ἀληθῆ γιγνώσκων ὑπάρχειν ὁ Ἐρασίστρατος οὔτ' ἐμνημόνευσεν οὔτ' ἀντεῖπεν οὔθ'
ὅλως ἀπεφήνατο, τίν' ἔχει δόξαν ὑπὲρ τῆς τῶν
οὔρων διακρίσεως.

Ἢ διὰ τί προειπὼν εὐθὺς κατ' ἀρχὰς τῶν καθ'
ὅλου λόγων, ὡς ὑπὲρ τῶν φυσικῶν ἐνεργειῶν ἐρεῖ,
πρῶτον τίνες τ' εἰσὶ καὶ πῶς γίγνονται καὶ διὰ
τίνων τόπων, ἐπὶ τῆς τῶν οὔρων διακρίσεως, ὅτι
μὲν διὰ νεφρῶν, ἀπεφήνατο, τὸ δ' ὅπως γίγνεται
παρέλιπε; μάτην οὖν ἡμᾶς καὶ περὶ τῆς πέψεως
ἐδίδαξεν, ὅπως γίγνεται, καὶ περὶ τῆς τοῦ χολώ-
δους περιττώματος διακρίσεως κατατρίβει. ἤρκει
γὰρ εἰπεῖν κἀνταῦθα τὰ μόρια, δι' ὧν γίγνεται, τὸ
δ' ὅπως παραλιπεῖν. ἀλλὰ περὶ μὲν ἐκείνων εἶχε
λέγειν, οὐ μόνον δι' ὧν ὀργάνων ἀλλὰ καὶ καθ'
ὅντινα γίγνεται τρόπον, ὥσπερ οἶμαι καὶ περὶ τῆς
ἀναδόσεως· οὐ γὰρ ἤρκεσεν εἰπεῖν αὐτῷ μόνον,
ὅτι διὰ φλεβῶν, ἀλλὰ καὶ πῶς ἐπεξῆλθεν, ὅτι τῇ
64 πρὸς ‖ τὸ κενούμενον ἀκολουθίᾳ· περὶ δὲ τῶν
οὔρων τῆς διακρίσεως, ὅτι μὲν διὰ νεφρῶν γίγνεται, γράφει, τὸ δ' ὅπως οὐκέτι προστίθησιν.
οὐδὲ γὰρ οἶμαι τῇ πρὸς τὸ κενούμενον ἀκολουθίᾳ
ἢν εἰπεῖν· οὕτω γὰρ ἂν οὐδεὶς ὑπ' ἰσχουρίας
ἀπέθανεν οὐδέποτε μὴ δυναμένου πλείονος ἐπιρ-

---

[1] This was Erasistratus's favourite principle, known in
Latin as the "horror vacui" and in English as "Nature's
abhorrence of a vacuum," although these terms are not an
exact translation of the Greek. τὸ κενούμενον probably means

natural tendency of a vacuum to become refilled,[1] when once we have granted the attractive faculty of the kidneys. Now, although Erasistratus knew that this faculty most certainly existed, he neither mentioned it nor denied it, nor did he make any statement as to his views on the secretion of urine.

Why did he give notice at the very beginning of his "General Principles" that he was going to speak about natural activities—firstly what they are, how they take place, and in what situations—and then, in the case of urinary secretion, declared that this took place through the kidneys, but left out its method of occurrence? It must, then, have been for no purpose that he told us how digestion occurs, or spends time upon the secretion of biliary super-fluities;[2] for in these cases also it would have been sufficient to have named the parts through which the function takes place, and to have omitted the method. On the contrary, in these cases he was able to tell us not merely through what organs, but also in what way it occurs—as he also did, I think, in the case of *anadosis*; for he was not satisfied with saying that this took place through the veins, but he also considered fully the method, which he held to be from the tendency of a vacuum to become refilled. Concerning the secretion of urine, however, he writes that this occurs through the kidneys, but does not add in what *way* it occurs. I do not think he could say that this was from the tendency of matter to fill a vacuum,[3] for, if this were so, nobody would have ever died of retention of urine, since no more can

the vacuum, not the *matter evacuated,* although Galen else-where uses κενόω in the latter (non-classical) sense, *e.g.* pp. 67, 215. Akolouthia is a *following-up,* a *sequence,* almost a *con-sequence.* ² *v.* p. 123. ³ *cf.* Book II., chap. i.

ῥυῆναί ποτε παρὰ τὸ κενούμενον· ἄλλης γὰρ
αἰτίας μηδεμιᾶς προστεθείσης, ἀλλὰ μόνης τῆς
πρὸς τὸ κενούμενον ἀκολουθίας ποδηγούσης τὸ
συνεχές, οὐκ ἐγχωρεῖ πλέον ἐπιρρυῆναί ποτε τοῦ
κενουμένου. ἀλλ᾽ οὐδ᾽ ἄλλην τινὰ προσθεῖναι
πιθανὴν αἰτίαν εἶχεν, ὡς ἐπὶ τῆς ἀναδόσεως τὴν
ἔκθλιψιν τῆς γαστρός. ἀλλ᾽ αὕτη γ᾽ ἐπὶ τοῦ
κατὰ τὴν κοίλην αἵματος ἀπωλώλει τελέως, οὐ
τῷ μήκει μόνον τῆς ἀποστάσεως ἐκλυθεῖσα, ἀλλὰ
καὶ τῷ τὴν καρδίαν ὑπερκειμένην ἐξαρπάζειν
αὐτῆς σφοδρῶς καθ᾽ ἑκάστην διαστολὴν οὐκ
ὀλίγον αἷμα.

Μόνη δή τις ἔτι καὶ πάντων ἔρημος ἀπελείπετο
τῶν σοφισμάτων ἐν τοῖς κάτω τῆς κοίλης ἡ πρὸς ‖
65 τὸ κενούμενον ἀκολουθία, διά τε τοὺς ἐπὶ ταῖς
ἰσχουρίαις ἀποθνήσκοντας ἀπολωλεκυῖα τὴν πι-
θανότητα καὶ διὰ τὴν τῶν νεφρῶν θέσιν οὐδὲν
ἧττον. εἰ μὲν γὰρ ἅπαν ἐπ᾽ αὐτοὺς ἐφέρετο τὸ
αἷμα, δεόντως ἄν τις ἅπαν ἔφασκεν αὐτὸ καθαίρεσθαι. νυνὶ δέ, οὐ γὰρ ὅλον ἀλλὰ τοσοῦτον
αὐτοῦ μέρος, ὅσον αἱ μέχρι νεφρῶν δέχονται
φλέβες, ἐπ᾽ αὐτοὺς ἔρχεται, μόνον ἐκεῖνο καθαρ-
θήσεται. καὶ τὸ μὲν ὀρρῶδες αὐτοῦ καὶ λεπτὸν
οἷον δι᾽ ἠθμῶν τινων τῶν νεφρῶν διαδύσεται· τὸ
δ᾽ αἱματῶδές τε καὶ παχὺ κατὰ τὰς φλέβας ὑπο-
μένον ἐμποδὼν στήσεται τῷ κατόπιν ἐπιρρέοντι.
παλινδρομεῖν οὖν αὐτὸ πρότερον ἐπὶ τὴν κοίλην
ἀναγκαῖον καὶ κενὰς οὕτως ἐργάζεσθαι τὰς ἐπὶ
τοὺς νεφροὺς ἰούσας φλέβας, αἳ δεύτερον οὐκέτι

---

[1] Vital factor necessary over and above the mechanical.
[2] cf. p. 119, note 2.   [3] pp. 91, 93.

flow into a vacuum than has run out. For, if no other factor comes into operation [1] save only this tendency by which a vacuum becomes refilled, no more could ever flow in than had been evacuated. Nor could he suggest any other plausible cause, such, for example, as the expression of nutriment by the stomach [2] which occurs in the process of anadosis ; this had been entirely disproved in the case of blood in the vena cava ; [3] it is excluded, not merely owing to the long distance, but also from the fact that the overlying heart, at each diastole, robs the vena cava by violence of a considerable quantity of blood.

In relation to the lower part of the vena cava [4] there would still remain, solitary and abandoned, the specious theory concerning the filling of a vacuum. This, however, is deprived of plausibility by the fact that people die of retention of urine, and also, no less, by the situation of the kidneys. For, if the whole of the blood were carried to the kidneys, one might properly maintain that it all undergoes purification there. But, as a matter of fact, the whole of it does not go to them, but only so much as can be contained in the veins going to the kidneys ; [5] this portion only, therefore, will be purified. Further, the thin serous part of this will pass through the kidneys as if through a sieve, while the thick sanguineous portion remaining in the veins will obstruct the blood flowing in from behind ; this will first, therefore, have to run back to the vena cava, and so to empty the veins going to the kidneys ; these veins will no longer be able to

[4] *i.e.* the part below the liver ; *cf.* p. 91, note 2.
[5] Renal veins.

παρακομιοῦσιν ἐπ᾽ αὐτοὺς ἀκάθαρτον αἷμα· κατ-
ειληφότος γὰρ αὐτὰς τοῦ προτέρου πάροδος
οὐδεμία λέλειπται. τίς οὖν ἡμῖν ἡ δύναμις ἀπά-
ξει πάλιν ὀπίσω τῶν νεφρῶν τὸ καθαρὸν αἷμα;
τίς δὲ τοῦτο μὲν διαδεξαμένη κελεύσει πάλιν πρὸς
τὸ κάτω μέρος ἰέναι τῆς κοίλης, ἑτέρῳ δ᾽ ἄνωθεν
66 ἐπιφερομένῳ προστάξει, πρὶν ‖ ἐπὶ τοὺς νεφροὺς
ἀπελθεῖν, μὴ φέρεσθαι κάτω;

Ταῦτ᾽ οὖν ἅπαντα συνιδὼν ὁ Ἐρασίστρατος
ἀποριῶν μεστὰ καὶ μίαν μόνην δόξαν εὔπορον
εὑρὼν ἐν ἅπασι τὴν τῆς ὁλκῆς, οὔτ᾽ ἀπορεῖσθαι
βουλόμενος οὔτε τὴν Ἱπποκράτους ἐθέλων λέγειν
ἄμεινον ὑπέλαβε σιωπητέον εἶναι περὶ τοῦ τρόπου τῆς διακρίσεως.

Ἀλλ᾽ εἰ κἀκεῖνος ἐσίγησεν, ἡμεῖς οὐ σιωπήσο-
μεν· ἴσμεν γάρ, ὡς οὐκ ἐνδέχεται παρελθόντα
τὴν Ἱπποκράτειον δόξαν, εἶθ᾽ ἕτερόν τι περὶ
νεφρῶν ἐνεργείας εἰπόντα μὴ οὐ καταγέλαστον
εἶναι παντάπασι. διὰ τοῦτ᾽ Ἐρασίστρατος μὲν
ἐσιώπησεν, Ἀσκληπιάδης δ᾽ ἐψεύσατο παραπλησίως οἰκέταις λάλοις μὲν τὰ πρόσθεν τοῦ βίου
καὶ πολλὰ πολλάκις ἐγκλήματα διαλυσαμένοις
ὑπὸ περιττῆς πανουργίας, ἐπ᾽ αὐτοφώρῳ δέ ποτε
κατειλημμένοις, εἶτ᾽ οὐδὲν ἐξευρίσκουσι σόφισμα
κἄπειτ᾽ ἐνταῦθα τοῦ μὲν αἰδημονεστέρου σιωπῶν-
τος, οἷον ἀποπληξίᾳ τινὶ κατειλημμένου, τοῦ δ᾽
ἀναισχυντοτέρου κρύπτοντος μὲν ἔθ᾽ ὑπὸ μάλης
τὸ ζητούμενον, ἐξομνυμένου δὲ καὶ μηδ᾽ ἑωρακέναι
πώποτε φάσκοντος. οὕτω γάρ τοι καὶ ὁ Ἀσκλη-
67 πιάδης ‖ ἐπιλειπόντων αὐτὸν τῶν τῆς πανουργίας
σοφισμάτων καὶ μήτε τῆς πρὸς τὸ λεπτομερὲς

conduct a second quantity of unpurified blood to the kidneys—occupied as they are by the blood which had preceded, there is no passage left. What power have we, then, which will draw back the purified blood from the kidneys? And what power, in the next place, will bid this blood retire to the lower part of the vena cava, and will enjoin on another quantity coming from above not to proceed downwards before turning off into the kidneys?

Now Erasistratus realized that all these ideas were open to many objections, and he could only find one idea which held good in all respects—namely, that of *attraction*. Since, therefore, he did not wish either to get into difficulties or to mention the view of Hippocrates, he deemed it better to say nothing at all as to the manner in which secretion occurs.

But even if he kept silence, I am not going to do so. For I know that if one passes over the Hippocratic view and makes some other pronouncement about the function of the kidneys, one cannot fail to make oneself utterly ridiculous. It was for this reason that Erasistratus kept silence and Asclepiades lied ; they are like slaves who have had plenty to say in the early part of their career, and have managed by excessive rascality to escape many and frequent accusations, but who, later, when caught in the act of thieving, cannot find any excuse ; the more modest one then keeps silence, as though thunderstruck, whilst the more shameless continues to hide the missing article beneath his arm and denies on oath that he has ever seen it. For it was in this way also that Asclepiades, when all subtle excuses had failed him and there was no longer any room for nonsense about " conveyance towards the

φορᾶς ἐχούσης ἔτι χώραν ἐνταυθοῖ ληρεῖσθαι
μήθ᾽ ὡς ὑπὸ τῶν νεφρῶν γεννᾶται τουτὶ τὸ περίττωμα, καθάπερ ὑπὸ τῶν ἐν ἥπατι πόρων ἡ χολή,
δυνατὸν ὂν εἰπόντα μὴ οὐ μέγιστον ὀφλεῖν γέλ-
ωτα, ἐξόμνυταί τε καὶ ψεύδεται φανερῶς, οὐ
διήκειν λέγων ἐπὶ τοὺς νεφροὺς τὸ οὖρον ἀλλ᾽
ἀτμοειδῶς εὐθὺς ἐκ τῶν κατὰ τὴν κοίλην¹ μερῶν
εἰς τὴν κύστιν ἀθροίζεσθαι.

Οὗτοι μὲν οὖν τοῖς ἐπ᾽ αὐτοφώρῳ κατειλημμένοις οἰκέταις ὁμοίως ἐκπλαγέντες ὁ μὲν ἐσιώπησεν, ὁ δ᾽ ἀναισχύντως ψεύδεται.

XVII

Τῶν δὲ νεωτέρων ὅσοι τοῖς τούτων ὀνόμασιν
ἑαυτοὺς ἐσέμνυναν Ἐρασιστρατείους τε καὶ Ἀσ-
κληπιαδείους ἐπονομάσαντες, ὁμοίως τοῖς ὑπὸ τοῦ
βελτίστου Μενάνδρου κατὰ τὰς κωμῳδίας εἰσαγο-
μένοις οἰκέταις, Δάοις τέ τισι καὶ Γέταις, οὐδὲν
ἡγουμένοις σφίσι πεπρᾶχθαι γενναῖον, εἰ μὴ τρὶς
ἐξαπατήσειαν τὸν δεσπότην, οὕτω καὶ αὐτοὶ κατὰ
πολλὴν σχολὴν ἀναίσχυντα σοφίσματα συνέθε-
σαν, οἱ μέν, ἵνα μηδ᾽ ὅλως ἐξελεγχθείη ποτ᾽ ‖
68 Ἀσκληπιάδης ψευδόμενος, οἱ δ᾽, ἵνα κακῶς εἴπω-
σιν, ἃ καλῶς ἐσιώπησεν Ἐρασίστρατος.

Ἀλλὰ τῶν μὲν Ἀσκληπιαδείων ἅλις. οἱ δ᾽
Ἐρασιστράτειοι λέγειν ἐπιχειροῦντες, ὅπως οἱ
νεφροὶ διηθοῦσι τὸ οὖρον, ἅπαντα δρῶσί τε καὶ

---

¹ cf. p. 87, note 3.
² κοίλην : the usual reading is κοιλίαν, which would make

rarefied part [of the air]," [1] and when it was impossible without incurring the greatest derision to say that this superfluity [*i.e.* the urine] is generated by the kidneys as is bile by the canals in the liver—he, then, I say, clearly lied when he swore that the urine does not reach the kidneys, and maintained that it passes, in the form of vapour, straight from the region of the vena cava,[2] to collect in the bladder.

Like slaves, then, caught in the act of stealing, these two are quite bewildered, and while the one says nothing, the other indulges in shameless lying.

XVII

Now such of the younger men as have dignified themselves with the names of these two authorities by taking the appellations "Erasistrateans" or "Asclepiadeans" are like the *Davi* and *Getae*—the slaves introduced by the excellent Menander into his comedies. As these slaves held that they had done nothing fine unless they had cheated their master three times, so also the men I am discussing have taken their time over the construction of impudent sophisms, the one party striving to prevent the lies of Asclepiades from ever being refuted, and the other saying stupidly what Erasistratus had the sense to keep silence about.

But enough about the Asclepiadeans. The Erasistrateans, in attempting to say how the kidneys let the urine through, will do anything or suffer anything

it "from the region of the alimentary canal." *cf.* p. 118, note 1.

πάσχουσι καὶ παντοῖοι γίγνονται πιθανὸν ἐξευρεῖν τι ζητοῦντες αἴτιον ὁλκῆς μὴ δεόμενον.

Οἱ μὲν δὴ πλησίον Ἐρασιστράτου τοῖς χρόνοις γενόμενοι τὰ μὲν ἄνω τῶν νεφρῶν μόρια καθαρὸν αἷμα λαμβάνειν φασί, τῷ δὲ βάρος ἔχειν τὸ ὑδατῶδες περίττωμα βρίθειν τε καὶ ὑπορρεῖν κάτω· διηθούμενον δ' ἐνταῦθα κατὰ τοὺς νεφροὺς αὐτοὺς χρηστὸν οὕτω γενόμενον ἅπασι τοῖς κάτω τῶν νεφρῶν ἐπιπέμπεσθαι τὸ αἷμα.

Καὶ μέχρι γέ τινος εὐδοκίμησεν ἥδε ἡ δόξα καὶ ἤκμασε καὶ ἀληθὴς ἐνομίσθη· χρόνῳ δ' ὕστερον καὶ αὐτοῖς τοῖς Ἐρασιστρατείοις ὕποπτος ἐφάνη καὶ τελευτῶντες ἀπέστησαν αὐτῆς. αἰτεῖσθαι γὰρ ἐδόκουν δύο ταῦτα μήτε συγχωρούμενα πρός τινος ἀλλ' οὐδ' ἀποδειχθῆναι δυνάμενα, πρῶτον μὲν τὸ βάρος τῆς ὀρρώδους ὑγρότητος ἐν τῇ
69 κοίλῃ ‖ φλεβὶ γεννώμενον, ὥσπερ οὐκ ἐξ ἀρχῆς ὑπάρχον, ὁπότ' ἐκ τῆς κοιλίας εἰς ἧπαρ ἀνεφέρετο. τί δὴ οὖν οὐκ εὐθὺς ἐν ἐκείνοις τοῖς χωρίοις ὑπέρρει κάτω; πῶς δ' ἄν τῳ δόξειεν εὐλόγως εἰρῆσθαι συντελεῖν εἰς τὴν ἀνάδοσιν ἡ ὑδατώδης ὑγρότης, εἴπερ οὕτως ἐστὶ βαρεῖα;

Δεύτερον δ' ἄτοπον, ὅτι κἂν κάτω συγχωρηθῇ φέρεσθαι πᾶσα καὶ μὴ κατ' ἄλλο χωρίον ἢ τὴν κοίλην φλέβα, τίνα τρόπον εἰς τοὺς νεφροὺς ἐμπεσεῖται, χαλεπόν, μᾶλλον δ' ἀδύνατον εἰπεῖν, μήτ' ἐν τοῖς κάτω μέρεσι κειμένων αὐτῶν τῆς φλεβὸς ἀλλ' ἐκ τῶν πλαγίων μήτ' ἐμφυομένης εἰς αὐτοὺς τῆς κοίλης ἀλλ' ἀπόφυσίν τινα μόνον

into each of them, as it also does into all the other parts.

What doctrine, then, took the place of this one when it was condemned ? One which to me seems far more foolish than the first, although it also flourished at one time. For they say, that if oil be mixed with water and poured upon the ground, each will take a different route, the one flowing this way and the other that, and that, therefore, it is not surprising that the watery fluid runs into the kidneys, while the blood falls downwards along the vena cava. Now this doctrine also stands already condemned. For why, of the countless veins which spring from the vena cava, should blood flow into all the others, and the serous fluid be diverted to those going to the kidneys? They have not answered the question which was asked ; they merely state what happens and imagine they have thereby assigned the reason.

Once again, then (the third cup to the Saviour !),[1] let us now speak of the worst doctrine of all, lately invented by Lycus of Macedonia,[2] but which is popular owing to its novelty. This Lycus, then, maintains, as though uttering an oracle from the inner sanctuary, that urine is *residual matter from the nutrition of the kidneys* ![3] Now, the amount of urine passed every day shows clearly that it is the whole of the fluid drunk which becomes urine, except for that which comes away with the dejections or passes off as sweat or insensible perspiration. This is most easily recognized in winter in those who are doing no work but are carousing, especially if the wine be thin and diffusible ;

[1] In a toast, the third cup was drunk to Zeus Sôtêr (the Saviour).

[2] An anatomist of the Alexandrian school.

[3] *cf.* nasal mucus, p. 90, note 1.

οὗτοι διὰ ταχέων ὀλίγου δεῖν, ὅσονπερ καὶ
πίνουσιν. ὅτι δὲ καὶ ὁ Ἐρασίστρατος οὕτως
ἐγίγνωσκεν, οἱ τὸ πρῶτον ἀνεγνωκότες αὐτοῦ
σύγγραμμα τῶν καθόλου λόγων ἐπίστανται. ὥσθ᾽
ὁ Λύκος οὔτ᾽ ἀληθῆ φαίνεται λέγων οὔτ᾽ Ἐρασιστράτεια, δῆλον δ᾽ ὡς οὐδ᾽ Ἀσκληπιάδεια, πολὺ
δὲ μᾶλλον οὐδ᾽ Ἱπποκράτεια. λευκῷ τοίνυν κατὰ
τὴν παροιμίαν ἔοικε κόρακι μήτ᾽ αὐτοῖς τοῖς κόρα-
ξιν ἀναμιχθῆναι δυναμένῳ διὰ τὴν χρόαν μήτε
ταῖς περιστεραῖς διὰ τὸ μέγεθος, ἀλλ᾽ οὔτι που
τούτου γ᾽ ἕνεκα παροπτέος· ἴσως γάρ τι λέγει
θαυμαστόν, ὃ μηδεὶς τῶν ἔμπροσθεν ἔγνω.

Τὸ μὲν οὖν ἅπαντα τὰ τρεφόμενα μόρια ποιεῖν
τι περίττωμα συγχωρούμενον, τὸ δὲ τοὺς νεφροὺς
μόνους, οὕτω σμικρὰ σώματα, χόας ὅλους τέτταρας ἢ καὶ πλείους ἴσχειν ἐνίοτε περιττώματος
οὔθ᾽ ὁμολογούμενον οὔτε λόγον ἔχον· τὸ γὰρ
ἑκάστου τῶν μειζόνων σπλάγχνων περίττωμα
πλεῖον ἀναγκαῖον ὑπάρχειν. οἷον αὐτίκα τὸ τοῦ
πνεύμονος, εἴπερ ἀνάλογον τῷ μεγέθει τοῦ
72 σπλάγχνου γίγνοιτο, πολλαπλά‖σιον ἔσται δήπου τοῦ κατὰ τοὺς νεφρούς, ὥσθ᾽ ὅλος μὲν ὁ
θώραξ ἐμπλησθήσεται, πνιγήσεται δ᾽ αὐτίκα τὸ
ζῷον. ἀλλ᾽ εἰ ἴσον φήσει τις γίγνεσθαι τὸ καθ᾽
ἕκαστον τῶν ἄλλων μορίων περίττωμα, διὰ ποίων
κύστεων ἐκκρίνεται; εἰ γὰρ οἱ νεφροὶ τοῖς κωθωνιζομένοις τρεῖς ἢ τέτταρας ἐνίοτε χόας ποιοῦσι
περιττώματος, ἑκάστου τῶν ἄλλων σπλάγχνων
πολλῷ πλείους ἔσονται καὶ πίθου τινὸς οὕτω
μεγίστου δεήσει τοῦ δεξομένου τὰ πάντων περιτ-

---

[1] " Sur l'Ensemble des Choses " (Daremberg).

these people rapidly pass almost the same quantity as they drink. And that even Erasistratus was aware of this is known to those who have read the first book of his "General Principles." [1] Thus Lycus is speaking neither good Erasistratism, nor good Asclepiadism, far less good Hippocratism. He is, therefore, as the saying is, like a white crow, which cannot mix with the genuine crows owing to its colour, nor with the pigeons owing to its size. For all this, however, he is not to be disregarded; he may, perhaps, be stating some wonderful truth, unknown to any of his predecessors.

Now it is agreed that all parts which are undergoing nutrition produce a certain amount of residue, but it is neither agreed nor is it likely, that the kidneys alone, small bodies as they are, could hold four whole *congii*,[2] and sometimes even more, of residual matter. For this surplus must necessarily be greater in quantity in each of the larger viscera; thus, for example, that of the lung, if it corresponds in amount to the size of the viscus, will obviously be many times more than that in the kidneys, and thus the whole of the thorax will become filled, and the animal will be at once suffocated. But if it be said that the residual matter is equal in amount in each of the other parts, where are the *bladders*, one may ask, through which it is excreted? For, if the kidneys produce in drinkers three and sometimes four *congii* of superfluous matter, that of each of the other viscera will be much more, and thus an enormous barrel will be needed to contain the waste products of them all.

[2] About twelve quarts. This is about five times as much as the average daily excretion, and could only be passed if a very large amount of wine were drunk.

τώματα. καίτοι πολλάκις, ὅσον ἔπιέ τις, ὀλίγου δεῖν οὔρησεν ἅπαν, ὡς ἂν ἐπὶ τοὺς νεφροὺς φερομένου τοῦ πόματος ἅπαντος.

Ἔοικεν οὖν ὁ τὸ τρίτον ἐξαπατῶν οὗτος οὐδὲν ἀνύειν ἀλλ᾽ εὐθὺς γεγονέναι κατάφωρος καὶ μένειν ἔτι τὸ ἐξ ἀρχῆς ἄπορον Ἐρασιστράτῳ τε καὶ τοῖς ἄλλοις ἅπασι πλὴν Ἱπποκράτους. διατρίβω δ᾽ ἑκὼν ἐν τῷ τόπῳ σαφῶς εἰδώς, ὅτι μηδὲν εἰπεῖν ἔχει μηδεὶς ἄλλος περὶ τῆς τῶν νεφρῶν ἐνεργείας, ἀλλ᾽ ἀναγκαῖον ἢ τῶν μαγείρων ἀμαθεστέρους φαίνεσθαι μηδ᾽ ὅτι διηθεῖται δι᾽ αὐτῶν τὸ οὖρον 73 ὁμολογοῦντας ἢ ‖ τοῦτο συγχωρήσαντας μηδὲν ἔτ᾽ ἔχειν εἰπεῖν ἕτερον αἴτιον τῆς διακρίσεως πλὴν τῆς ὁλκῆς.

Ἀλλ᾽ εἰ μὴ τῶν οὔρων ἡ φορὰ τῇ πρὸς τὸ κενούμενον ἀκολουθίᾳ γίγνεται, δῆλον, ὡς οὐδ᾽ ἡ τοῦ αἵματος οὐδ᾽ ἡ τῆς χολῆς ἢ εἴπερ ἐκείνων καὶ τούτου· πάντα γὰρ ὡσαύτως ἀναγκαῖον ἐπιτελεῖσθαι καὶ κατ᾽ αὐτὸν τὸν Ἐρασίστρατον.

Εἰρήσεται δ᾽ ἐπὶ πλέον ὑπὲρ αὐτῶν ἐν τῷ μετὰ ταῦτα γράμματι.

Yet one often urinates practically the same quantity as one has drunk, which would show that the whole of what one drinks goes to the kidneys.

Thus the author of this third piece of trickery would appear to have achieved nothing, but to have been at once detected, and there still remains the original difficulty which was insoluble by Erasistratus and by all others except Hippocrates. I dwell purposely on this topic, knowing well that nobody else has anything to say about the function of the kidneys, but that either we must prove more foolish than the very butchers [1] if we do not agree that the urine passes through the kidneys; or, if one acknowledges this, that then one cannot possibly give any other reason for the secretion than the principle of attraction.

Now, if the movement of urine does not depend on the tendency of a vacuum to become refilled,[2] it is clear that neither does that of the blood nor that of the bile; or if that of these latter does so, then so also does that of the former. For they must all be accomplished in one and the same way, even according to Erasistratus himself.

This matter, however, will be discussed more fully in the book following this.

[1] cf. p. 51.   [2] Horror vacui. Note analogical reasoning; cf. p. 289, note 1.

BOOK II

B

I

74 Ὅτι μὲν οὖν ἀναγκαῖόν ἐστιν οὐκ Ἐρασιστράτῳ μόνον ἀλλὰ καὶ τοῖς ἄλλοις ἅπασιν, ὅσοι
μέλλουσι περὶ διακρίσεως οὔρων ἐρεῖν τι χρηστόν, ὁμολογῆσαι δύναμίν τιν' ὑπάρχειν τοῖς
νεφροῖς ἕλκουσαν εἰς ἑαυτοὺς ποιότητα τοιαύτην,
οἵα ἐν τοῖς οὔροις ἐστί, διὰ τοῦ πρόσθεν ἐπιδέδεικται γράμματος, ἀναμιμνησκόντων ἅμ' αὐτῷ
καὶ τοῦθ' ἡμῶν, ὡς οὐκ ἄλλως μὲν εἰς τὴν κύστιν
φέρεται τὰ οὖρα διὰ τῶν νεφρῶν, ἄλλως δ' εἰς
ἅπαντα τοῦ ζῴου τὰ μόρια τὸ αἷμα, κατ' ἄλλον
δέ τινα τρόπον ἡ ξανθὴ χολὴ διακρίνεται.   δειχ-
75 θείσης γὰρ ἐναργῶς ἐφ' ἑνὸς ‖ οὑτινοσοῦν ὀργάνου
τῆς ἑλκτικῆς τε καὶ ἐπισπαστικῆς ὀνομαζομένης
δυνάμεως οὐδὲν ἔτι χαλεπὸν ἐπὶ τὰ λοιπὰ μεταφέρειν αὐτήν· οὐ γὰρ δὴ τοῖς μὲν νεφροῖς ἡ φύσις
ἔδωκέ τινα τοιαύτην δύναμιν, οὐχὶ δέ γε καὶ τοῖς
τὸ χολῶδες ὑγρὸν ἕλκουσιν ἀγγείοις οὐδὲ τούτοις
μέν, οὐκέτι δὲ καὶ τῶν ἄλλων μορίων ἑκάστῳ.
καὶ μὴν εἰ τοῦτ' ἀληθές ἐστι, θαυμάζειν χρὴ τοῦ
Ἐρασιστράτου ψευδεῖς οὕτω λόγους ὑπὲρ ἀνα-

---

[1] cf. p. 89.   [2] This term is nowadays limited to the
drawing action of a blister. cf. p. 223.

116

BOOK II

I

In the previous book we demonstrated that not
only Erasistratus, but also all others who would say
anything to the purpose about urinary secretion,
must acknowledge that the kidneys possess some
faculty which attracts to them this particular quality
existing in the urine.[1] Besides this we drew attention to the fact that the urine is not carried through
the kidneys into the bladder by one method, the
blood into parts of the animal by another, and the
yellow bile separated out on yet another principle.
For when once there has been demonstrated in
any one organ, the drawing, or so-called *epispastic*[2]
faculty, there is then no difficulty in transferring
it to the rest. Certainly Nature did not give a
power such as this to the kidneys without giving
it also to the vessels which abstract the biliary
fluid,[3] nor did she give it to the latter without
also giving it to each of the other parts. And,
assuredly, if this is true, we must marvel that
Erasistratus should make statements concerning the
delivery of nutriment from the food-canal[4] which are

---

[3] The radicles of the hepatic ducts in the liver were supposed to be the active agents in extracting bile from the
blood. *cf.* pp. 145–149.   [4] *Anadosis*; *cf.* p. 13, note 5.

GALEN

δόσεως τροφῆς εἰπόντος, ὡς μηδ' Ἀσκληπιάδην
λαθεῖν. καίτοι γ' οἴεται παντὸς μᾶλλον ἀληθὲς
ὑπάρχειν, ὡς, εἴπερ ἐκ τῶν φλεβῶν ἀπορρέοι
τι, δυοῖν θάτερον ἢ κενὸς ἔσται τόπος ἀθρόως ἢ
τὸ συνεχὲς ἐπιρρυήσεται τὴν βάσιν ἀναπληροῦν
τοῦ κενουμένου. ἀλλ' ὅ γ' Ἀσκληπιάδης οὐ
δυοῖν θάτερόν φησιν, ἀλλὰ τριῶν ἕν τι χρῆναι
λέγειν ἐπὶ τοῖς κενουμένοις ἀγγείοις ἕπεσθαι ἢ
κενὸν ἀθρόως τόπον ἢ τὸ συνεχὲς ἀκολουθήσειν ἢ
συσταλήσεσθαι τὸ ἀγγεῖον. ἐπὶ μὲν γὰρ τῶν
καλάμων καὶ τῶν αὐλίσκων τῶν εἰς τὸ ὕδωρ
καθιεμένων ἀληθὲς εἰπεῖν, ὅτι κενουμένου τοῦ
76 περιεχομένου κατὰ τὴν ‖ εὐρυχωρίαν αὐτῶν ἀέρος
ἢ κενὸς ἀθρόως ἔσται τόπος ἢ ἀκολουθήσει τὸ
συνεχές· ἐπὶ δὲ τῶν φλεβῶν οὐκέτ' ἐγχωρεῖ, δυναμένου δὴ τοῦ χιτῶνος αὐτῶν εἰς ἑαυτὸν συνιζάνειν
καὶ διὰ τοῦτο καταπίπτειν εἰς τὴν ἐντὸς εὐρυχωρίαν. οὕτω μὲν δὴ ψευδὴς ἡ περὶ τῆς πρὸς
τὸ κενούμενον ἀκολουθίας οὐκ ἀπόδειξις μὰ Δί'
εἴποιμ' ἂν ἀλλ' ὑπόθεσις Ἐρασιστράτειος.

Καθ' ἕτερον δ' αὖ τρόπον, εἰ καὶ ἀληθὴς εἴη,
περιττή, τῆς μὲν κοιλίας ἐνθλίβειν ταῖς φλεψὶ
δυναμένης, ὡς αὐτὸς ὑπέθετο, τῶν φλεβῶν δ' αὖ
περιστέλλεσθαι τῷ ἐνυπάρχοντι καὶ προωθεῖν
αὐτό. τά τε γὰρ ἄλλα καὶ πλῆθος οὐκ ἂν ἐν τῷ
σώματι γένοιτο, τῇ πρὸς τὸ κενούμενον ἀκολουθίᾳ
μόνῃ τῆς ἀναδόσεως ἐπιτελουμένης. εἰ μὲν οὖν
ἡ τῆς γαστρὸς ἔνθλιψις ἐκλύεται προϊοῦσα καὶ

so false as to be detected even by Asclepiades. Now, Erasistratus considers it absolutely certain that, if anything flows from the veins, one of two things must happen: either a completely empty space will result, or the contiguous quantum of fluid will run in and take the place of that which has been evacuated. Asclepiades, however, holds that not one of two, but one of three things must be said to result in the emptied vessels: either there will be an entirely empty space, or the contiguous portion will flow in, or the vessel will contract. For whereas, in the case of reeds and tubes it is true to say that, if these be submerged in water, and are emptied of the air which they contain in their lumens, then either a completely empty space will be left, or the contiguous portion will move onwards; in the case of veins this no longer holds, since their coats can collapse and so fall in upon the interior cavity. It may be seen, then, how false this hypothesis—by Zeus, I cannot call it a demonstration!—of Erasistratus is.

And, from another point of view, even if it were true, it is superfluous, if the stomach [1] has the power of compressing the veins, as he himself supposed, and the veins again of contracting upon their contents and propelling them forwards.[2] For, apart from other considerations, no *plethora* [3] would ever take place in the body, if delivery of nutriment resulted merely from the tendency of a vacuum to become refilled. Now, if the compression of the stomach becomes weaker the further it goes, and cannot reach to an

from which nutriment was believed to be absorbed by the mesenteric veins; *cf.* p. 309, note 2.

[2] *cf.* p. 100, note 2; p. 167, note 2.

[3] A characteristic "lesion" in Erasistratus's pathology.

μέχρι παντὸς ἀδυνατός ἐστιν ἐξικνεῖσθαι καὶ διὰ
τοῦτ' ἄλλης τινὸς δεῖ μηχανῆς εἰς τὴν πάντη
φορὰν τοῦ αἵματος, ἀναγκαία μὲν ἡ πρὸς τὸ
κενούμενον ἀκολουθία προσεξεύρηται· πλῆθος δ'
77 ἐν οὐδενὶ τῶν μεθ' ἧπαρ ἔσται ‖ μορίων, ἤ, εἴπερ
ἄρα, περὶ τὴν καρδίαν τε καὶ τὸν πνεύμονα. μόνη
γὰρ αὕτη τῶν μεθ' ἧπαρ εἰς τὴν δεξιὰν αὑτῆς
κοιλίαν ἕλκει τὴν τροφήν, εἶτα διὰ τῆς φλεβὸς
τῆς ἀρτηριώδους ἐκπέμπει τῷ πνεύμονι· τῶν γὰρ
ἄλλων οὐδὲν οὐδ' αὐτὸς ὁ Ἐρασίστρατος ἐκ καρδίας βούλεται τρέφεσθαι διὰ τὴν τῶν ὑμένων
ἐπίφυσιν. εἰ δέ γ', ἵνα πλῆθος γένηται, φυλάξομεν
ἄχρι παντὸς τὴν ῥώμην τῆς κατὰ τὴν κοιλίαν
ἐνθλίψεως, οὐδὲν ἔτι δεόμεθα τῆς πρὸς τὸ κενούμενον ἀκολουθίας, μάλιστ' εἰ καὶ τὴν τῶν φλεβῶν
συνυποθοίμεθα περιστολήν, ὡς αὖ καὶ τοῦτ' αὐτῷ
πάλιν ἀρέσκει τῷ Ἐρασιστράτῳ.

II

Ἀναμνηστέον οὖν αὖθις αὐτόν, κἂν μὴ βού-
ληται, τῶν νεφρῶν καὶ λεκτέον, ὡς ἔλεγχος
οὗτοι φανερώτατος ἁπάντων τῶν ἀποχωρούντων
τῆς ὁλκῆς· οὐδεὶς γὰρ οὐδὲν οὔτ' εἶπε πιθανόν,
ἀλλ' οὐδ' ἐξευρεῖν εἶχε κατ' οὐδένα τρόπον, ὡς

---

[1] A certain subordinate place allowed to the horror vacui.
[2] i.e. the parts to which the veins convey blood after it
leaves the liver—second stage of anadosis; cf. p. 91, note 2 ;
p. 13, note 5.

indefinite distance, and if, therefore, there is need
of some other mechanism to explain why the blood
is conveyed in all directions, then the principle of
the refilling of a vacuum may be looked on as a
necessary addition;[1] there will not, however, be a
plethora in any of the parts coming after the liver,[2]
or, if there be, it will be in the region of the
heart and lungs; for the heart alone of the parts
which come after the liver draws the nutriment
into its right ventricle, thereafter sending it through
the *arterioid vein*[3] to the lungs (for Erasistratus
himself will have it that, owing to the membranous
excrescences,[4] no other parts save the lungs receive
nourishment from the heart). If, however, in order
to explain how plethora comes about, we suppose
the force of compression by the stomach to persist
indefinitely, we have no further need of the principle
of the refilling of a vacuum, especially if we assume
contraction of the veins in addition—as is, again,
agreeable to Erasistratus himself.

II

LET me draw his attention, then, once again, even
if he does not wish it, to the kidneys, and let me
state that these confute in the very clearest manner
such people as object to the principle of *attraction*.
Nobody has ever said anything plausible, nor, as we
previously showed, has anyone been able to discover,

[3] What we now call the pulmonary artery. Galen believed
that the right ventricle existed for the purpose of sending
nutrient blood to the lungs.
[4] Lit. owing to the ongrowth (*epiphysis*) of membranes;
he means the tricuspid valve; *cf.* p. 314, note 2; p. 321, note 4.

ἔμπροσθεν ἐδείκνυμεν, ἕτερον αἴτιον οὔρων διακρίσεως, ἀλλ' ἀναγκαῖον ἢ μαίνεσθαι δοκεῖν, εἰ
78 φήσαιμεν ἀτμοει‖δῶς εἰς τὴν κύστιν ἰέναι τὸ οὖρον
ἢ ἀσχημονεῖν τῆς πρὸς τὸ κενούμενον ἀκολουθίας
μνημονεύοντας, ληρώδους μὲν οὔσης κἀπὶ τοῦ
αἵματος, ἀδυνάτου δὲ καὶ ἠλιθίου παντάπασιν
ἐπὶ τῶν οὔρων.

Ἓν μὲν δὴ τοῦτο σφάλμα τῶν ἀποστάντων τῆς
ὁλκῆς· ἕτερον δὲ τὸ περὶ τῆς κατὰ τὴν ξανθὴν
χολὴν διακρίσεως. οὐδὲ γὰρ οὐδ' ἐκεῖ παραρρέ-
οντος τοῦ αἵματος τὰ στόματα τῶν χοληδόχων
ἀγγείων ἀκριβῶς διακριθήσεται τὸ χολῶδες
περίττωμα. καὶ μὴ διακρινέσθω, φασίν, ἀλλὰ
συναναφερέσθω τῷ αἵματι πάντῃ τοῦ σώματος.
ἀλλ', ὦ σοφώτατοι, προνοητικὴν τοῦ ζῴου καὶ
τεχνικὴν αὐτὸς ὁ Ἐρασίστρατος ὑπέθετο τὴν
φύσιν. ἀλλὰ καὶ τὸ χολῶδες ὑγρὸν ἄχρηστον
εἶναι παντάπασι τοῖς ζῴοις ἔφασκεν. οὐ συμβαίνει δ' ἀλλήλοις ἄμφω ταῦτα. πῶς γὰρ ἂν
ἔτι προνοεῖσθαι τοῦ ζῴου δόξειεν ἐπιτρέπουσα
συναναφέρεσθαι τῷ αἵματι μοχθηρὸν οὕτω χυμόν;

Ἀλλὰ ταῦτα μὲν σμικρά· τὸ δὲ μέγιστον καὶ
σαφέστατον πάλιν ἐνταῦθ' ἁμάρτημα καὶ δὴ
φράσω. εἴπερ γὰρ δι' οὐδὲν ἀλλ' ἢ ὅτι παχύτερον
79 μέν ἐστι τὸ αἷμα, λεπτοτέρα δ' ἡ ‖ ξανθὴ χολὴ
καὶ τὰ μὲν τῶν φλεβῶν εὐρύτερα στόματα, τὰ

by any means, any other cause for the secretion of urine ; we necessarily appear mad if we maintain that the urine passes into the kidneys in the form of vapour, and we certainly cut a poor figure when we talk about the tendency of a vacuum to become refilled ;[1] this idea is foolish in the case of blood, and impossible, nay, perfectly nonsensical, in the case of the urine.[2]

This, then, is one blunder made by those who dissociate themselves from the principle of attraction. Another is that which they make about the *secretion of yellow bile.* For in this case, too, it is not a fact that when the blood runs past the mouths [stomata] of the bile-ducts there will be a thorough separation out [secretion] of biliary waste-matter. " Well," say they, " let us suppose that it is not secreted but carried with the blood all over the body." But, you sapient folk, Erasistratus himself supposed that Nature took thought for the animals' future, and was workmanlike in her method ; and at the same time he maintained that the biliary fluid was useless in every way for the animals. Now these two things are incompatible. For how could Nature be still looked on as exercising forethought for the animal when she allowed a noxious humour such as this to be carried off and distributed with the blood ? . . .

This, however, is a small matter. I shall again point out here the greatest and most obvious error. For if the yellow bile adjusts itself to the narrower vessels and stomata, and the blood to the wider ones, for no other reason than that blood is thicker and bile thinner, and that the stomata of the veins are

[1] Horror vacui.   [2] But Erasistratus had never upheld this in the case of urinary secretion. *cf.* p. 99.

δὲ τῶν χοληδόχων ἀγγείων στενότερα, διὰ τοῦθ'
ἡ μὲν χολὴ τοῖς στενοτέροις ἀγγείοις τε καὶ
στόμασιν ἐναρμόττει, τὸ δ' αἷμα τοῖς εὐρυτέροις,
δῆλον, ὡς καὶ τὸ ὑδατῶδες τοῦτο καὶ ὀρρῶδες
περίττωμα τοσούτῳ πρότερον εἰσρυήσεται τοῖς
χοληδόχοις ἀγγείοις, ὅσῳ λεπτότερόν ἐστι τῆς
χολῆς. πῶς οὖν οὐκ εἰσρεῖ; ὅτι παχύτερόν ἐστι
νὴ Δία τὸ οὖρον τῆς χολῆς· τοῦτο γὰρ ἐτόλμησέ
τις εἰπεῖν τῶν καθ' ἡμᾶς Ἐρασιστρατείων
ἀποστὰς δηλονότι τῶν αἰσθήσεων, αἷς ἐπίστευσεν
ἐπί τε τῆς χολῆς καὶ τοῦ αἵματος. εἴτε γὰρ
ὅτι μᾶλλον ἡ χολὴ τοῦ αἵματος ῥεῖ, διὰ τοῦτο
λεπτοτέραν αὐτὴν ἡμῖν ἐστι νομιστέον, εἴθ' ὅτι
δι' ὀθόνης ἢ ῥάκους ἤ τινος ἠθμοῦ ῥᾷον διεξέρχεται
καὶ ταύτης τὸ ὀρρῶδες περίττωμα, κατὰ ταῦτα
τὰ γνωρίσματα παχυτέρα τῆς ὑδατώδους ὑγρότητος καὶ αὕτη γενήσεται. πάλιν γὰρ οὐδ'
ἐνταῦθα λόγος οὐδείς ἐστιν, ὃς ἀποδείξει λεπτό-
τέραν τὴν χολὴν τῶν ὀρρωδῶν περιττωμάτων.

Ἀλλ' ὅταν τις ἀναισχυντῇ περιπλέκων τε καὶ
80 μήπω καταπεπτωκέναι συγχωρῶν, || ὅμοιος ἔσται
τοῖς ἰδιώταις τῶν παλαιστῶν, οἳ καταβληθέντες
ὑπὸ τῶν παλαιστρικῶν καὶ κατὰ τῆς γῆς ὕπτιοι
κείμενοι τοσούτου δέουσι τὸ πτῶμα γνωρίζειν,
ὥστε καὶ κρατοῦσι τῶν αὐχένων αὐτοὺς τοὺς
καταβαλόντας οὐκ ἐῶντες ἀπαλλάττεσθαι, κἂν
τούτῳ νικᾶν ὑπολαμβάνουσι.

---

[1] This was the characteristically "anatomical" explanation
of bile-secretion made by Erasistratus. cf. p. 170, note 2.

wider and those of the bile-ducts narrower,[1] then it is clear that this watery and serous superfluity,[2] too, will run out into the bile-ducts quicker than does the bile, exactly in proportion as it is thinner than the bile! How is it, then, that it does not run out? "Because," it may be said, "urine is thicker than bile!" This was what one of our Erasistrateans ventured to say, herein clearly disregarding the evidence of his senses, although he had trusted these in the case of the bile and blood. For, if it be that we are to look on bile as thinner than blood because it runs more, then, since the serous residue [2] passes through fine linen or lint or a sieve more easily even than does bile, by these tokens bile must also be thicker than the watery fluid. For here, again, there is no argument which will demonstrate that bile is thinner than the serous superfluities.

But when a man shamelessly goes on using circumlocutions, and never acknowledges when he has had a fall, he is like the amateur wrestlers, who, when they have been overthrown by the experts and are lying on their backs on the ground, so far from recognizing their fall, actually seize their victorious adversaries by the necks and prevent them from getting away, thus supposing themselves to be the winners!

Why, then, says Galen, does not urine, rather than bile, enter the bile-ducts? [2] Urine, or, more exactly, blood-serum.

GALEN

III

Λῆρος οὖν μακρὸς ἅπασα πόρων ὑπόθεσις εἰς φυσικὴν ἐνέργειαν. εἰ μὴ γὰρ δύναμίς τις σύμφυτος ἑκάστῳ τῶν ὀργάνων ὑπὸ τῆς φύσεως εὐθὺς ἐξ ἀρχῆς δοθείη, διαρκεῖν οὐ δυνήσεται τὰ ζῷα, μὴ ὅτι τοσοῦτον ἀριθμὸν ἐτῶν ἀλλ' οὐδ' ἡμερῶν ὀλιγίστων· ἀνεπιτρόπευτα γὰρ ἐάσαντες αὐτὰ καὶ τέχνης καὶ προνοίας ἔρημα μόναις ταῖς τῶν ὑλῶν οἰακιζόμενα ῥοπαῖς, οὐδαμοῦ δυνάμεως οὐδεμιᾶς τῆς μὲν ἑλκούσης τὸ προσῆκον ἑαυτῇ, τῆς δ' ἀπωθούσης τὸ ἀλλότριον, τῆς δ' ἀλλοιούσης τε καὶ προσφυούσης τὸ θρέψον, οὐκ οἶδ' ὅπως οὐκ ἂν εἴημεν καταγέλαστοι περί τε τῶν φυσικῶν ἐνεργειῶν διαλεγόμενοι καὶ πολὺ μᾶλλον ἔτι περὶ 81 τῶν ψυχικῶν καὶ ‖ συμπάσης γε τῆς ζωῆς.

Οὐδὲ γὰρ ζῆν οὐδὲ διαμένειν οὐδενὶ τῶν ζῴων οὐδ' εἰς ἐλάχιστον χρόνον ἔσται δυνατόν, εἰ τοσαῦτα κεκτημένον ἐν ἑαυτῷ μόρια καὶ οὕτω διαφέροντα μήθ' ἑλκτικῇ τῶν οἰκείων χρήσεται δυνάμει μήτ' ἀποκριτικῇ τῶν ἀλλοτρίων μήτ' ἀλλοιωτικῇ τῶν θρεψόντων. καὶ μὴν εἰ ταύτας ἔχοιμεν, οὐδὲν ἔτι πόρων μικρῶν ἢ μεγάλων ἐξ ὑποθέσεως ἀναποδείκτου λαμβανομένων εἰς οὔρου καὶ χολῆς διάκρισιν δεόμεθα καί τινος ἐπικαίρου θέσεως, ἐν ᾧ μόνῳ σωφρονεῖν ἔοικεν ὁ Ἐρασίστρατος ἅπαντα καλῶς τεθῆναί τε καὶ διαπλασ-

---

[1] Or ducts, canals, conduits, *i.e. morphological* factors.
[2] Or artistic skill, "artistry." *cf.* Book I., chap. xii.
[3] "Only"; *cf.* Introd., p. xxviii.
[4] Note how Galen, although he has not yet clearly differ-

126

III

THUS, every hypothesis of *channels*[1] as an explanation of natural functioning is perfect nonsense. For, if there were not *an inborn faculty* given by Nature to each one of the organs at the very beginning, then animals could not continue to live even for a few days, far less for the number of years which they actually do. For let us suppose they were under no guardianship, lacking in creative ingenuity[2] and forethought; let us suppose they were steered only by material forces,[3] and not by any special *faculties* (the one attracting what is proper to it, another rejecting what is foreign, and yet another causing alteration and adhesion of the matter destined to nourish it); if we suppose this, I am sure it would be ridiculous for us to discuss natural, or, still more, psychical, activities—or, in fact, life as a whole.[4]

For there is not a single animal which could live or endure for the shortest time if, possessing within itself so many different parts, it did not employ faculties which were attractive of what is appropriate, eliminative of what is foreign, and alterative of what is destined for nutrition. On the other hand, if we have these faculties, we no longer need *channels*, little or big, resting on an unproven hypothesis, for explaining the secretion of urine and bile, and the conception of some *favourable situation* (in which point alone Erasistratus shows some common sense, since he does regard all the parts of the body as

entiated physiological from physical processes (both are "natural") yet separates them definitely from the psychical. *cf.* p. 2, footnote. A *psychical* function or activity is, in Latin, *actio animalis* (from *anima = psyche*).

θῆναι τὰ μόρια τοῦ σώματος ὑπὸ τῆς φύσεως
οἰόμενος.

Ἀλλ᾽ εἰ παρακολουθήσειεν ἑαυτῷ φύσιν ὀνο-
μάζοντι τεχνικήν, εὐθὺς μὲν ἐξ ἀρχῆς ἅπαντα
καλῶς διαπλάσασάν τε καὶ διαθεῖσαν τοῦ ζῴου
τὰ μόρια, μετὰ δὲ τὴν τοιαύτην ἐνέργειαν, ὡς
οὐδὲν ἔλειπεν, ἔτι προαγαγοῦσαν εἰς φῶς αὐτὸ
σύν τισι δυνάμεσιν, ὧν ἄνευ ζῆν οὐκ ἠδύνατο, καὶ
μετὰ ταῦτα κατὰ βραχὺ προσαυξήσασαν ἄχρι
τοῦ πρέποντος μεγέθους, οὐκ οἶδα πῶς ὑπομένει
82 πόρων σμικρότησιν ‖ ἢ μεγέθεσιν ἢ τισιν ἄλλαις
οὕτω ληρώδεσιν ὑποθέσεσι φυσικὰς ἐνεργείας
ἐπιτρέπειν. ἡ γὰρ διαπλάττουσα τὰ μόρια φύσις
ἐκείνη καὶ κατὰ βραχὺ προσαύξουσα πάντως
δήπου δι᾽ ὅλων αὐτῶν ἐκτέταται· καὶ γὰρ ὅλα δι᾽
ὅλων οὐκ ἔξωθεν μόνον αὐτὰ διαπλάττει τε καὶ
τρέφει καὶ προσαύξει. Πραξιτέλης μὲν γὰρ ἢ
Φειδίας ἤ τις ἄλλος ἀγαλματοποιὸς ἔξωθεν μόνον
ἐκόσμουν τὰς ὕλας, καθὰ καὶ ψαύειν αὐτῶν ἠδύ-
ναντο, τὸ βάθος δ᾽ ἀκόσμητον καὶ ἀργὸν καὶ
ἄτεχνον καὶ ἀπρονόητον ἀπέλιπον, ὡς ἂν μὴ
δυνάμενοι κατελθεῖν εἰς αὐτὸ καὶ καταδῦναι καὶ
θιγεῖν ἁπάντων τῆς ὕλης τῶν μερῶν. ἡ φύσις δ᾽
οὐχ οὕτως, ἀλλὰ τὸ μὲν ὀστοῦ μέρος ἅπαν ὀστοῦν
ἀποτελεῖ, τὸ δὲ σαρκὸς σάρκα, τὸ δὲ πιμελῆς
πιμελὴν καὶ τῶν ἄλλων ἕκαστον· οὐδὲν γάρ ἐστιν
ἄψαυστον αὐτῇ μέρος οὐδ᾽ ἀνεξέργαστον οὐδ᾽
ἀκόσμητον. ἀλλὰ τὸν μὲν κηρὸν ὁ Φειδίας οὐκ
ἠδύνατο ποιεῖν ἐλέφαντα καὶ χρυσόν, ἀλλ᾽ οὐδὲ
τὸν χρυσὸν κηρόν· ἕκαστον γὰρ αὐτῶν μένον, οἷον
ἦν ἐξ ἀρχῆς, ἔξωθεν μόνον ἠμφιεσμένον εἶδός τι

having been well and truly placed and shaped by Nature).

But let us suppose he remained true to his own statement that Nature is "artistic"—this Nature which, at the beginning, well and truly shaped and disposed all the parts of the animal,[1] and, after carrying out this function (for she left nothing undone), brought it forward to the light of day, endowed with certain faculties necessary for its very existence, and, thereafter, gradually increased it until it reached its due size. If he argued consistently on this principle, I fail to see how he can continue to refer natural functions to the smallness or largeness of canals, or to any other similarly absurd hypothesis. For this Nature which shapes and gradually adds to the parts is most certainly extended throughout their whole substance. Yes indeed, she shapes and nourishes and increases them through and through, not on the outside only. For Praxiteles and Phidias and all the other statuaries used merely to decorate their material on the outside, in so far as they were able to touch it; but its inner parts they left unembellished, unwrought, unaffected by art or forethought, since they were unable to penetrate therein and to reach and handle all portions of the material. It is not so, however, with Nature. Every part of a bone she makes bone, every part of the flesh she makes flesh, and so with fat and all the rest; there is no part which she has not touched, elaborated, and embellished. Phidias, on the other hand, could not turn wax into ivory and gold, nor yet gold into wax: for each of these remains as it was at the commencement, and becomes a perfect statue

---

[1] The stage of organogenesis or *diaplasis*; *cf.* p. 25, note 4.

83 καὶ σχῆμα τεχνικόν, ἄγαλμα τέλειον ‖ γέγονεν. ἡ
φύσις δ᾽ οὐδεμιᾶς ἔτι φυλάττει τῶν ὑλῶν τὴν
ἀρχαίαν ἰδέαν· αἷμα γὰρ ἂν ἦν οὕτως ἅπαντα
τοῦ ζῴου τὰ μόρια, τὸ παρὰ τῆς κυούσης ἐπιρρέον τῷ σπέρματι, δίκην κηροῦ τινος ὕλη μία καὶ
μονοειδὴς ὑποβεβλημένη τῷ τεχνίτῃ. γίγνεται
δ᾽ ἐξ αὐτῆς οὐδὲν τῶν τοῦ ζῴου μορίων οὔτ᾽ ἐρυθρὸν οὕτως οὔθ᾽ ὑγρόν. ὀστοῦν γὰρ καὶ ἀρτηρία
καὶ φλὲψ καὶ νεῦρον καὶ χόνδρος καὶ πιμελὴ καὶ
ἀδὴν καὶ ὑμὴν καὶ μυελὸς ἄναιμα μέν, ἐξ αἵματος
δὲ γέγονε.

Τίνος ἀλλοιώσαντος καὶ τίνος πήξαντος καὶ
τίνος διαπλάσαντος ἐδεόμην ἄν μοι τὸν Ἐρασίστρατον αὐτὸν ἀποκρίνασθαι. πάντως γὰρ ἂν
εἶπεν ἤτοι τὴν φύσιν ἢ τὸ σπέρμα, ταὐτὸν μὲν
λέγων καθ᾽ ἑκάτερον, διαφόροις δ᾽ ἐπινοίαις ἑρμη-
νεύων· ὃ γὰρ ἦν πρότερον σπέρμα, τοῦθ᾽, ὅταν
ἄρξηται φύειν τε καὶ διαπλάττειν τὸ ζῷον, φύσις
τις γίγνεται. καθάπερ γὰρ ὁ Φειδίας εἶχε μὲν τὰς
δυνάμεις τῆς τέχνης καὶ πρὶν ψαύειν τῆς ὕλης,
ἐνήργει δ᾽ αὐταῖς περὶ τὴν ὕλην—ἅπασα γὰρ
δύναμις ἀργεῖ ἀποροῦσα τῆς οἰκείας ὕλης—, οὕτω
84 καὶ τὸ σπέρμα τὰς μὲν ‖ δυνάμεις οἴκοθεν ἐκέκτητο,
τὰς δ᾽ ἐνεργείας οὐκ ἐκ τῆς ὕλης ἔλαβεν, ἀλλὰ
περὶ τὴν ὕλην ἐπεδείξατο.

Καὶ μὴν εἰ πολλῷ μὲν ἐπικλύζοιτο τῷ αἵματι
τὸ σπέρμα, διαφθείροιτ᾽ ἄν· εἰ δ᾽ ὅλως ἀποροίη

---

[1] The spermatozoon now becomes an " organism " proper.
[2] Galen attributed to the *sperma* or semen what we should

130

simply by being clothed externally in a form and artificial shape. But Nature does not preserve the original character of any kind of matter; if she did so, then all parts of the animal would be blood—that blood, namely, which flows to the semen from the impregnated female and which is, so to speak, like the statuary's wax, a single uniform matter, subjected to the artificer. From this blood there arises no part of the animal which is as red and moist [as blood is], for bone, artery, vein, nerve, cartilage, fat, gland, membrane, and marrow are not blood, though they arise from it.

I would then ask Erasistratus himself to inform me what the altering, coagulating, and shaping agent is. He would doubtless say, "Either Nature or the semen," meaning the same thing in both cases, but explaining it by different devices. For that which was previously semen, when it begins to procreate and to shape the animal, becomes, so to say, a special *nature*.[1] For in the same way that Phidias possessed the faculties of his art even before touching his material, and then activated these in connection with this material (for every faculty remains inoperative in the absence of its proper material), so it is with the semen: its faculties it possessed from the beginning,[2] while its activities it does not receive from its material, but it manifests them in connection therewith.

And, of course, if it were to be overwhelmed with a great quantity of blood, it would perish, while if it were to be entirely deprived of blood

to the fertilized ovum: to him the maternal contribution is purely passive—mere food for the sperm. The epoch-making Ovum Theory was not developed till the seventeenth century. *cf.* p. 19, note 3.

παντάπασιν ἀργοῦν, οὐκ ἂν γένοιτο φύσις. ἵν'
οὖν μήτε φθείρηται καὶ γίγνηται φύσις ἀντὶ
σπέρματος, ὀλίγον ἐπιρρεῖν ἀναγκαῖον αὐτῷ τοῦ
αἵματος, μᾶλλον δ' οὐκ ὀλίγον λέγειν χρή, ἀλλὰ
σύμμετρον τῷ πλήθει τοῦ σπέρματος. τίς οὖν
ὁ μετρῶν αὐτοῦ τὸ ποσὸν τῆς ἐπιρροῆς; τίς ὁ
κωλύων ἰέναι πλέον; τίς ὁ προτρέπων, ἵν' ἐνδεέστερον μὴ ἴῃ; τίνα ζητήσομεν ἐνταῦθα τρίτον
ἐπιστάτην τοῦ ζῴου τῆς γενέσεως, ὃς χορηγήσει
τῷ σπέρματι τὸ σύμμετρον αἷμα; τί ἂν εἶπεν
Ἐρασίστρατος, εἰ ζῶν ταῦτ' ἠρωτήθη; τὸ σπέρμα
αὐτὸ δηλονότι· τοῦτο γάρ ἐστιν ὁ τεχνίτης ὁ ἀνα-
λογῶν τῷ Φειδίᾳ, τὸ δ' αἷμα τῷ κηρῷ προσέοικεν.

Οὔκουν πρέπει τὸν κηρὸν αὐτὸν ἑαυτῷ τὸ
μέτρον ἐξευρίσκειν, ἀλλὰ τὸν Φειδίαν. ἕλξει δὴ
τοσοῦτον αἵματος ὁ τεχνίτης εἰς ἑαυτόν, ὁπόσου
85 δεῖται. ἀλλ' ἐν‖ταῦθα χρὴ προσέχειν ἤδη τὸν
νοῦν καὶ σκοπεῖν, μή πως λάθωμεν τῷ σπέρματι
λογισμόν τινα καὶ νοῦν χαρισάμενοι· οὕτω γὰρ
ἂν οὔτε σπέρμα ποιήσαιμεν οὔτε φύσιν ἀλλ' ἤδη
ζῷον αὐτό. καὶ μὴν εἰ φυλάξομεν ἀμφότερα,
τήν θ' ὁλκὴν τοῦ συμμέτρου καὶ τὸ χωρὶς
λογισμοῦ, δύναμίν τινα, καθάπερ ἡ λίθος ἑλκτικὴν εἶχε τοῦ σιδήρου, καὶ τῷ σπέρματι φήσομεν
ὑπάρχειν αἵματος ἐπισπαστικήν. ἠναγκάσθημεν
οὖν πάλιν κἀνταῦθα, καθάπερ ἤδη πολλάκις
ἔμπροσθεν, ἑλκτικήν τινα δύναμιν ὁμολογῆσαι
κατὰ τὸ σπέρμα.

---

[1] i.e. we should be talking psychology, not biology; cf.
stomach, p. 307, note 3.
[2] Attraction now described not merely as *qualitative* but
also as *quantitative*. cf. p. 85, note 3.

it would remain inoperative and would not turn into a *nature*. Therefore, in order that it may not perish, but may become a *nature* in place of semen, there must be an afflux to it of a little blood—or, rather, one should not say a little, but a quantity commensurate with that of the semen. What is it then that measures the quantity of this afflux? What prevents more from coming? What ensures against a deficiency? What is this third overseer of animal generation that we are to look for, which will furnish the semen with à due amount of blood? What would Erasistratus have said if he had been alive, and had been asked this question? Obviously, the semen itself. This, in fact, is the artificer analogous with Phidias, whilst the blood corresponds to the statuary's wax.

Now, it is not for the wax to discover for itself how much of it is required; that is the business of Phidias. Accordingly the artificer will draw to itself as much blood as it needs. Here, however, we must pay attention and take care not unwittingly to credit the semen with reason and intelligence; if we were to do this, we would be making neither semen nor a nature, but an actual living animal.[1] And if we retain these two principles—that of proportionate attraction[2] and that of the non-participation of intelligence—we shall ascribe to the semen a faculty for attracting blood similar to that possessed by the lodestone for iron.[3] Here, then, again, in the case of the semen, as in so many previous instances, we have been compelled to acknowledge some kind of attractive faculty.

[3] He still tends either to biologize physics, or to physicize biology—whichever way we prefer to look at it. *cf.* Book I., chap. xiv.

Τί δ' ἦν τὸ σπέρμα; ἡ ἀρχὴ τοῦ ζῴου δηλονότι ἡ δραστική· ἡ γὰρ ὑλικὴ τὸ καταμήνιόν ἐστιν. εἶτ' αὐτῆς τῆς ἀρχῆς πρώτῃ ταύτῃ τῇ δυνάμει χρωμένης, ἵνα γένηται τῶν ὑπ' αὐτῆς τι δεδημιουργημένων, ἄμοιρον εἶναι τῆς οἰκείας δυνάμεως οὐκ ἐνδέχεται. πῶς οὖν Ἐρασίστρατος αὐτὴν οὐκ οἶδεν, εἰ δὴ πρώτη μὲν αὕτη τοῦ σπέρματος ἐνέργια τὸ σύμμετρον αἵματος ἐπισπᾶσθαι πρὸς ἑαυτό; σύμμετρον δ' ἂν εἴη τὸ λεπτὸν οὕτω καὶ ἀτμῶδες, ὥστ' εὐθὺς εἰς πᾶν μόριον ἑλκόμενον τοῦ σπέρματος δροσοειδῶς μηδαμοῦ τὴν || ἑαυτοῦ παρεμφαίνειν ἰδέαν. οὕτω γὰρ αὐτοῦ καὶ κρατήσει ῥᾳδίως τὸ σπέρμα καὶ ταχέως ἐξομοιώσει καὶ τροφὴν ἑαυτῷ ποιήσεται κἄπειτ' οἶμαι δεύτερον ἐπισπάσεται καὶ τρίτον, ὡς ὄγκον ἑαυτῷ καὶ πλῆθος ἀξιόλογον ἐργάσασθαι τραφέντι. καὶ μὴν ἤδη καὶ ἡ ἀλλοιωτικὴ δύναμις ἐξεύρηται μηδ' αὐτὴ πρὸς Ἐρασιστράτου γεγραμμένη. τρίτη δ' ἂν ἡ διαπλαστικὴ φανείη, καθ' ἣν πρῶτον μὲν οἷον ἐπίπαγόν τινα λεπτὸν ὑμένα περιτίθησιν ἑαυτῷ τὸ σπέρμα, τὸν ὑφ' Ἱπποκράτους ἐπὶ τῆς ἐκταίας γονῆς, ἣν ἐκπεσεῖν ἔλεγε τῆς μουσουργοῦ, τῷ τῶν ὠῶν εἰκασθέντα χιτῶνι· μετὰ δὲ τοῦτον ἤδη καὶ τἆλλ', ὅσα πρὸς ἐκείνου λέγεται διὰ τοῦ περὶ φύσιος παιδίου συγγράμματος.

Ἀλλ' εἰ τῶν διαπλασθέντων ἕκαστον οὕτω μείνειε σμικρόν, ὡς ἐξ ἀρχῆς ἐγένετο, τί ἂν εἴη πλέον; αὐξάνεσθαι τοίνυν αὐτὰ χρή. πῶς οὖν

---

[1] Aristotelian and Stoic duality of an active and a passive principle.

[2] Note that early embryonic development is described as a process of *nutrition.* *cf.* p. 130, note 2.

And what is the semen? Clearly the active principle of the animal, the material principle being the menstrual blood.[1] Next, seeing that the active principle employs this faculty primarily, therefore, in order that any one of the things fashioned by it may come into existence, it [the principle] must necessarily be possessed of its own faculty. How, then, was Erasistratus unaware of it, if the primary function of the semen be to draw to itself a due proportion of blood? Now, this fluid would be in due proportion if it were so thin and vaporous, that, as soon as it was drawn like dew into every part of the semen, it would everywhere cease to display its own particular character; for so the semen will easily dominate and quickly assimilate it—in fact, will use it as food. It will then, I imagine, draw to itself a second and a third quantum, and thus by feeding it acquires for itself considerable bulk and quantity.[2] In fact, *the alterative faculty* has now been discovered as well, although about this also Erasistratus has not written a word. And, thirdly the *shaping*[3] faculty will become evident, by virtue of which the semen firstly surrounds itself with a thin membrane like a kind of superficial condensation; this is what was described by Hippocrates in the sixth-day birth, which, according to his statement, fell from the singing-girl and resembled the pellicle of an egg. And following this all the other stages will occur, such as are described by him in his work "On the Child's Nature."

But if each of the parts formed were to remain as small as when it first came into existence, of what use would that be? They have, then, to grow.

[3] On the *alterative* and *shaping* faculties *cf.* p. 18, note 1.

αὐξηθήσεται; πάντη διατεινόμενα θ ἅμα καὶ
τρεφόμενα. καί μοι τῶν ἔμπροσθεν εἰρημένων
ἐπὶ τῆς κύστεως, ἣν οἱ παῖδες ἐμφυσῶντες ἔτρι-
87 βον, ἀναμνησθεὶς μαθήσῃ μᾶλλον ‖ κἀκ τῶν νῦν
ῥηθησομένων.

Ἐννόησον γὰρ δὴ τὴν καρδίαν οὕτω μὲν μικρὰν
εἶναι κατ' ἀρχάς, ὡς κέγχρου μηδὲν διαφέρειν ἤ,
εἰ βούλει, κυάμου, καὶ ζήτησον, ὅπως ἂν ἄλλως
αὕτη γένοιτο μεγάλη χωρὶς τοῦ πάντη διατεινο-
μένην τρέφεσθαι δι' ὅλης ἑαυτῆς, ὡς ὀλίγῳ πρόσθεν ἐδείκνυτο τὸ σπέρμα τρεφόμενον. ἀλλ' οὐδὲ
τοῦτ' Ἐρασίστρατος οἶδεν ὁ τὴν τέχνην τῆς
φύσεως ὑμνῶν, ἀλλ' οὕτως αὐξάνεσθαι τὰ ζῷα
νομίζει καθάπερ τινὰ κρησέραν ἢ σειρὰν ἢ σάκκον
ἢ τάλαρον, ὧν ἑκάστῳ κατὰ τὸ πέρας ἐπιπλεκο-
μένων ὁμοίων ἑτέρων τοῖς ἐξ ἀρχῆς αὐτὰ συντι-
θεῖσιν ἡ πρόσθεσις γίγνεται.

Ἀλλὰ τοῦτό γ' οὐκ αὔξησίς ἐστιν ἀλλὰ γένεσις, ὦ σοφώτατε· γίγνεται γὰρ ὁ θύλακος ἔτι καὶ
ὁ σάκκος καὶ θοἰμάτιον καὶ ἡ οἰκία καὶ τὸ πλοῖον
καὶ τῶν ἄλλων ἕκαστον, ὅταν μηδέπω τὸ προσῆκον εἶδος, οὗ χάριν ὑπὸ τοῦ τεχνίτου δημιουρ-
γεῖται, συμπεπληρωμένον ᾖ. πότ' οὖν αὐξάνεται;
ὅταν ἤδη τέλειος ὢν ὁ τάλαρος, ὡς ἔχειν πυθμένα
τέ τινα καὶ στόμα καὶ οἷον γαστέρα καὶ τὰ
τούτων μεταξύ, μείζων ἅπασι τούτοις γένηται.
88 καὶ πῶς ‖ ἔσται τοῦτο; φήσει τις. πῶς δ' ἄλλως
ἢ εἰ ζῷον ἐξαίφνης ἢ φυτὸν ὁ τάλαρος ἡμῖν
γένοιτο; μόνων γὰρ τῶν ζώντων ἡ αὔξησις. σὺ
δ' ἴσως οἴει τὴν οἰκίαν οἰκοδομουμένην αὐξάνε-

Now, how will they grow? By becoming extended in all directions and at the same time receiving nourishment. And if you will recall what I previously said about the bladder which the children blew up and rubbed,[1] you will also understand my meaning better as expressed in what I am now about to say.

Imagine the heart to be, at the beginning, so small as to differ in no respect from a millet-seed, or, if you will, a bean ; and consider how otherwise it is to become large than by being extended in all directions and acquiring nourishment throughout its whole substance, in the way that, as I showed a short while ago, the semen is nourished. But even this was unknown to Erasistratus—the man who sings the artistic skill of Nature ! He imagines that animals grow like webs, ropes, sacks, or baskets, each of which has, woven on to its end or margin, other material similar to that of which it was originally composed.

But this, most sapient sir, is not growth, but genesis ! For a bag, sack, garment, house, ship, or the like is said to be still coming into existence [undergoing genesis] so long as the appropriate form for the sake of which it is being constructed by the artificer is still incomplete. Then, when does it grow ? Only when the basket, being complete, with a bottom, a mouth, and a belly, as it were, as well as the intermediate parts, now becomes larger in all these respects. "And how can this happen ?" someone will ask. Only by our basket suddenly becoming an animal or a plant ; for growth belongs to living things alone. Possibly you imagine that a house *grows* when it is being built, or a basket when being

[1] pp. 27–29.

σθαι καὶ τὸν τάλαρον πλεκόμενον καὶ θοἰμάτιον
ὑφαινόμενον. ἀλλ' οὐχ ὧδ' ἔχει· τοῦ μὲν γὰρ
ἤδη συμπεπληρωμένου κατὰ τὸ εἶδος ἡ αὔξησις,
τοῦ δ' ἔτι γιγνομένου ἡ εἰς τὸ εἶδος ὁδὸς οὐκ
αὔξησις ἀλλὰ γένεσις ὀνομάζεται· αὐξάνεται μὲν
γὰρ τὸ ὄν, γίγνεται δὲ τὸ οὐκ ὄν.

IV

Καὶ ταῦτ' Ἐρασίστρατος οὐκ οἶδεν, ὃν οὐδὲν
λανθάνει, εἴπερ ὅλως ἀληθεύουσιν οἱ ἀπ' αὐτοῦ
φάσκοντες ὡμιληκέναι τοῖς ἐκ τοῦ περιπάτου
φιλοσόφοις αὐτόν. ἄχρι μὲν οὖν τοῦ τὴν φύσιν
ὑμνεῖν ὡς τεχνικὴν κἀγὼ γνωρίζω τὰ τοῦ περιπάτου δόγματα, τῶν δ' ἄλλων οὐδὲν οὐδ' ἐγγύς.
εἰ γάρ τις ὁμιλήσειε τοῖς Ἀριστοτέλους καὶ
Θεοφράστου γράμμασι, τῆς Ἱπποκράτους ἂν
αὐτὰ δόξειε φυσιολογίας ὑπομνήματα συγκεῖσθαι,
89 τὸ θερμὸν καὶ τὸ ψυχρὸν ‖ καὶ τὸ ξηρὸν καὶ τὸ
ὑγρὸν εἰς ἄλληλα δρῶντα καὶ πάσχοντα καὶ
τούτων αὐτῶν δραστικώτατον μὲν τὸ θερμόν,
δεύτερον δὲ τῇ δυνάμει τὸ ψυχρὸν Ἱπποκράτους
ταῦτα σύμπαντα πρώτου, δευτέρου δ' Ἀριστοτέλους εἰπόντος. τρέφεσθαι δὲ δι' ὅλων αὐτῶν
τὰ τρεφόμενα καὶ κεράννυσθαι δι' ὅλων τὰ
κεραννύμενα καὶ ἀλλοιοῦσθαι δι' ὅλων τὰ ἀλλοι-
ούμενα, καὶ ταῦθ' Ἱπποκράτειά θ' ἅμα καὶ
Ἀριστοτέλεια. καὶ τὴν πέψιν ἀλλοίωσίν τιν'

---

[1] cf. Introduction, p. xxvi.   [2] cf. p. 15.

plaited, or a garment when being woven? It is not so, however. Growth belongs to that which has already been completed in respect to its form, whereas the process by which that which is still *becoming* attains its form is termed not growth but genesis. That which *is*, grows, while that which *is not*, becomes.

IV

THIS also was unknown to Erasistratus, whom nothing escaped, if his followers speak in any way truly in maintaining that he was familiar with the Peripatetic philosophers. Now, in so far as he acclaims Nature as being an artist in construction, even I recognize the Peripatetic teachings, but in other respects he does not come near them. For if anyone will make himself acquainted with the writings of Aristotle and Theophrastus, these will appear to him to consist of commentaries on the Nature-lore [physiology][1] of Hippocrates—according to which the principles of heat, cold, dryness and moisture act upon and are acted upon by one another, the hot principle being the most active, and the cold coming next to it in power; all this was stated in the first place by Hippocrates and secondly by Aristotle.[2] Further, it is at once the Hippocratic and the Aristotelian teaching that the parts which are being nourished receive that nourishment throughout their whole substance, and that, similarly, processes of *mingling* and *alteration* involve the entire substance.[3] Moreover, that digestion is a species of

[3] For definitions of *alteration* and *mingling* (*crasis*, "temperament") *cf.* Book I., chaps. ii. and iii.

ὑπάρχειν καὶ μεταβολὴν τοῦ τρέφοντος εἰς τὴν
οἰκείαν τοῦ τρεφομένου ποιότητα, τὴν δ' ἐξαι-
μάτωσιν ἀλλοίωσιν εἶναι καὶ τὴν θρέψιν ὡσαύτως
καὶ τὴν αὔξησιν ἐκ τῆς πάντη διατάσεως καὶ
θρέψεως γίγνεσθαι, τὴν δ' ἀλλοίωσιν ὑπὸ τοῦ
θερμοῦ μάλιστα συντελεῖσθαι καὶ διὰ τοῦτο καὶ
τὴν πέψιν καὶ τὴν θρέψιν καὶ τὴν τῶν χυμῶν
ἁπάντων γένεσιν, ἤδη δὲ καὶ τοῖς περιττώμασι
τὰς ποιότητας ὑπὸ τῆς ἐμφύτου θερμασίας ἐγγί-
γνεσθαι, ταῦτα σύμπαντα καὶ πρὸς τούτοις ἕτερα
πολλὰ τά τε τῶν προειρημένων δυνάμεων καὶ
90 τὰ ‖ τῶν νοσημάτων τῆς γενέσεως καὶ τὰ τῶν
ἰαμάτων τῆς εὑρέσεως Ἱπποκράτης μὲν πρῶτος
ἁπάντων ὧν ἴσμεν ὀρθῶς εἶπεν, Ἀριστοτέλης δὲ
δεύτερος ὀρθῶς ἐξηγήσατο. καὶ μὴν εἰ ταῦτα
σύμπαντα τοῖς ἐκ τοῦ περιπάτου δοκεῖ, καθάπερ
οὖν δοκεῖ, μηδὲν δ' αὐτῶν ἀρέσκει τῷ Ἐρασιστράτῳ, τί ποτε βούλεται τοῖς Ἐρασιστρατείοις ἡ
πρὸς τοὺς φιλοσόφους ἐκείνους τοῦ τῆς αἱρέσεως
αὐτῶν ἡγεμόνος ὁμιλία; θαυμάζουσι μὲν γὰρ
αὐτὸν ὡς θεὸν καὶ πάντ' ἀληθεύειν νομίζουσιν.
εἰ δ' οὕτως ἔχει ταῦτα, πάμπολυ δήπου τῆς
ἀληθείας ἐσφάλθαι χρὴ νομίζειν τοὺς ἐκ τοῦ
περιπάτου φιλοσόφους, οἷς μηδὲν ὧν Ἐρασίστρατος ὑπελάμβανεν ἀρέσκει. καὶ μὴν ὥσπερ
τιν' εὐγένειαν αὐτῷ τῆς φυσιολογίας τὴν πρὸς
τοὺς ἄνδρας ἐκείνους συνουσίαν ἐκπορίζουσι.

Πάλιν οὖν ἀναστρέψωμεν τὸν λόγον ἑτέρως ἢ
ὡς ὀλίγῳ πρόσθεν ἐτύχομεν εἰπόντες. εἴπερ γὰρ
οἱ ἐκ τοῦ περιπάτου καλῶς ἐφυσιολόγησαν,
οὐδὲν ἂν εἴη ληρωδέστερον Ἐρασιστράτου καὶ
δίδωμι τοῖς Ἐρασιστρατείοις αὐτοῖς τὴν αἵρεσιν·

alteration—a transmutation of the nutriment into the proper quality of the thing receiving it; that blood-production also is an alteration, and nutrition as well; that growth results from extension in all directions, combined with nutrition; that alteration is effected mainly by the warm principle, and that therefore digestion, nutrition, and the generation of the various humours, as well as the qualities of the surplus substances, result from the *innate heat*; [1] all these and many other points besides in regard to the aforesaid faculties, the origin of diseases, and the discovery of remedies, were correctly stated first by Hippocrates of all writers whom we know, and were in the second place correctly expounded by Aristotle. Now, if all these views meet with the approval of the Peripatetics, as they undoubtedly do, and if none of them satisfy Erasistratus, what can the Erasistrateans possibly mean by claiming that their leader was associated with these philosophers? The fact is, they revere him as a god, and think that everything he says is true. If this be so, then we must suppose the Peripatetics to have strayed very far from truth, since they approve of none of the ideas of Erasistratus. And, indeed, the disciples of the latter produce his connection with the Peripatetics in order to furnish his Nature-lore with a respectable pedigree.

Now, let us reverse our argument and put it in a different way from that which we have just employed. For if the Peripatetics were correct in their teaching about Nature, there could be nothing more absurd than the contentions of Erasistratus. And, I will leave it to the Erasistrateans themselves to decide;

[1] *i.e.* are associated with oxidation? *cf.* p. 41, note 3.

91 ἢ γὰρ τὸν πρότερον λόγον ἢ τοῦτον ‖ προσήσονται. λέγει δ᾽ ὁ μὲν πρότερος οὐδὲν ὀρθῶς ἐγνωκέναι περὶ φύσεως τοὺς περιπατητικούς, ὁ δὲ δεύτερος Ἐρασίστρατον. ἐμὸν μὲν οὖν ὑπομνῆσαι τῶν δογμάτων τὴν μάχην, ἐκείνων δ᾽ ἡ αἵρεσις.

Ἀλλ᾽ οὐκ ἂν ἀποσταῖεν τοῦ θαυμάζειν Ἐρασίστρατον· οὐκοῦν σιωπάτωσαν περὶ τῶν ἐκ τοῦ περιπάτου φιλοσόφων. παμπόλλων γὰρ ὄντων δογμάτων φυσικῶν περί τε γένεσιν καὶ φθορὰν τῶν ζῴων καὶ ὑγίειαν καὶ νόσους καὶ τὰς θεραπείας αὐτῶν ἓν μόνον εὑρεθήσεται ταὐτὸν Ἐρασιστράτῳ κἀκείνοις τοῖς ἀνδράσι, τό τινος ἕνεκα πάντα ποιεῖν τὴν φύσιν καὶ μάτην μηδέν.

Ἀλλὰ καὶ αὐτὸ τοῦτο μέχρι λόγου κοινόν, ἔργῳ δὲ μυριάκις Ἐρασίστρατος αὐτὸ διαφθείρει· μάτην μὲν γὰρ ὁ σπλὴν ἐγένετο, μάτην δὲ τὸ ἐπίπλοον, μάτην δ᾽ αἱ εἰς τοὺς νεφροὺς ἀρτηρίαι καταφυόμεναι, σχεδὸν ἁπασῶν τῶν ἀπὸ τῆς μεγάλης ἀρτηρίας ἀποβλαστανουσῶν οὖσαι μέγισται, μάτην δ᾽ ἄλλα μυρία κατά γε τὸν Ἐρασιστράτειον λόγον· ἅπερ εἰ μὲν οὐδ᾽ ὅλως γιγνώσκει, βραχεῖ μαγείρου σοφώτερός ἐστιν ἐν ταῖς ἀνατομαῖς, εἰ δ᾽ εἰδὼς οὐ λέγει τὴν χρείαν 92 αὐτῶν, οἴεται ‖ δηλονότι παραπλησίως τῷ σπληνὶ μάτην αὐτὰ γεγονέναι. καίτοι τί ταῦτ᾽ ἐπεξέρχομαι τῆς περὶ χρείας μορίων πραγματείας ὄντα μελλούσης ἡμῖν ἰδίᾳ περαίνεσθαι;

---

[1] "Useless" organs ; cf. p. 56, note 2.   For fallacy of Erasistratus's view on the spleen v. p. 205.

they must either advance the one proposition or the other. According to the former one the Peripatetics had no accurate acquaintance with Nature, and according to the second, Erasistratus. It is my task, then, to point out the opposition between the two doctrines, and theirs to make the choice. . . .

But they certainly will not abandon their reverence for Erasistratus. Very well, then; let them stop talking about the Peripatetic philosophers. For among the numerous physiological teachings regarding the genesis and destruction of animals, their health, their diseases, and the methods of treating these, there will be found one only which is common to Erasistratus and the Peripatetics—namely, the view that Nature does everything for some purpose, and nothing in vain.

But even as regards this doctrine their agreement is only verbal; in practice Erasistratus makes havoc of it a thousand times over. For, according to him, the spleen was made for no purpose, as also the omentum; similarly, too, the arteries which are inserted into kidneys [1]—although these are practically the largest of all those that spring from the great artery [aorta]! And to judge by the Erasistratean argument, there must be countless other useless structures; for, if he knows nothing at all about these structures, he has little more anatomical knowledge than a butcher, while, if he is acquainted with them and yet does not state their use, he clearly imagines that they were made for no purpose, like the spleen. Why, however, should I discuss these structures fully, belonging as they do to the treatise "On the Use of Parts," which I am personally about to complete?

Πάλιν οὖν ἀναλάβωμεν τὸν αὐτὸν λόγον εἰπόντες τέ τι βραχὺ πρὸς τοὺς Ἐρασιστρατείους ἔτι τῶν ἐφεξῆς ἐχώμεθα. δοκοῦσι γάρ μοι μηδὲν ἀνεγνωκέναι τῶν Ἀριστοτέλους οὗτοι συγγραμμάτων, ἀλλ' ἄλλων ἀκούοντες, ὡς δεινὸς ἦν περὶ φύσιν ὁ ἄνθρωπος καὶ ὡς οἱ ἀπὸ τῆς στοᾶς κατ' ἴχνη τῆς ἐκείνου φυσιολογίας βαδίζουσιν, εἶθ' εὑρόντες ἔν τι τῶν περιφερομένων δογμάτων κοινὸν αὐτῷ πρὸς Ἐρασίστρατον ἀναπλάσαι τινὰ συνουσίαν αὐτοῦ πρὸς ἐκείνους τοὺς ἄνδρας. ἀλλ' ὅτι μὲν τῆς Ἀριστοτέλους φυσιολογίας οὐδὲν Ἐρασιστράτῳ μέτεστιν, ὁ κατάλογος τῶν προειρημένων ἐνδείκνυται δογμάτων, ἃ πρώτου μὲν Ἱπποκράτους ἦν, δευτέρου δ' Ἀριστοτέλους, τρίτων δὲ τῶν Στωϊκῶν,[1] ἑνὸς μόνου μετατιθεμένου τοῦ τὰς ποιότητας εἶναι σώματα.[3]

Τάχα δ' ἂν τῆς λογικῆς ἕνεκα θεωρίας ὡμιληκέναι φαῖεν τὸν Ἐρασίστρατον τοῖς ἐκ τοῦ περιπάτου[2] φιλοσόφοις, οὐκ εἰδότες, ὡς ἐκεῖνοι
93 μὲν ψευ‖δεῖς καὶ ἀπεράντους οὐκ ἔγραψαν λόγους, τὰ δ' Ἐρασιστράτεια βιβλία παμπόλλους ἔχει τοὺς τοιούτους.

Τάχ' ἂν οὖν ἤδη τις θαυμάζοι καὶ διαποροίη, τί παθὼν ὁ Ἐρασίστρατος εἰς τοσοῦτον τῶν Ἱπποκράτους δογμάτων ἀπετράπετο καὶ διὰ τί τῶν ἐν ἥπατι πόρων τῶν χοληδόχων, ἅλις γὰρ ἤδη νεφρῶν, ἀφελόμενος τὴν ἑλκτικὴν δύναμιν ἐπίκαιρον αἰτιᾶται θέσιν καὶ στομάτων

---

[1] The Stoics.   [2] The Peripatetics (Aristotelians).
[3] Aristotle regarded the *qualitative* differences apprehended by our senses (the cold, the warm, the moist, and the dry) as fundamental, while the Stoics held the four corporeal elements

Let us, then, sum up again this same argument, and, having said a few words more in answer to the Erasistrateans, proceed to our next topic. The fact is, these people seem to me to have read none of Aristotle's writings, but to have heard from others how great an authority he was on "Nature," and that those of the Porch[1] follow in the steps of his Nature-lore; apparently they then discovered a single one of the current ideas which is common to Aristotle and Erasistratus, and made up some story of a connection between Erasistratus and these people.[2] That Erasistratus, however, has no share in the Nature-lore of Aristotle is shown by an enumeration of the aforesaid doctrines, which emanated first from Hippocrates, secondly from Aristotle, thirdly from the Stoics (with a single modification, namely, that for them the *qualities* are *bodies*).[3]

Perhaps, however, they will maintain that it was in the matter of *logic* that Erasistratus associated himself with the Peripatetic philosophers? Here they show ignorance of the fact that these philosophers never brought forward false or inconclusive arguments, while the Erasistratean books are full of them.

So perhaps somebody may already be asking, in some surprise, what possessed Erasistratus that he turned so completely from the doctrines of Hippocrates, and why it is that he takes away the attractive faculty from the biliary[4] passages in the liver—for we have sufficiently discussed the kidneys —alleging [as the cause of bile-secretion] a favourable situation, the narrowness of vessels, and *a*

(earth, air, fire, and water) to be still more fundamental. *cf.* p. 8, note 3.   [4] Lit. bile-receiving (choledochous).

στενότητα καὶ χώραν τινὰ κοινήν, εἰς ἣν παρ-
άγουσι μὲν αἱ ἀπὸ τῶν πυλῶν τὸ ἀκάθαρτον αἷμα,
μεταλαμβάνουσι δὲ πρότεροι μὲν οἱ πόροι τὴν
χολήν, δεύτεραι δ᾽ αἱ ἀπὸ τῆς κοίλης φλεβὸς
τὸ καθαρὸν αἷμα. πρὸς γὰρ τῷ μηδὲν ἂν βλα-
βῆναι τὴν ὁλκὴν εἰπὼν ἄλλων μυρίων ἔμελλεν
ἀμφισβητουμένων ἀπαλλάξεσθαι λόγων.

V

Ὡς νῦν γε πόλεμος οὐ σμικρός ἐστι τοῖς
Ἐρασιστρατείοις οὐ πρὸς τοὺς ἄλλους μόνον
ἀλλὰ καὶ πρὸς ἀλλήλους, οὐκ ἔχουσιν, ὅπως
ἐξηγήσωνται τὴν ἐκ τοῦ πρώτου τῶν καθόλου
94 λόγων λέξιν, ἐν ᾗ φησιν· "Εἰς τὸ ‖ αὐτὸ δ᾽ ἀνεστομωμένων ἑτέρων δύο ἀγγείων τῶν τ᾽ ἐπὶ τὴν
χοληδόχον τεινόντων καὶ τῶν ἐπὶ τὴν κοίλην
φλέβα συμβαίνει τῆς ἀναφερομένης ἐκ τῆς
κοιλίας τροφῆς τὰ ἐναρμόζοντα ἑκατέροις τῶν
στομάτων εἰς ἑκάτερα τῶν ἀγγείων μεταλαμβάνεσθαι καὶ τὰ μὲν ἐπὶ τὴν χοληδόχον
φέρεσθαι, τὰ δ᾽ ἐπὶ τὴν κοίλην φλέβα περαιοῦ-
σθαι." τὸ γὰρ "εἰς τὸ αὐτὸ ἀνεστομωμένων,"
ὃ κατ᾽ ἀρχὰς τῆς λέξεως γέγραπται, τί ποτε χρὴ
νοῆσαι, χαλεπὸν εἰπεῖν. ἤτοι γὰρ οὕτως εἰς
ταὐτόν, ὥστε τῷ τῆς ἐν τοῖς σιμοῖς φλεβὸς
πέρατι συνάπτειν δύο ἕτερα πέρατα, τό τ᾽ ἐν τοῖς

---

[1] *Jecoris portae*, the transverse fissure, by which the portal
vein enters the liver.

common space into which the veins from the gate-way [of the liver][1] conduct the unpurified blood, and from which, in the first place, the [biliary] passages take over the bile, and secondly, the [branches] of the vena cava take over the purified blood. For it would not only have done him no harm to have mentioned the idea of *attraction*, but he would there-by have been able to get rid of countless other dis-puted questions.

V

At the actual moment, however, the Erasi-strateans are engaged in a considerable battle, not only with others but also amongst themselves, and so they cannot explain the passage from the first book of the " General Principles," in which Erasistratus says, " Since there are two kinds of vessels opening [2] at the same place, the one kind extending to the gall-bladder and the other to the vena cava, the result is that, of the nutriment carried up from the alimentary canal, that part which fits both kinds of stomata is received into both kinds of vessels, some being carried into the gall-bladder, and the rest passing over into the vena cava." For it is difficult to say what we are to understand by the words " opening at the same place " which are written at the beginning of this passage. Either they mean there is a *junction* [3] between the termination of the vein which is on the concave surface of the liver[4] and two other vascular terminations (that of the vessel on the convex surface of the liver [5]

[2] Lit. " anastomosing."   [3] More literally, " synapse."
[4] The portal vein.   [5] The hepatic vein or veins.

κυρτοῖς καὶ τὸ τοῦ χοληδόχου πόρου, ἤ, εἰ μὴ
οὕτω, χώραν τινὰ κοινὴν ἐπινοῆσαι χρὴ τῶν
τριῶν ἀγγείων οἷον δεξαμενήν τινα, πληρουμένην
μὲν ὑπὸ τῆς κάτω φλεβός, ἐκκενουμένην δ᾽ εἴς τε
τοὺς χοληδόχους πόρους καὶ τὰς τῆς κοίλης
ἀποσχίδας· καθ᾽ ἑκατέραν δὲ τῶν ἐξηγήσεων
ἄτοπα πολλά, περὶ ὧν εἰ πάντων λέγοιμι, λάθοιμ᾽
ἂν ἐμαυτὸν ἐξηγήσεις Ἐρασιστράτου γράφων,
οὐχ, ὅπερ ἐξ ἀρχῆς προὐθέμην, περαίνων. κοινὸν
δ᾽ ἀμφοτέραις ταῖς ἐξηγήσεσιν ἄτοπον τὸ μὴ ‖
95 καθαίρεσθαι πᾶν τὸ αἷμα. χρὴ γὰρ ὡς εἰς
ἠθμόν τινα τὸ χοληδόχον ἀγγεῖον ἐμπίπτειν
αὐτό, οὐ παρέρχεσθαι καὶ παρρεῖν ὠκέως εἰς
τὸ μεῖζον στόμα τῇ ῥύμῃ τῆς ἀναδόσεως φερό-
μενον.

Ἆρ᾽ οὖν ἐν τούτοις μόνον ἀπορίαις ἀφύκτοις ὁ
Ἐρασιστράτου λόγος ἐνέχεται μὴ βουληθέντος
χρήσασθαι ταῖς ἑλκτικαῖς δυνάμεσιν εἰς μηδέν, ἢ
σφοδρότατα μὲν ἐν τούτοις καὶ σαφῶς οὕτως, ὡς
ἂν μηδὲ παῖδα λαθεῖν;

VI

Εἰ δ᾽ ἐπισκοποῖτό τις ἐπιμελῶς, οὐδ᾽ ὁ περὶ
θρέψεως αὐτοῦ λόγος, ὃν ἐν τῷ δευτέρῳ τῶν
καθόλου λόγων διεξέρχεται, τὰς αὐτὰς ἀπορίας
ἐκφεύγει. τῇ γὰρ πρὸς τὸ κενούμενον ἀκολουθίᾳ
συγχωρηθέντος ἑνὸς λήμματος, ὡς πρόσθεν
ἐδείκνυμεν, ἐπέραινέ τι περὶ φλεβῶν μόνων καὶ
τοῦ κατ᾽ αὐτὰς αἵματος. ἐκρέοντος γάρ τινος

---

[1] The portal vein. [2] cf. p. 120, note 1.

and that of the bile-duct), or, if not, then we must
suppose that there is, as it were, a common space
for all three vessels, which becomes filled from the
lower vein,[1] and empties itself both into the bile-
duct and into the branches of the vena cava.   Now,
there are many difficulties in both of these explanations, but if I were to state them all, I should find
myself inadvertently writing an exposition of the
teaching of Erasistratus, instead of carrying out my
original undertaking.   There is, however, one difficulty common to both these explanations, namely,
that the whole of the blood does not become
purified.   For it ought to fall into the bile-duct as
into a kind of sieve, instead of going (running, in
fact, rapidly) past it, into the larger stoma, by virtue
of the impulse of *anadosis.*

Are these, then, the only inevitable difficulties in
which the argument of Erasistratus becomes involved
through his disinclination to make any use of the
attractive faculty, or is it that the difficulty is greatest
here, and also so obvious that even a child could not
avoid seeing it?

VI

AND if one looks carefully into the matter one
will find that even Erasistratus's reasoning on the
subject of *nutrition*, which he takes up in the
second book of his " General Principles," fails to
escape this same difficulty.   For, having conceded
one premise to the principle that matter tends to fill
a vacuum, as we previously showed, he was only
able to draw a conclusion in the case of the veins
and their contained blood.[2]   That is to say, when

κατὰ τὰ στόματ' αὐτῶν καὶ διαφορουμένου καὶ
μήτ' ἀθρόως τόπου κενοῦ δυναμένου γενέσθαι
μήτε τῶν φλεβῶν συμπεσεῖν, τοῦτο γὰρ ἦν τὸ
παραλειπόμενον, ἀναγκαῖον ἦν ἕπεσθαι τὸ συνεχὲς
96 ἀναπληροῦν τοῦ κενου‖μένου τὴν βάσιν. αἱ μὲν
δὴ φλέβες ἡμῖν οὕτω θρέψονται τοῦ περιεχομένου
κατ' αὐτὰς αἵματος ἀπολαύουσαι· τὰ δὲ νεῦρα
πῶς; οὐ γὰρ δὴ κἂν τούτοις ἐστὶν αἷμα. πρό-
χειρον μὲν γὰρ ἦν εἰπεῖν, ἕλκοντα παρὰ τῶν
φλεβῶν· ἀλλ' οὐ βούλεται. τί ποτ' οὖν κἀνταῦθα ἐπιτεχνᾶται; φλέβας ἔχειν ἐν ἑαυτῷ
καὶ ἀρτηρίας τὸ νεῦρον ὥσπερ τινὰ σειρὰν ἐκ
τριῶν ἱμάντων διαφερόντων τῇ φύσει πεπλεγ-
μένην. ᾠήθη γὰρ ἐκ ταύτης τῆς ὑποθέσεως
ἐκφεύξεσθαι τῷ λόγῳ τὴν ὁλκήν· οὐ γὰρ ἂν ἔτι
δεήσεσθαι τὸ νεῦρον ἐν ἑαυτῷ περιέχον αἵματος
ἀγγεῖον ἐπιρρύτου τινὸς ἔξωθεν ἐκ τῆς παρακειμένης φλεβὸς τῆς ἀληθινῆς αἵματος ἑτέρου,
ἀλλ' ἱκανὸν αὐτῷ πρὸς τὴν θρέψιν ἔσεσθαι
τὸ κατεψευσμένον ἀγγεῖον ἐκεῖνο τὸ λόγῳ θεω-
ρητόν.

Ἀλλὰ κἀνταῦθα πάλιν αὐτὸν ὁμοία τις ἀπορία
διεδέξατο. τουτὶ γὰρ τὸ σμικρὸν ἀγγεῖον ἑαυτὸ
μὲν θρέψει, τὸ παρακείμενον μέντοι νεῦρον ἐκεῖνο
τὸ ἁπλοῦν ἢ τὴν ἀρτηρίαν οὐχ οἷόν τ' ἔσται
τρέφειν ἄνευ τοῦ σύμφυτόν τιν' ὑπάρχειν αὐτοῖς
97 ὁλκὴν τῆς τροφῆς. ‖ τῇ μὲν γὰρ πρὸς τὸ κενούμενον ἀκολουθίᾳ πῶς ἂν ἔτι δύναιτο τὴν τροφὴν
ἐπισπᾶσθαι τὸ ἁπλοῦν νεῦρον, ὥσπερ αἱ φλέβες

---

[1] cf. p. 272, note 1.
[2] i.e. one might assume an *attraction*.

blood is running away through the stomata of
the veins, and is being dispersed, then, since an
absolutely empty space cannot result, and the veins
cannot collapse (for this was what he overlooked),
it was therefore shown to be necessary that the
adjoining quantum of fluid should flow in and fill
the place of the fluid evacuated. It is in this way
that we may suppose the veins to be nourished;
they get the benefit of the blood which they contain.
But how about the nerves?[1] For they do not also
contain blood. One might obviously say that they
draw their supply from the veins.[2] But Erasistratus
will not have it so. What further contrivance, then,
does he suppose? He says that a nerve has within
itself veins and arteries, like a rope woven by
Nature out of three different strands. By means of
this hypothesis he imagined that his theory would
escape from the idea of *attraction*. For if the nerve
contain within itself a blood-vessel it will no
longer need the adventitious flow of other blood
from the real vein lying adjacent; this fictitious
vessel, perceptible only in theory,[3] will suffice it for
nourishment.

But this, again, is succeeded by another similar
difficulty. For this small vessel will nourish itself,
but it will not be able to nourish this adjacent
simple nerve or artery, unless these possess some
innate proclivity for attracting nutriment. For how
could the *nerve*, being simple, attract its nourishment,
as do the composite veins, by virtue of the tendency

---

[3] *i.e.* visible to the mind's eye as distinguished from
the bodily eye. *cf.* p. 21, note 4. *Theoreton* without quali-
fication means merely *visible*, not *theoretic*. *cf.* p. 205,
note 1.

αἱ σύνθετοι; κοιλότης μὲν γάρ τίς ἐστιν ἐν αὐτῷ
κατ' αὐτόν, ἀλλ' οὐχ αἵματος αὕτη γ' ἀλλὰ
πνεύματος ψυχικοῦ μεστή. δεόμεθα δ' ἡμεῖς
οὐκ εἰς τὴν κοιλότητα ταύτην εἰσάγειν τῷ λόγῳ
τὴν τροφὴν ἀλλ' εἰς τὸ περιέχον αὐτὴν ἀγγεῖον,
εἴτ' οὖν τρέφεσθαι μόνον εἴτε καὶ αὔξεσθαι δέοιτο.
πῶς οὖν εἰσάξομεν; οὕτω γάρ ἐστι σμικρὸν ἐκεῖνο
τὸ ἁπλοῦν ἀγγεῖον καὶ μέντοι καὶ τῶν ἄλλων
ἑκάτερον, ὥστ', εἰ τῇ λεπτοτάτῃ βελόνῃ νύξειάς
τι μέρος, ἅμα διαιρήσεις τὰ τρία. τόπος οὖν
αἰσθητὸς ἀθρόως κενὸς οὐκ ἄν ποτ' ἐν αὐτῷ
γένοιτο· λόγῳ δὲ θεωρητὸς τόπος κενούμενος οὐκ
ἦν ἀναγκαστικὸς τῆς τοῦ συνεχοῦς ἀκολουθίας.

Ἠβουλόμην δ' αὖ πάλιν μοι κἀνταῦθα τὸν
Ἐρασίστρατον αὐτὸν ἀποκρίνασθαι περὶ τοῦ
στοιχειώδους ἐκείνου νεύρου τοῦ σμικροῦ, πότερον
ἕν τι καὶ συνεχὲς ἀκριβῶς ἐστιν ἢ ἐκ πολλῶν
καὶ σμικρῶν σωμάτων, ὧν Ἐπίκουρος καὶ Λεύ-
98 κιππος καὶ Δημόκριτος ὑπέθεντο, σύγ‖κειται.
καὶ γὰρ καὶ περὶ τούτου τοὺς Ἐρασιστρατείους
ὁρῶ διαφερομένους. οἱ μὲν γὰρ ἕν τι καὶ συνεχὲς
αὐτὸ νομίζουσιν ἢ οὐκ ἂν ἁπλοῦν εἰρῆσθαι πρὸς
αὐτοῦ φασι· τινὲς δὲ καὶ τοῦτο διαλύειν εἰς ἕτερα
στοιχειώδη τολμῶσιν. ἀλλ' εἰ μὲν ἕν τι καὶ
συνεχές ἐστι, τὸ κενούμενον ἐξ αὐτοῦ κατὰ τὴν
ἄδηλον ὑπὸ τῶν ἰατρῶν ὀνομαζομένην διαπνοὴν

---

[1] According to the Pneumatist school, certain of whose
ideas were accepted by Erasistratus, the air, breath, pneuma,
or spirit was brought by inspiration into the left side of the
heart, where it was converted into natural, vital, and
psychic pneuma ; the latter then went to the brain, whence
it was distributed through the nervous system ; practically

of a vacuum to become refilled? For, although according to Erasistratus, it contains within itself a cavity of sorts, this is not occupied with blood, but with *psychic pneuma*,[1] and we are required to imagine the nutriment introduced, not into this cavity, but into the vessel containing it, whether it needs merely to be nourished, or to grow as well. How, then, are we to imagine it introduced? For this simple vessel [*i.e.* nerve] is so small—as are also the other two—that if you prick it at any part with the finest needle you will tear the whole three of them at once. Thus there could never be in it a perceptible space entirely empty. And an emptied space which merely existed in theory could not compel the adjacent fluid to come and fill it.

At this point, again, I should like Erasistratus himself to answer regarding this small elementary nerve, whether it is actually one and definitely continuous, or whether it consists of many small bodies, such as those assumed by Epicurus, Leucippus, and Democritus.[2] For I see that the Erasistrateans are at variance on this subject. Some of them consider it one and continuous, for otherwise, as they say, he would not have called it *simple*; and some venture to resolve it into yet other elementary bodies. But if it be one and continuous, then what is evacuated from it in the so-called *insensible transpiration* of the

this teaching involved the idea of a *psyche*, or conscious vital principle. "Psychic pneuma" is in Latin *spiritus animalis* (*anima = psyche*); *cf.* p. 126, note 4. Introduction, p. xxxiv.

[2] Observe that Erasistratus's "simple nerve" may be almost looked on as an anticipation of the *cell*. The question Galen now asks is whether this vessel is a "unit mass of living matter," or merely an agglomeration of *atoms* subject to mechanical law. *cf.* Galen's "fibres," p. 329.

οὐδεμίαν ἐν ἑαυτῷ καταλείψει χώραν κενήν.
οὕτω γὰρ οὐχ ἓν ἀλλὰ πολλὰ γενήσεται, διειργό-
μενα δήπου ταῖς κεναῖς χώραις. εἰ δ' ἐκ πολλῶν
σύγκειται, τῇ κηπαίᾳ κατὰ τὴν παροιμίαν πρὸς
Ἀσκληπιάδην ἀπεχωρήσαμεν ἄναρμά τινα στοιχεῖα τιθέμενοι. πάλιν οὖν ἄτεχνος ἡμῖν ἡ φύσις
λεγέσθω· τοῖς γὰρ τοιούτοις στοιχείοις ἐξ ἀνάγκης τοῦθ' ἕπεται.

Διὸ δή μοι καὶ δοκοῦσιν ἀμαθῶς πάνυ τὴν εἰς
τὰ τοιαῦτα στοιχεῖα τῶν ἁπλῶν ἀγγείων εἰσάγειν
διάλυσιν ἔνιοι τῶν Ἐρασιστρατείων. ἐμοὶ γοῦν
οὐδὲν διαφέρει. καθ' ἑκατέρους γὰρ ἄτοπος ὁ
τῆς θρέψεως ἔσται λόγος, ἐκείνοις τοῖς ἁπλοῖς
ἀγγείοις τοῖς σμικροῖς τοῖς συντιθεῖσι τὰ μεγάλα ‖
99 τε καὶ αἰσθητὰ νεῦρα κατὰ μὲν τοὺς συνεχῆ
φυλάττοντας αὐτὰ μὴ δυναμένης γενέσθαι τῆς
πρὸς τὸ κενούμενον ἀκολουθίας, ὅτι μηδὲν ἐν τῷ
συνεχεῖ γίγνεται κενόν, κἂν ἀπορρέῃ τι· συνέρ-
χεται γὰρ πρὸς ἄλληλα τὰ καταλειπόμενα μόρια,
καθάπερ ἐπὶ τοῦ ὕδατος ὁρᾶται, καὶ πάλιν ἓν
γίγνεται πάντα τὴν χώραν τοῦ διαφορηθέντος
αὐτὰ καταλαμβάνοντα· κατὰ δὲ τοὺς ἑτέρους,
ὅτι τῶν στοιχείων ἐκείνων οὐδὲν δεῖται τῆς πρὸς
τὸ κενούμενον ἀκολουθίας. ἐπὶ γὰρ τῶν αἰσθητῶν
μόνων, οὐκ ἐπὶ τῶν λόγῳ θεωρητῶν ἔχει δύναμιν,
ὡς αὐτὸς ὁ Ἐρασίστρατος ὁμολογεῖ διαρρήδην,
οὐ περὶ τοῦ τοιούτου κενοῦ φάσκων ἑκάστοτε
ποιεῖσθαι τὸν λόγον, ὃ κατὰ βραχὺ παρέσπαρται
τοῖς σώμασιν, ἀλλὰ περὶ τοῦ σαφοῦς καὶ αἰσθητοῦ
καὶ ἀθρόου καὶ μεγάλου καὶ ἐναργοῦς καὶ ὅπως
ἂν ἄλλως ὀνομάζειν ἐθέλῃς. Ἐρασίστρατος μὲν
γὰρ αὐτὸς αἰσθητὸν ἀθρόως οὔ φησι δύνασθαι

physicians will leave no empty space in it; otherwise it would not be one body but many, separated by empty spaces. But if it consists of many bodies, then we have "escaped by the back door," as the saying is, to Asclepiades, seeing that we have postulated certain *inharmonious elements*. Once again, then, we must call Nature "inartistic"; for this necessarily follows the assumption of such elements.

For this reason some of the Erasistrateans seem to me to have done very foolishly in reducing the simple vessels to elements such as these. Yet it makes no difference to me, since the theory of both parties regarding nutrition will be shown to be absurd. For in these minute simple vessels constituting the large perceptible nerves, it is impossible, according to the theory of those who would keep the former continuous, that any "refilling of a vacuum" should take place, since no vacuum can occur in a continuum even if anything does run away; for the parts left come together (as is seen in the case of water) and again become one, taking up the whole space of that which previously separated them. Nor will any "refilling" occur if we accept the argument of the other Erasistrateans, since none of their *elements* need it. For this principle only holds of things which are perceptible, and not of those which exist merely in theory; this Erasistratus expressly acknowledges, for he states that it is not a vacuum such as this, interspersed in small portions among the corpuscles, that his various treatises deal with, but a vacuum which is clear, perceptible, complete in itself, large in size, evident, or however else one cares to term it (for, what Erasistratus himself says is, that " there cannot be a

γενέσθαι κενόν· ἐγὼ δ' ἐκ περιουσίας εὐπορήσας
ὀνομάτων ταὐτὸν δηλοῦν ἔν γε τῷ νῦν προκειμένῳ
λόγῳ δυναμένων καὶ τἆλλα προσέθηκα.

100 Κάλλιον οὖν μοι δοκεῖ καὶ ‖ ἡμᾶς τι συνεισενέγ-
κασθαι τοῖς Ἐρασιστρατείοις, ἐπειδὴ κατὰ τοῦτο
γεγόναμεν, καὶ συμβουλεῦσαι τοῖς τὸ πρῶτον
ἐκεῖνο καὶ ἁπλοῦν ὑπ' Ἐρασιστράτου καλούμενον
ἀγγεῖον εἰς ἕτερ' ἄττα σώματα στοιχειώδη
διαλύουσιν ἀποστῆναι τῆς ὑπολήψεως, ὡς πρὸς
τῷ μηδὲν ἔχειν πλέον ἔτι καὶ διαφερομένοις
Ἐρασιστράτῳ. ὅτι μὲν οὖν οὐδὲν ἔχει πλέον,
ἐπιδέδεικται σαφῶς· οὐδὲ γὰρ ἠδυνήθη διαφυγεῖν
τὴν περὶ τῆς θρέψεως ἀπορίαν ἡ ὑπόθεσις· ὅτι δ'
οὐδ' Ἐρασιστράτῳ σύμφωνός ἐστιν, ὃ ἐκεῖνος
ἁπλοῦν καὶ πρῶτον ὀνομάζει, σύνθετον ἀπο-
φαίνουσα, καὶ τὴν τῆς φύσεως τέχνην ἀναιροῦσα,
πρόδηλον καὶ τοῦτ' εἶναί μοι δοκεῖ. εἰ μὴ γὰρ
κἂν τοῖς ἁπλοῖς τούτοις ἕνωσίν τινα τῆς οὐσίας
ἀπολείψομεν, ἀλλ' εἰς ἄναρμα καὶ ἀμέριστα
καταβησόμεθα στοιχεῖα, παντάπασιν ἀναιρήσομεν
τῆς φύσεως τὴν τέχνην, ὥσπερ καὶ πάντες οἱ ἐκ
ταύτης ὁρμώμενοι τῆς ὑποθέσεως ἰατροὶ καὶ
φιλόσοφοι. δευτέρα γὰρ τῶν τοῦ ζῴου μορίων
κατὰ τὴν τοιαύτην ὑπόθεσιν ἡ φύσις, οὐ πρώτη
101 γίγνεται. διαπλάττειν δὲ ‖ καὶ δημιουργεῖν οὐ
τοῦ δευτέρου γεγονότος, ἀλλὰ τοῦ προϋπάρχοντός
ἐστιν· ὥστ' ἀναγκαῖόν ἐστιν εὐθὺς ἐκ σπερμάτων
ὑποθέσθαι τὰς δυνάμεις τῆς φύσεως, αἷς δια-

---

[1] cf. Book I., chap. xii.
[2] i.e. in biology we must begin with living substance—
with something which is specifically alive—here with the
"unit mass of living matter." cf. p. 73, note 3.

perceptible space which is entirely empty "; while I, for my part, being abundantly equipped with terms which are equally elucidatory, at least in relation to the present topic of discussion, have added them as well).

Thus it seems to me better that we also should help the Erasistrateans with some contribution, since we are on the subject, and should advise those who reduce the vessel called *primary* and *simple* by Erasistratus into other elementary bodies to give up their opinion; for not only do they gain nothing by it, but they are also at variance with Erasistratus in this matter. That they gain nothing by it has been clearly demonstrated; for this hypothesis could not escape the difficulty regarding *nutrition*. And it also seems perfectly evident to me that this hypothesis is not in consonance with the view of Erasistratus, when it declares that what he calls simple and primary is composite, and when it destroys the principle of Nature's artistic skill.[1] For, if we do not grant a certain *unity of substance*[2] to these simple structures as well, and if we arrive eventually at inharmonious and indivisible elements,[3] we shall most assuredly deprive Nature of her artistic skill, as do all the physicians and philosophers who start from this hypothesis. For, according to such a hypothesis, Nature does not precede, but is secondary to the *parts* of the animal.[4] Now, it is not the province of what comes secondarily, but of what pre-exists, to shape and to construct. Thus we must necessarily suppose that the faculties of Nature, by which she

[3] " Ad elementa quae nec coalescere possunt nec in partes dividi " (Linacre). On the two contrasted schools *cf.* p. 45.
[4] *cf. loc. cit.*

πλάττει τε καὶ αὐξάνει καὶ τρέφει τὸ ζῷον·
ἀλλ' ἐκείνων τῶν σωμάτων τῶν ἀνάρμων καὶ
ἀμερῶν οὐδὲν ἐν ἑαυτῷ διαπλαστικὴν ἔχει δύναμιν ἢ αὐξητικὴν ἢ θρεπτικὴν ἢ ὅλως τεχνικήν·
ἀπαθὲς γὰρ καὶ ἀμετάβλητον ὑπόκειται. τῶν δ'
εἰρημένων οὐδὲν ἄνευ μεταβολῆς καὶ ἀλλοιώσεως
καὶ τῆς δι' ὅλων κράσεως γίγνεται, καθάπερ καὶ
διὰ τῶν ἔμπροσθεν ἐνεδειξάμεθα. καὶ διὰ ταύτην
τὴν ἀνάγκην οὐκ ἔχοντες, ὅπως τὰ ἀκόλουθα τοῖς
στοιχείοις, οἷς ὑπέθεντο, φυλάττοιεν, οἱ ἀπὸ τῶν
τοιούτων αἱρέσεων ἅπαντες ἄτεχνον ἠναγκάσθησαν ἀποφήνασθαι τὴν φύσιν. καίτοι ταῦτά γ'
οὐ παρ' ἡμῶν ἐχρῆν μανθάνειν τοὺς Ἐρασιστρατείους, ἀλλὰ παρ' αὐτῶν τῶν φιλοσόφων,
οἷς μάλιστα δοκεῖ πρῶτον ἐπισκοπεῖσθαι τὰ
στοιχεῖα τῶν ὄντων ἁπάντων.

Οὔκουν οὐδ' Ἐρασίστρατον ἄν τις ὀρθῶς ἄχρι
τοσαύτης ἀμαθίας νομίζοι προήκειν, ὡς μηδὲ
102 ταύτην γνωρίσαι δυνηθῆναι τὴν ἀκολου‖θίαν,
ἀλλ' ἅμα μὲν ὑποθέσθαι τεχνικὴν τὴν φύσιν,
ἅμα δ' εἰς ἀπαθῆ καὶ ἄναρμα καὶ ἀμετάβλητα
στοιχεῖα καταθραῦσαι τὴν οὐσίαν. καὶ μὴν
εἰ δώσει τιν' ἐν τοῖς στοιχείοις ἀλλοίωσίν τε
καὶ μεταβολὴν καὶ ἕνωσιν καὶ συνέχειαν, ἓν
ἀσύνθετον αὐτῷ τὸ ἁπλοῦν ἀγγεῖον ἐκεῖνο,
καθάπερ καὶ αὐτὸς ὀνομάζει, γενήσεται. ἀλλ'
ἡ μὲν ἁπλῆ φλὲψ ἐξ αὑτῆς τραφήσεται, τὸ
νεῦρον δὲ καὶ ἡ ἀρτηρία παρὰ τῆς φλεβός.

---

[1] "Auxetic." cf. p. 26, note 1.
[2] "At corporum quae nec una committi nec dividi possunt nullum in se formatricem, auctricem, nutricem, aut

shapes the animal, and makes it grow and receive nourishment, are present from the seed onwards; whereas none of these inharmonious and non-partite corpuscles contains within itself any formative, incremental,[1] nutritive, or, in a word, any artistic power; it is, by hypothesis, unimpressionable and untransformable,[2] whereas, as we have previously shown,[3] none of the processes mentioned takes place without transformation, alteration, and complete intermixture. And, owing to this necessity, those who belong to these sects are unable to follow out the consequences of their supposed elements, and they are all therefore forced to declare Nature devoid of art. It is not from us, however, that the Erasistrateans should have learnt this, but from those very philosophers who lay most stress on a preliminary investigation into the elements of all existing things.

Now, one can hardly be right in supposing that Erasistratus could reach such a pitch of foolishness as to be incapable of recognizing the logical consequences of this theory, and that, while assuming Nature to be artistically creative, he would at the same time break up substance into insensible, inharmonious, and untransformable elements. If, however, he will grant that there occurs in the elements a process of alteration and transformation, and that there exists in them unity and continuity, then that *simple vessel* of his (as he himself names it) will turn out to be single and uncompounded. And the simple vein will receive nourishment from itself, and the nerve and artery from the vein. How, and in what

in summa artificem facultatem habet; quippe quod impatibile esse immutibileque praesumitur" (Linacre).

[3] Book I., chaps. v.–xi.

πῶς καὶ τίνα τρόπον; ἐν τούτῳ γὰρ δὴ καὶ
πρόσθεν γενόμενοι τῷ λόγῳ τῆς τῶν Ἐρασιστρατείων διαφωνίας ἐμνημονεύσαμεν, ἐπεδείξαμεν δὲ
καὶ καθ' ἑκατέρους μὲν ἄπορον εἶναι τὴν τῶν
ἁπλῶν ἐκείνων ἀγγείων θρέψιν, ἀλλὰ καὶ κρῖναι
τὴν μάχην αὐτῶν οὐκ ὠκνήσαμεν καὶ τιμῆσαι τὸν
Ἐρασίστρατον εἰς τὴν βελτίονα μεταστήσαντες
αἵρεσιν.

Αὖθις οὖν ἐπὶ τὴν ἐν ἁπλοῦν ἡνωμένον ἑαυτῷ
πάντη τὸ στοιχειῶδες ἐκεῖνο νεῦρον ὑποτιθε-
μένην αἵρεσιν ὁ λόγος μεταβὰς ἐπισκοπείσθω,
πῶς τραφήσεται· τὸ γὰρ εὑρεθὲν ἐνταῦθα κοινὸν
ἂν ἤδη καὶ τῆς Ἱπποκράτους αἱρέσεως γένοιτο.

103   Κάλλιον δ' ἄν μοι δοκῶ τὸ ζητού‖μενον ἐπὶ
τῶν νενοσηκότων καὶ σφόδρα καταλελεπτυσ-
μένων βασανισθῆναι. πάντα γὰρ τούτοις ἐναργῶς φαίνεται τὰ μόρια τοῦ σώματος ἄτροφα
καὶ λεπτὰ καὶ πολλῆς προσθήκης τε καὶ ἀναθρέψεως δεόμενα. καὶ τοίνυν καὶ τὸ νεῦρον τοῦτο
τὸ αἰσθητόν, ἐφ' οὗπερ ἐξ ἀρχῆς ἐποιησάμην
τὸν λόγον, ἰσχνὸν μὲν ἱκανῶς γέγονε, δεῖται δὲ
θρέψεως. ἔχει δ' ἐν ἑαυτῷ μέρη πάμπολλα
μὲν ἐκεῖνα τὰ πρῶτα καὶ ἀόρατα νεῦρα τὰ σμικρὰ
καί τινας ἀρτηρίας ἁπλᾶς ὀλίγας καὶ φλέβας
ὁμοίως. ἅπαντ' οὖν αὐτοῦ τὰ νεῦρα τὰ στοιχειώδη καταλελέπτυνται δηλονότι καὶ αὐτά, ἤ,
εἰ μηδ' ἐκεῖνα, οὐδὲ τὸ ὅλον. καὶ τοίνυν καὶ
θρέψεως οὐ τὸ μὲν ὅλον δεῖται νεῦρον, ἕκαστον δ'
ἐκείνων οὐ δεῖται. καὶ μὴν εἰ δεῖται μὲν ἀναθρέψεως, οὐδὲν δ' ἡ πρὸς τὸ κενούμενον ἀκολουθία

way? For, when we were at this point before, we drew attention to the disagreement among the Erasistrateans,[1] and we showed that the nutrition of these simple vessels was impracticable according to the teachings of both parties, although we did not hesitate to adjudicate in their quarrel and to do Erasistratus the honour of placing him in the better sect.[2]

Let our argument, then, be transferred again to the doctrine which assumes this *elementary nerve*[3] to be a single, simple, and entirely unified structure, and let us consider how it is to be nourished; for what is discovered here will at once be found to be common also to the school of Hippocrates.

It seems to me that our enquiry can be most rigorously pursued in subjects who are suffering from illness and have become very emaciated, since in these people all parts of the body are obviously atrophied and thin, and in need of additional substance and feeding-up; for the same reason the ordinary *perceptible* nerve, regarding which we originally began this discussion, has become thin, and requires nourishment. Now, this contains within itself various parts, namely, a great many of these primary, invisible, minute nerves, a few simple arteries, and similarly also veins. Thus, all its elementary nerves have themselves also obviously become emaciated; for, if they had not, neither would the nerve as a whole; and of course, in such a case, the whole nerve cannot require nourishment without each of these requiring it too. Now, if on the one hand they stand in need of feeding-up, and if on the

---

[1] *cf.* p. 153.
[2] On account of his idea of a simple tissue not susceptible of further analysis.   [3] Or " cell "; *cf.* p. 153, note 2.

GALEN

βοηθεῖν αὐτοῖς δύναται διά τε τὰς ἔμπροσθεν
εἰρημένας ἀπορίας καὶ διὰ τὴν ὑπόγυιον ἰσχνό-
τητα, καθάπερ δείξω, ζητητέον ἡμῖν ἐστιν ἑτέραν
αἰτίαν θρέψεως.

Πῶς οὖν ἡ πρὸς τὸ κενούμενον ἀκολουθία
τρέφειν ἀδύνατός ἐστι τὸν οὕτω διακείμενον;
104 ὅτι τοσοῦτον ἀκολουθεῖν ‖ ἀναγκάζει τῶν συν-
εχῶν, ὅσον ἀπορρεῖ. τοῦτο δ᾽ ἐπὶ μὲν τῶν
εὐεκτούντων ἱκανόν ἐστιν εἰς τὴν θρέψιν, ἴσα
γὰρ ἐπ᾽ αὐτῶν εἶναι χρὴ τοῖς ἀπορρέουσι τὰ
προστιθέμενα· ἐπὶ δὲ τῶν ἐσχάτως ἰσχνῶν καὶ
πολλῆς ἀναθρέψεως δεομένων εἰ μὴ πολλαπλά-
σιον εἴη τὸ προστιθέμενον τοῦ κενουμένου, τὴν
ἐξ ἀρχῆς ἕξιν ἀναλαβεῖν οὐκ ἄν ποτε δύναιντο.
δῆλον οὖν, ὡς ἕλκειν αὐτὰ δεήσει τοσούτῳ
πλεῖον, ὅσῳ καὶ δεῖται πλείονος. Ἐρασίστρατος
δὲ κἀνταῦθα πρότερον ποιήσας τὸ δεύτερον οὐκ
οἶδ᾽ ὅπως οὐκ αἰσθάνεται. διότι γάρ, φησί,
πολλὴ πρόσθεσις εἰς ἀνάθρεψιν γίγνεται τοῖς
νενοσηκόσι, διὰ τοῦτο καὶ ἡ πρὸς ταύτην ἀκολουθία πολλή. πῶς δ᾽ ἂν πολλὴ πρόσθεσις γένοιτο
μὴ προηγουμένης ἀναδόσεως δαψιλοῦς; εἰ δὲ
τὴν διὰ τῶν φλεβῶν φορὰν τῆς τροφῆς ἀνάδοσιν
καλεῖ, τὴν δ᾽ εἰς ἕκαστον τῶν ἁπλῶν καὶ ἀοράτων
ἐκείνων νεύρων καὶ ἀρτηριῶν μετάληψιν οὐκ
ἀνάδοσιν ἀλλὰ διάδοσιν, ὥς τινες ὀνομάζειν
105 ἠξίωσαν, εἶτα ‖ τὴν διὰ τῶν φλεβῶν μόνῃ τῇ

---

[1] The *horror vacui.*
[2] *Prosthesis* of nutriment ; *cf.* p. 39, note 6.

other the principle of the refilling of a vacuum[1] can give them no help—both by reason of the difficulties previously mentioned and the actual thinness, as I shall show—we must then seek another cause for nutrition.

How is it, then, that the tendency of a vacuum to become refilled is unable to afford nourishment to one in such a condition ? Because its rule is that only so much of the contiguous matter should succeed as has flowed away. Now this is sufficient for nourishment in the case of those who are in good condition, for, in them, what is *presented* [2] must be equal to what has flowed away. But in the case of those who are very emaciated and who need a great restoration of nutrition, unless what was presented were many times greater than what has been emptied out, they would never be able to regain their original habit. It is clear, therefore, that these parts will have to exert a greater amount of *attraction*, in so far as their requirements are greater. And I fail to understand how Erasistratus does not perceive that here again he is putting the cart before the horse. Because, in the case of the sick, there must be a large amount of *presentation* [2] in order to feed them up, he argues that the factor of "re-filling"[1] must play an equally large part. And how could much *presentation* take place if it were not preceded by an abundant *delivery* [3] of nutriment? And if he calls the conveyance of food through the veins delivery, and its assumption by each of these simple and visible nerves and arteries not delivery but *distribution*,[4] as some people have thought fit to name it, and then ascribes conveyance

[3] *Anadosis*, "absorption"; *cf.* p. 13, note 5.   [4] Lit. *diadosis*.

GALEN

πρὸς τὸ κενούμενον ἀκολουθίᾳ φησὶ γίγνεσθαι,
τὴν εἰς τὰ λόγῳ θεωρητὰ μετάληψιν ἡμῖν ἐξηγη-
σάσθω. ὅτι μὲν γὰρ οὐκέτ' ἐπὶ τούτων ἡ πρὸς
τὸ κενούμενον ἀκολουθίᾳ λέγεσθαι δύναται καὶ
μάλιστ' ἐπὶ τῶν ἐσχάτως ἰσχνῶν, ἀποδέδεικται.
τί δέ φησιν ἐπ' αὐτῶν ἐν τῷ δευτέρῳ τῶν
καθόλου λόγων ὁ Ἐρασίστρατος, ἄξιον ἐπακοῦ-
σαι τῆς λέξεως· "Τοῖς δ' ἐσχάτοις τε καὶ ἁπλοῖς,
λεπτοῖς τε καὶ στενοῖς οὖσιν, ἐκ τῶν παρακειμένων ἀγγείων ἡ πρόσθεσις συμβαίνει εἰς τὰ
κενώματα τῶν ἀπενεχθέντων κατὰ τὰ πλάγια
τῶν ἀγγείων ἑλκομένης τῆς τροφῆς καὶ κατα-
χωριζομένης." ἐκ ταύτης τῆς λέξεως πρῶτον
μὲν τὸ κατὰ τὰ πλάγια προσίεμαί τε καὶ
ἀποδέχομαι· κατὰ μὲν γὰρ αὐτὸ τὸ στόμα τὸ
ἁπλοῦν νεῦρον οὐκ ἂν δύναιτο δεχόμενον τὴν
τροφὴν οὕτως εἰς ὅλον ἑαυτὸ διανέμειν· ἀνάκειται
γὰρ ἐκεῖνο τῷ ψυχικῷ πνεύματι· κατὰ δὲ τὸ
πλάγιον ἐκ τῆς παρακειμένης φλεβὸς τῆς ἁπλῆς
ἐγχωρεῖ λαβεῖν αὐτό. δεύτερον δ' ἀποδέχομαι
τῶν ἐκ τῆς Ἐρασιστράτου λέξεως ὀνομάτων τὸ
106 γεγραμμένον ἐφεξῆς τῷ κατὰ τὰ πλάγια. ‖ τί
γάρ φησι; "Κατὰ τὰ πλάγια τῶν ἀγγείων ἑλκομένης τῆς τροφῆς." ὅτι μὲν οὖν ἕλκεται, καὶ ἡμεῖς
ὁμολογοῦμεν, ὅτι δ' οὐ τῇ πρὸς τὸ κενούμενον
ἀκολουθίᾳ, δέδεικται πρόσθεν.

VII

Ἐξεύρωμεν οὖν κοινῇ, πῶς ἕλκεται. πῶς δ'
ἄλλως ἢ ὡς ὁ σίδηρος ὑπὸ τῆς ἡρακλείας λίθου

---

[1] i.e. let him explain the *diadosis*.

164

through the veins to the principle of vacuum-refilling alone, let him explain to us the assumption of food by the hypothetical elements.[1] For it has been shown that at least in relation to these there is no question of the refilling of a vacuum being in operation, and especially where the parts are very attenuated. It is worth while listening to what Erasistratus says about these cases in the second book of his "General Principles": "In the ultimate simple [vessels], which are thin and narrow, presentation takes place from the adjacent vessels, the nutriment being attracted through the sides of the vessels and deposited in the empty spaces left by the matter which has been carried away." Now, in this statement firstly I admit and accept the words "through the sides." For, if the simple nerve were actually to take in the food through its mouth, it could not distribute it through its whole substance ; for the mouth is dedicated to the psychic pneuma.[2] It can, however, take it in through its sides from the adjacent simple vein. Secondly, I also accept in Erasistratus's statement the expression which precedes "through the sides." What does this say ? " The nutriment being attracted through the sides of the vessels." Now I, too, agree that it is attracted, but it has been previously shown that this is not through the tendency of evacuated matter to be replaced.

VII

LET us, then, consider together how it is attracted. How else than in the way that iron is attracted by

---

[2] "Spiritus animalis" ; *cf.* p. 152, note 1. The nutriment was for the *walls* of the vessels, not for their cavities. *cf.* p. 319, note 3.

δύναμιν ἐχούσης ἑλκτικὴν τοιαύτης ποιότητος;
ἀλλ᾽ εἰ τὴν μὲν ἀρχὴν τῆς ἀναδόσεως ἡ τῆς
κοιλίας ἔνθλιψις παρέχεται, τὴν δὲ μετὰ ταῦτα
φορὰν ἅπασαν αἵ τε φλέβες περιστελλόμεναι καὶ
προωθοῦσαι καὶ τῶν τρεφομένων ἕκαστον ἐπισπώ-
μενον εἰς ἑαυτό, τῆς πρὸς τὸ κενούμενον ἀκολουθίας ἀποστάντες, ὡς οὐ πρεπούσης ἀνδρὶ
τεχνικὴν ὑποθεμένῳ τὴν φύσιν, οὕτως ἂν ἤδη
καὶ τὴν ἀντιλογίαν εἴημεν πεφευγότες τὴν
Ἀσκληπιάδου μὴ δυνάμενοί γε λύειν αὐτήν. τὸ
γὰρ εἰς τὴν ἀπόδειξιν παραλαμβανόμενον λῆμμα
τὸ διεζευγμένον οὐκ ἐκ δυοῖν ἀλλ᾽ ἐκ τριῶν ἐστι
κατά γε τὴν ἀλήθειαν διεζευγμένον. εἰ μὲν οὖν
107 ὡς ἐκ δυοῖν αὐτῷ χρη‖σαίμεθα, ψεῦδος ἔσται τι
τῶν εἰς τὴν ἀπόδειξιν παρειλημμένων· εἰ δ᾽ ὡς ἐκ
τριῶν, ἀπέραντος ὁ λόγος γενήσεται.

VIII

Καὶ ταῦτ᾽ οὐκ ἐχρῆν ἀγνοεῖν τὸν Ἐρασίστρατον, εἴπερ κἂν ὄναρ ποτὲ τοῖς ἐκ τοῦ περιπάτου
συνέτυχεν, ὥσπερ οὖν οὐδὲ τὰ περὶ τῆς γενέσεως
τῶν χυμῶν, ὑπὲρ ὧν οὐδὲν ἔχων εἰπεῖν οὐδὲ
μέχρι τοῦ μετρίου πιθανὸν οἴεται παρακρούεσθαι
σκηπτόμενος, ὡς οὐδὲ χρήσιμος ὅλως ἐστὶν ἡ
τῶν τοιούτων ἐπίσκεψις. εἶτ᾽, ὦ πρὸς θεῶν,
ὅπως μὲν τὰ σιτία κατὰ τὴν γαστέρα πέττεται
χρήσιμον ἐπίστασθαι, πῶς δ᾽ ἐν ταῖς φλεψὶν ἡ

the lodestone, the latter having a faculty attractive
of this particular quality [existing in iron]?[1]   But if
the beginning of anadosis depends on the squeezing
action of the stomach,[2] and the whole movement
thereafter on the peristalsis and propulsive action of
the veins, as well as on the traction exerted by each
of the parts which are undergoing nourishment, then
we can abandon the principle of replacement of
evacuated matter, as not being suitable for a man
who assumes Nature to be a skilled artist; thus we
shall also have avoided the contradiction of Asclepiades[3] though we cannot refute it: for the disjunctive
argument used for the purposes of demonstration
is, in reality, disjunctive not of two but of three
alternatives; now, if we treat the disjunction as a
disjunction of two alternatives, one of the two
propositions assumed in constructing our proof must
be false; and if as a disjunctive of three alternatives,
no conclusion will be arrived at.

VIII

Now Erasistratus ought not to have been ignorant
of this if he had ever had anything to do with the
Peripatetics—even in a dream.   Nor, similarly, should
he have been unacquainted with the genesis of the
humours, about which, not having even anything
moderately plausible to say, he thinks to deceive us
by the excuse that the consideration of such matters
is not the least useful.   Then, in Heaven's name, is it
useful to know how food is digested in the stomach,
but unnecessary to know how *bile* comes into existence

---

[1] Specific attraction ; *cf.* Book I., chap. xiv.
[2] *cf.* p. 100, note 2.   [3] In Book II., chap. i.

χολὴ γίγνεται, περιττόν; καὶ τῆς κενώσεως ἄρα
φροντιστέον αὐτῆς μόνης, ἀμελητέον δὲ τῆς
γενέσεως; ὥσπερ οὐκ ἄμεινον ὑπάρχον μακρῷ
τὸ κωλύειν εὐθὺς ἐξ ἀρχῆς γεννᾶσθαι πλείονα
τοῦ πράγματ' ἔχειν ἐκκενοῦντας. θαυμαστὸν δὲ
καὶ τὸ διαπορεῖν, εἴτ' ἐν τῷ σώματι τὴν γένεσιν
αὐτῆς ὑποθετέον εἴτ' εὐθὺς ἔξωθεν ἐν τοῖς σιτίοις
περιέχεσθαι φατέον. εἰ γὰρ δὴ τοῦτο καλῶς
ἠπόρηται, τί οὐχὶ καὶ περὶ τοῦ αἵματος ἐπισκε-
108 ψόμεθα, πότερον ἐν τῷ σώματι ‖ λαμβάνει τὴν
γένεσιν ἢ τοῖς σιτίοις παρέσπαρται, καθάπερ οἱ
τὰς ὁμοιομερείας ὑποτιθέμενοί φασι; καὶ μὴν
πολλῷ γ' ἦν χρησιμώτερον ζητεῖσθαι, ποῖα τῶν
σιτίων ὁμολογεῖ τῇ τῆς αἱματώσεως ἐνεργείᾳ καὶ
ποῖα διαφέρεται, τοῦ ζητεῖν, τίνα μὲν τῇ τῆς
γαστρὸς ἐνεργείᾳ νικᾶται ῥᾳδίως, τίνα δ' ἀντιβαίνει καὶ μάχεται. τούτων μὲν γὰρ ἡ ἔκλεξις
εἰς πέψιν μόνην, ἐκείνων δ' εἰς αἵματος χρηστοῦ
διαφέρει γένεσιν. οὐδὲ γὰρ ἴσον ἐστὶν ἢ μὴ
καλῶς ἐν τῇ γαστρὶ χυλωθῆναι τὴν τροφὴν ἢ
μὴ χρηστὸν αἷμα γεννηθῆναι. πῶς δ' οὐκ αἰδεῖται
τὰς μὲν τῆς πέψεως ἀποτυχίας διαιρούμενος,
ὡς πολλαί τ' εἰσὶ καὶ κατὰ πολλὰς γίγνονται
προφάσεις, ὑπὲρ δὲ τῶν τῆς αἱματώσεως σφαλ-
μάτων οὐδ' ἄχρι ῥήματος ἑνὸς οὐδ' ἄχρι συλλαβῆς
μιᾶς φθεγξάμενος; καὶ μὴν εὑρίσκεταί γε καὶ
παχὺ καὶ λεπτὸν ἐν ταῖς φλεψὶν αἷμα καὶ τοῖς
μὲν ἐρυθρότερον, τοῖς δὲ ξανθότερον, τοῖς δὲ
μελάντερον, τοῖς δὲ φλεγματωδέστερον. εἰ δ' ὅτι

---

[1] Prevention better than cure.
[2] e.g. Anaxagoras; cf. p. 7, note 5; p. 20, note 3.
[3] Lit. haematosis.　　　　[4] cf. p. 174, note 4.

in the veins? Are we to pay attention merely to the evacuation of this humour, and not to its genesis? As though it were not far better to prevent its excessive development from the beginning than to give ourselves all the trouble of expelling it![1] And it is a strange thing to be entirely unaware as to whether its genesis is to be looked on as taking place in the body, or whether it comes from without and is contained in the food. For, if it was right to raise this problem, why should we not make investigations concerning the *blood* as well—whether it takes its origin in the body, or is distributed through the food as is maintained by those who postulate *homœomeries*?[2] Assuredly it would be much more useful to investigate what kinds of food are suited, and what kinds unsuited, to the process of blood-production[3] rather than to enquire into what articles of diet are easily mastered by the activity of the stomach, and what resist and contend with it. For the choice of the latter bears reference merely to digestion, while that of the former is of importance in regard to the generation of useful blood. For it is not equally important whether the aliment be imperfectly chylified[4] in the stomach or whether it fail to be turned into useful blood. Why is Erasistratus not ashamed to distinguish all the various kinds of digestive failure and all the occasions which give rise to them, whilst in reference to the errors of blood-production he does not utter a single word—nay, not a syllable? Now, there is certainly to be found in the veins both thick and thin blood; in some people it is redder, in others yellower, in some blacker, in others more of the nature of phlegm. And one who realizes that it

καὶ δυσῶδες οὐχ ἕνα τρόπον ἀλλ᾿ ἐν πολλαῖς
109 πάνυ διαφοραῖς ἀρρήτοις μὲν λόγῳ, σα‖φεστάταις
δ᾿ αἰσθήσεσι φαίνεται γιγνόμενον, εἰδείη τις, οὐκ
ἂν οἶμαι μετρίως ἔτι καταγνώσεσθαι τῆς Ἐρασιστράτου ῥαθυμίας αὐτὸν οὕτω γ᾿ ἀναγκαίαν
εἰς τὰ ἔργα τῆς τέχνης θεωρίαν παραλιπόντος.

Ἐναργῆ γὰρ δὴ καὶ τὰ περὶ τῶν ὑδέρων ἁμαρ-
τήματα τῇ ῥαθυμίᾳ ταύτῃ κατὰ λόγον ἠκολουθη-
κότα. τό τε γὰρ τῇ στενοχωρίᾳ τῶν ὁδῶν
κωλύεσθαι νομίζειν πρόσω τοῦ ἥπατος ἰέναι τὸ
αἷμα καὶ μηδέποτ᾿ ἂν ἄλλως ὕδερον δύνασθαι
συστῆναι πῶς οὐκ ἐσχάτην ἐνδείκνυται ῥαθυμίαν;
τό τε μὴ διὰ τὸν σπλῆνα μηδὲ δι᾿ ἄλλο τι μόριον,
ἀλλ᾿ ἀεὶ διὰ τὸν ἐν τῷ ἥπατι σκίρρον ὕδερον
οἴεσθαι γίγνεσθαι τελέως ἀργοῦ τὴν διάνοιαν
ἀνθρώπου καὶ μηδενὶ τῶν ὁσημέραι γιγνομένων
παρακολουθοῦντος. ἐπὶ μέν γε χρονίαις αἱμορ-
ροῖσιν ἐπισχεθείσαις ἢ διὰ κένωσιν ἄμετρον εἰς
ψῦξιν ἐσχάτην ἀγαγούσαις τὸν ἄνθρωπον οὐχ
ἅπαξ οὐδὲ δὶς ἀλλὰ πολλάκις ἤδη τεθεάμεθα
συστάντας ὑδέρους, ὥσπερ γε καὶ γυναιξὶν ἥ τε
τῆς ἐφ᾿ ἑκάστῳ μηνὶ καθάρσεως ἀπώλεια παν-
τελὴς καὶ ἄμετρος κένωσις, ὅταν αἱμορραγήσωσί
110 ποθ᾿ αἱ μῆτραι σφοδρῶς, ἐπεκαλέσαντο πολ‖λάκις
ὕδερον καί τισιν αὐτῶν καὶ ὁ γυναικεῖος ὀνομα-
ζόμενος ῥοῦς εἰς τοῦτ᾿ ἐτελεύτησε τὸ πάθος, ἵνα

---

[1] Erasistratus held the spleen to be useless. cf. p. 143.

[2] Induration : Gk. skirros, Lat. scirrhus. The condition is
now commonly known by Laënnec's term cirrhosis, from
Gk. kirros, meaning yellow or tawny. Here again we have an
example of Erasistratus's bias towards anatomical or structural
rather than functional explanations of disease. cf. p. 124, note 1.

may smell offensively not in one way only, but in a
great many different respects (which cannot be put
into words, although perfectly appreciable to the
senses), would, I imagine, condemn in no measured
terms the carelessness of Erasistratus in omitting
a consideration so essential to the practice of our
art.

Thus it is clear what errors in regard to the
subject of *dropsies* logically follow this carelessness.
For, does it not show the most extreme carelessness
to suppose that the blood is prevented from going
forward into the liver owing to the *narrowness of the
passages,* and that dropsy can never occur in any
other way? For, to imagine that dropsy is never
caused by the spleen[1] or any other part, but always by
induration of the liver,[2] is the standpoint of a man
whose intelligence is perfectly·torpid and who is
quite out of touch with things that happen every
day. For, not merely once or twice, but frequently,
we have observed dropsy produced by chronic
haemorrhoids which have been suppressed,[3] or
which, through immoderate bleeding, have given
the patient a severe chill; similarly, in women, the
complete disappearance of the monthly discharge,[4]
or an undue evacuation such as is caused by
violent bleeding from the womb, often provoke
dropsy; and in some of them the so-called female
flux ends in this disorder. I leave out of account

---

[3] On the risks which were supposed to attend the checking
of habitual bleeding from piles *cf.* Celsus (*De Re Med.* VI.
xviii. 9), " Atque in quibusdam parum tuto supprimitur, qui
sanguinis profluvio imbecilliores non fiunt; habent enim
purgationem hanc, non morbum." (*i.e.* the habit was to be
looked on as a periodical cleansing, not as a disease.)

[4] Lit. *catharsis.*·

τοὺς ἀπὸ τῶν κενεώνων ἀρχομένους ἢ ἄλλου τινὸς τῶν ἐπικαίρων μορίων ὑδέρους παραλίπω, σαφῶς μὲν καὶ αὐτοὺς ἐξελέγχοντας τὴν Ἐρασιστράτειον ὑπόληψιν, ἀλλ' οὐχ οὕτως ἐναργῶς ὡς οἱ διὰ κατάψυξιν σφοδρὰν τῆς ὅλης ἕξεως ἀποτελούμενοι. πρώτη γὰρ αὕτη γενέσεως ὑδέρων αἰτία διὰ τὴν ἀποτυχίαν τῆς αἱματώσεως γιγνομένη τρόπον ὁμοιότατον ταῖς ἐπὶ τῇ τῶν σιτίων ἀπεψίᾳ διαρροίαις. οὐ μὴν ἐσκίρρωταί γε κατὰ τοὺς τοιούτους ὑδέρους οὐδ' ἄλλο τι σπλάγχνον οὐδὲ τὸ ἧπαρ.

Ἀλλ' Ἐρασίστρατος ὁ σοφὸς ὑπεριδὼν καὶ καταφρονήσας, ὧν οὔθ' Ἱπποκράτης οὔτε Διοκλῆς οὔτε Πραξαγόρας οὔτε Φιλιστίων ἀλλ' οὐδὲ τῶν ἀρίστων φιλοσόφων οὐδεὶς κατεφρόνησεν οὔτε Πλάτων οὔτ' Ἀριστοτέλης οὔτε Θεόφραστος, ὅλας ἐνεργείας ὑπερβαίνει καθάπερ τι σμικρὸν καὶ τὸ τυχὸν τῆς τέχνης παραλιπὼν μέρος οὐδ'
111 ἀντειπεῖν ἀξιώσας, εἶτ' ὀρθῶς εἴτε καὶ μὴ ‖ σύμπαντες οὗτοι θερμῷ καὶ ψυχρῷ καὶ ξηρῷ καὶ ὑγρῷ, τοῖς μὲν ὡς δρῶσι, τοῖς δ' ὡς πάσχουσι, τὰ κατὰ τὸ σῶμα τῶν ζῴων ἁπάντων διοικεῖσθαί φασι καὶ ὡς τὸ θερμὸν ἐν αὐτοῖς εἴς τε τὰς ἄλλας ἐνεργείας καὶ μάλιστ' εἰς τὴν τῶν χυμῶν γένεσιν τὸ πλεῖστον δύναται. ἀλλὰ τὸ μὲν μὴ πείθεσθαι τοσούτοις τε καὶ τηλικούτοις ἀνδράσι καὶ πλέον αὐτῶν οἴεσθαί τι γιγνώσκειν ἀνεμέσητον, τὸ δὲ μήτ' ἀντιλογίας ἀξιῶσαι μήτε μνήμης οὕτως ἔνδοξον δόγμα θαυμαστήν τινα τὴν ὑπεροψίαν ἐνδείκνυται.

---

[1] Apparently some form of anaemia.

the dropsy which begins in the flanks or in any other susceptible part; this clearly confutes Erasistratus's assumption, although not so obviously as does that kind of dropsy which is brought about by an excessive chilling of the whole constitution; this, which is the primary reason for the occurrence of dropsy, results from a failure of blood-production,[1] very much like the diarrhoea which follows imperfect digestion of food; certainly in this kind of dropsy neither the liver nor any other viscus becomes indurated.

The learned Erasistratus, however, overlooks—nay, despises—what neither Hippocrates, Diocles, Praxagoras, nor Philistion[2] despised, nor indeed any of the best philosophers, whether Plato, Aristotle, or Theophrastus; he passes by whole functions as though it were but a trifling and casual department of medicine which he was neglecting, without deigning to argue whether or not these authorities are right in saying that the bodily parts of all animals are governed by the Warm, the Cold, the Dry and the Moist, the one pair being active and the other passive, and that among these the Warm has most power in connection with all functions, but especially with the genesis of the humours.[3] Now, one cannot be blamed for not agreeing with all these great men, nor for imagining that one knows more than they; but not to consider such distinguished teaching worthy either of contradiction or even mention shows an extraordinary arrogance.

[2] Philistion of Locri, a contemporary of Plato, was one of the chief representatives of the Sicilian school of medicine. For Diocles and Praxagoras see p. 51, note 1.

[3] cf. Book I., chap. iii.

GALEN

Καὶ μὴν σμικρότατός ἐστι τὴν γνώμην καὶ
ταπεινὸς ἐσχάτως ἐν ἁπάσαις ταῖς ἀντιλογίαις ἐν
μὲν τοῖς περὶ τῆς πέψεως λόγοις τοῖς σήπεσθαι
τὰ σιτία νομίζουσι φιλοτίμως ἀντιλέγων, ἐν δὲ
τοῖς περὶ τῆς ἀναδόσεως τοῖς διὰ τὴν παράθεσιν
τῶν ἀρτηριῶν ἀναδίδοσθαι τὸ διὰ τῶν φλεβῶν
αἷμα νομίζουσιν, ἐν δὲ τοῖς περὶ τῆς ἀναπνοῆς
τοῖς περιωθεῖσθαι τὸν ἀέρα φάσκουσιν. οὐκ
ὤκνησε δ᾽ οὐδὲ τοῖς ἀτμοειδῶς εἰς τὴν κύστιν
ἰέναι τὰ οὖρα νομίζουσιν ἀντειπεῖν οὐδὲ τοῖς εἰς ‖
112 τὸν πνεύμονα φέρεσθαι τὸ ποτόν. οὕτως ἐν ἅπασι
τὰς χειρίστας ἐπιλεγόμενος δόξας ἀγάλλεται δια-
τρίβων ἐπὶ πλέον ἐν ταῖς ἀντιλογίαις· ἐπὶ δὲ τῆς
τοῦ αἵματος γενέσεως οὐδὲν ἀτιμοτέρας οὔσης τῆς
ἐν τῇ γαστρὶ χυλώσεως τῶν σιτίων οὔτ᾽ ἀντειπεῖν
τινι τῶν πρεσβυτέρων ἠξίωσεν οὔτ᾽ αὐτὸς εἰση-
γήσασθαί τιν᾽ ἑτέραν γνώμην ἐτόλμησεν, ὁ περὶ
πασῶν τῶν φυσικῶν ἐνεργειῶν ἐν ἀρχῇ τῶν καθόλου λόγων ὑποσχόμενος ἐρεῖν, ὅπως τε γίγνονται
καὶ δι᾽ ὧντινων τοῦ ζῴου μορίων. ἢ τῆς μὲν
πέττειν τὰ σιτία πεφυκυίας δυνάμεως ἀρρωστού-
σης ἀπεπτήσει τὸ ζῷον, τῆς δ᾽ αἱματούσης τὰ
πεφθέντα οὐδὲν ἔσται πάθημα τὸ παράπαν, ἀλλ᾽
ἀδαμαντίνη τις ἡμῖν αὕτη μόνη καὶ ἀπαθής ἐστιν·
ἢ ἄλλο τι τῆς ἀρρωστίας αὐτῆς ἔκγονον ὑπάρξει

---

[1] Gk. *pepsis* ; otherwise rendered *coction*.
[2] *cf.* p. 13, note 5.   [3] *e.g.* Asclepiades.
[4] Lit. *chylosis*; *cf.* p. 238, note 2.
[5] That is to say, the haematopoietic function deserves

Now, Erasistratus is thoroughly small-minded and petty to the last degree in all his disputations—when, for instance, in his treatise "On Digestion," [1] he argues jealously with those who consider that this is a process of putrefaction of the food; and, in his work "On Anadosis," [2] with those who think that the anadosis of blood through the veins results from the contiguity of the arteries; also, in his work "On Respiration," with those who maintain that the air is forced along by contraction. Nay, he did not even hesitate to contradict those who maintain that the urine passes into the bladder in a vaporous state,[3] as also those who say that imbibed fluids are carried into the lung. Thus he delights to choose always the most valueless doctrines, and to spend his time more and more in contradicting these; whereas on the subject of the *origin of blood* (which is in no way less important than the chylification [4] of food in the stomach) he did not deign to dispute with any of the ancients, nor did he himself venture to bring forward any other opinion, despite the fact that at the beginning of his treatise on "General Principles" he undertook to say how all the various natural functions take place, and through what parts of the animal! Now, is it possible that, when the faculty which naturally digests food is weak, the animal's digestion fails, whereas the faculty which turns the digested food into blood cannot suffer any kind of impairment?[5] Are we to suppose this latter faculty alone to be as tough as steel and unaffected by circumstances? Or is it that weakness of this faculty will result in some-

consideration as much as the digestive processes which precede it.

καὶ οὐχ ὕδερος; δῆλος οὖν ἐναργῶς ἐστιν ὁ Ἐρασίστρατος ἐξ ὧν ἐν μὲν τοῖς ἄλλοις οὐδὲ ταῖς
φαυλοτάταις δόξαις ἀντιλέγειν ὤκνησεν, ἐνταυθοῖ
δ᾽ οὔτ᾽ ἀντειπεῖν τοῖς πρόσθεν οὔτ᾽ αὐτὸς εἰπεῖν
τι καινὸν ἐτόλμησε, τὸ σφάλμα τῆς ἑαυτοῦ γνωρί-
ζων αἱρέσεως.

Τί γὰρ ἂν καὶ λέγειν ἔσχεν ὑπὲρ αἵματος ‖
113 ἄνθρωπος εἰς μηδὲν τῷ συμφύτῳ θερμῷ χρώ-
μενος; τί δὲ περὶ ξανθῆς χολῆς ἢ μελαίνης ἢ
φλέγματος; ὅτι νὴ Δία δυνατόν ἐστιν ἀναμεμιγ-
μένην τοῖς σιτίοις εὐθὺς ἔξωθεν παραγίγνεσθαι
τὴν χολήν. λέγει γοῦν ὧδέ πως αὐτοῖς ὀνόμασι·
" Πότερον δ᾽ ἐν τῇ περὶ τὴν κοιλίαν κατεργασίᾳ
τῆς τροφῆς γεννᾶται τοιαύτη ὑγρασία ἢ μεμιγ-
μένη τοῖς ἔξωθεν προσφερομένοις παραγίγνεται,
οὐδὲν χρήσιμον πρὸς ἰατρικὴν ἐπεσκέφθαι." καὶ
μήν, ὦ γενναιότατε, καὶ κενοῦσθαι χρῆναι φάσ-
κεις ἐκ τοῦ ζῴου τὸν χυμὸν τοῦτον καὶ μεγάλως
λυπεῖν, εἰ μὴ κενωθείη. πῶς οὖν οὐδὲν ἐξ αὐτοῦ
χρηστὸν ὑπολαμβάνων γίγνεσθαι τολμᾷς ἄχρηστον λέγειν εἰς ἰατρικὴν εἶναι τὴν περὶ τῆς γενέσεως αὐτοῦ σκέψιν;

Ὑποκείσθω γὰρ ἐν μὲν τοῖς σιτίοις περιέχεσθαι, μὴ διακρίνεσθαι δ᾽ ἀκριβῶς ἐν ἥπατι·
ταῦτα γὰρ ἀμφότερα νομίζεις εἶναι δυνατά. καὶ
μὴν οὐ σμικρὸν ἐνταῦθα τὸ διαφέρον ἢ ἐλαχίστην
ἢ παμπόλλην χολὴν ἐν ἑαυτοῖς περιέχοντα
προσάρασθαι σιτία. τὰ μὲν γὰρ ἀκίνδυνα, τὰ δὲ
παμπόλλην περιέχοντα τῷ μὴ δύνασθαι πᾶσαν

---

[1] *i.e.* Erasistratus could obviously say nothing about any
of the humours or their origins, since he had not postulated

thing else than dropsy? The fact, therefore, that
Erasistratus, in regard to other matters, did not
hesitate to attack even the most trivial views, whilst
in this case he neither dared to contradict his
predecessors nor to advance any new view of his
own, proves plainly that he recognized the fallacy
of his own way of thinking.[1]

For what could a man possibly say about blood
who had no use for *innate heat*? What could he say
about yellow or black bile, or phlegm? Well, of
course, he might say that the bile could come
directly from without, mingled with the food! Thus
Erasistratus practically says so in the following
words: "It is of no value in practical medicine to
find out whether a fluid of this kind[2] arises from
the elaboration of food in the stomach-region, or
whether it reaches the body because it is mixed with
the food taken in from outside." But, my very good
Sir, you most certainly maintain also that this
humour has to be evacuated from the animal, and
that it causes great pain if it be not evacuated.
How, then, if you suppose that no good comes from
the bile, do you venture to say that an investigation
into its origin is of no value in medicine?

Well, let us suppose that it is contained in the
food, and not specifically secreted in the liver (for
you hold these two things possible). In this case,
it will certainly make a considerable difference
whether the ingested food contains a minimum or
a maximum of bile; for the one kind is harmless,
whereas that containing a large quantity of bile,
owing to the fact that it cannot be properly purified[3]

the four qualities (particularly the Warm—that is, innate
heat).   [2] *i.e.* bile.   [3] *i.e.* deprived of its bile.

114 αὐτὴν ἐν ‖ ἥπατι καθαρθῆναι καλῶς αἴτια κατα-
στήσεται τῶν τ᾽ ἄλλων παθῶν, ὧν αὐτὸς ὁ
Ἐρασίστρατος ἐπὶ πλήθει χολῆς γίγνεσθαί φησι,
καὶ τῶν ἰκτέρων οὐχ ἥκιστα. πῶς οὖν οὐκ
ἀναγκαιότατον ἰατρῷ γιγνώσκειν, πρῶτον μέν,
ὡς ἐν τοῖς σιτίοις αὐτοῖς ἔξωθεν ἡ χολὴ περιέχεται, δεύτερον δ᾽, ὡς τὸ μὲν τεῦτλον, εἰ τύχοι,
παμπόλλην, ὁ δ᾽ ἄρτος ἐλαχίστην καὶ τὸ μὲν
ἔλαιον πλείστην, ὁ δ᾽ οἶνος ὀλιγίστην ἕκαστόν
τε τῶν ἄλλων ἄνισον τῷ πλήθει περιέχει τὴν
χολήν; πῶς γὰρ οὐκ ἂν εἴη γελοιότατος, ὃς ἂν
ἑκὼν αἱρῆται τὰ πλείονα χολὴν ἐν ἑαυτοῖς περιέχοντα πρὸ τῶν ἐναντίων;

Τί δ᾽ εἰ μὴ περιέχεται μὲν ἐν τοῖς σιτίοις ἡ
χολή, γίγνεται δ᾽ ἐν τοῖς τῶν ζῴων σώμασιν; ἢ
οὐχὶ καὶ κατὰ τοῦτο χρήσιμον ἐπίστασθαι, τίνι
μὲν καταστάσει σώματος ἕπεται πλείων αὐτῆς ἡ
γένεσις, τίνι δ᾽ ἐλάττων; ἀλλοιοῦν γὰρ δήπου καὶ
μεταβάλλειν οἷοί τ᾽ ἐσμὲν καὶ τρέπειν ἐπὶ τὸ
βέλτιον ἀεὶ τὰς μοχθηρὰς καταστάσεις τοῦ σώματος. ἀλλ᾽ εἰ μὴ γιγνώσκοιμεν, καθότι μοχθηραὶ
καὶ ὅπῃ τῆς δεούσης ἐξίστανται, πῶς ἂν αὐτὰς
115 ἐπανάγειν οἷοί τ᾽ εἴημεν ἐπὶ τὸ ‖ κρεῖττον;

Οὔκουν ἄχρηστόν ἐστιν εἰς τὰς ἰάσεις, ὡς
Ἐρασίστρατός φησιν, ἐπίστασθαι τἀληθὲς αὐτὸ
περὶ γενέσεως χολῆς. οὐ μὴν οὐδ᾽ ἀδύνατον οὐδ᾽
ἀσαφὲς ἐξευρεῖν, ὅτι μὴ τῷ πλείστην ἐν ἑαυτῷ
περιέχειν τὸ μέλι τὴν ξανθὴν χολὴν ἀλλ᾽ ἐν τῷ
σώματι μεταβαλλόμενον εἰς αὐτὴν ἀλλοιοῦταί τε
καὶ τρέπεται. πικρόν τε γὰρ ἂν ἦν γευομένοις,
εἰ χολὴν ἔξωθεν εὐθὺς ἐν ἑαυτῷ περιεῖχεν ἅπασί
τ᾽ ἂν ὡσαύτως τοῖς ἀνθρώποις ἴσον αὐτῆς ἐγέννα

in the liver, will result in the various affections—
particularly jaundice—which Erasistratus himself
states to occur where there is much bile. Surely,
then, it is most essential for the physician to know
in the first place, that the bile is contained in the
food itself from outside, and, secondly, that for
example, beet contains a great deal of bile, and
bread very little, while olive oil contains most, and
wine least of all, and all the other articles of diet
different quantities. Would it not be absurd for
any one to choose voluntarily those articles which
contain more bile, rather than those containing less?

What, however, if the bile is not contained in the
food, but comes into existence in the animal's body?
Will it not also be useful to know what *state of the
body* is followed by a greater, and what by a smaller
occurrence of bile?[1] For obviously it is in our
power to alter and transmute morbid states of the
body—in fact, to give them a turn for the better.
But if we did not know in what respect they were
morbid or in what way they diverged from the
normal, how should we be able to ameliorate them?

Therefore it is not useless in treatment, as
Erasistratus says, to know the actual truth about
the genesis of bile. Certainly it is not impossible,
or even difficult to discover that the reason why
honey produces yellow bile is not that it contains a
large quantity of this within itself, but because it
[the honey] undergoes change, becoming *altered*
and transmuted into bile. For it would be bitter
to the taste if it contained bile from the outset,
and it would produce an equal quantity of bile

[1] Here it is rather the living organism we consider than
the particular food that is put into it.

τὸ πλῆθος. ἀλλ' οὐχ ὧδ' ἔχει τἀληθές. ἐν μὲν
γὰρ τοῖς ἀκμάζουσι καὶ μάλιστ' εἰ φύσει θερμό-
τεροι καὶ βίον εἶεν βιοῦντες ταλαίπωρον, ἅπαν
εἰς ξανθὴν χολὴν μεταβάλλει τὸ μέλι· τοῖς
γέρουσι δ' ἱκανῶς ἐστιν ἐπιτήδειον, ὡς ἂν οὐκ εἰς
χολὴν ἀλλ' εἰς αἷμα τὴν ἀλλοίωσιν ἐν ἐκείνοις
λαμβάνον. Ἐρασίστρατος δὲ πρὸς τῷ μηδὲν
τούτων γιγνώσκειν οὐδὲ περὶ τὴν διαίρεσιν τοῦ
λόγου σωφρονεῖ, πότερον ἐν τοῖς σιτίοις ἡ χολὴ
περιέχεται εὐθὺς ἐξ ἀρχῆς ἢ κατὰ τὴν ἐν τῇ
κοιλίᾳ κατεργασίαν ἐγένετο, μηδὲν εἶναι χρήσι-
116 μον εἰς ἰατρικὴν ἐπεσκέφθαι λέγων. ἐχρῆν ‖ γὰρ
δήπου προσθεῖναί τι καὶ περὶ τῆς ἐν ἥπατι καὶ
φλεψὶ γενέσεως αὐτῆς, ἐν τοῖσδε τοῖς ὀργάνοις
γεννᾶσθαι τὴν χολὴν ἅμα τῷ αἵματι τῶν παλαιῶν
ἰατρῶν τε καὶ φιλοσόφων ἀποφηναμένων. ἀλλὰ
τοῖς εὐθὺς ἐξ ἀρχῆς σφαλεῖσι καὶ διαμαρτάνουσι
τῆς ὀρθῆς ὁδοῦ τοιαῦτά τε ληρεῖν ἀναγκαῖόν ἐστι
καὶ προσέτι τῶν χρησιμωτάτων εἰς τὴν τέχνην
παραλιπεῖν τὴν ζήτησιν.

Ἡδέως δ' ἂν ἐνταῦθα τοῦ λόγου γεγονὼς
ἠρόμην τοὺς ὁμιλῆσαι φάσκοντας αὐτὸν ἐπὶ
πλεῖστον τοῖς ἐκ τοῦ περιπάτου φιλοσόφοις, εἰ
γιγνώσκουσιν, ὅσα περὶ τοῦ κεκρᾶσθαι τὰ σώμαθ'
ἡμῶν ἐκ θερμοῦ καὶ ψυχροῦ καὶ ξηροῦ καὶ ὑγροῦ
πρὸς Ἀριστοτέλους εἴρηταί τε καὶ ἀποδέδεικται,
καὶ ὡς τὸ θερμὸν ἐν αὐτοῖς ἐστι τὸ δραστικώτατον
καὶ ὡς τῶν ζῴων ὅσα μὲν θερμότερα φύσει, ταῦτα
πάντως ἔναιμα, τὰ δ' ἐπὶ πλέον ψυχρότερα
πάντως ἄναιμα καὶ διὰ τοῦτο τοῦ χειμῶνος ἀργὰ

in every person who took it. The facts, however, are not so.[1] For in those who are in the prime of life, especially if they are warm by nature and are leading a life of toil, the honey changes entirely into yellow bile. Old people, however, it suits well enough, inasmuch as the alteration which it undergoes is not into bile, but into blood. Erasistratus, however, in addition to knowing nothing about this, shows no intelligence even in the division of his argument; he says that it is of no practical importance to investigate whether the bile is contained in the food from the beginning or comes into existence as a result of gastric digestion. He ought surely to have added something about its genesis in liver and veins, seeing that the old physicians and philosophers declare that it along with the blood is generated in these organs. But it is inevitable that people who, from the very outset, go astray, and wander from the right road, should talk such nonsense, and should, over and above this, neglect to search for the factors of most practical importance in medicine.

Having come to this point in the argument, I should like to ask those who declare that Erasistratus was very familiar with the Peripatetics, whether they know what Aristotle stated and demonstrated with regard to our bodies being compounded out of the Warm, the Cold, the Dry and the Moist, and how he says that among these the Warm is the most active, and that those animals which are by nature warmest have abundance of blood, whilst those that are colder are entirely lacking in blood, and consequently in winter lie idle and motionless, lurking

[1] Supreme importance of the "soil." *cf.* Introduction, pp. xii. and xxxi.

καὶ ἀκίνητα κεῖται φωλεύοντα δίκην νεκρῶν.
εἴρηται δὲ καὶ περὶ τῆς χροιᾶς τοῦ αἵματος οὐκ
Ἀριστοτέλει μόνον, ἀλλὰ καὶ Πλάτωνι. καὶ
117 ἡμεῖς νῦν, ὅπερ ἤδη καὶ πρόσθεν εἶπον, ‖ οὐ τὰ
καλῶς ἀποδεδειγμένα τοῖς παλαιοῖς λέγειν προὐθέμεθα, μήτε τῇ γνώμῃ μήτε τῇ λέξει τοὺς ἄνδρας
ἐκείνους ὑπερβαλέσθαι δυνάμενοι· τὰ δ' ἤτοι
χωρὶς ἀποδείξεως ὡς ἐναργῆ πρὸς αὐτῶν εἰρημένα
διὰ τὸ μηδ' ὑπονοῆσαι μοχθηροὺς οὕτως ἔσεσθαί
τινας σοφιστάς, οἳ καταφρονήσουσι τῆς ἐν αὐτοῖς
ἀληθείας, ἢ καὶ παραλελειμμένα τελέως ὑπ'
ἐκείνων ἀξιοῦμεν εὑρίσκειν τε καὶ ἀποδεικνύναι.

Περὶ δὲ τῆς τῶν χυμῶν γενέσεως οὐκ οἶδ', εἰ
ἔχει τις ἕτερον προσθεῖναι σοφώτερον ὧν Ἱπποκράτης εἶπε καὶ Ἀριστοτέλης καὶ Πραξαγόρας
καὶ Φιλότιμος καὶ ἄλλοι πολλοὶ τῶν παλαιῶν.
ἀποδέδεικται γὰρ ἐκείνοις τοῖς ἀνδράσιν ἀλλοιου-
μένης τῆς τροφῆς ἐν ταῖς φλεψὶν ὑπὸ τῆς ἐμφύτου
θερμασίας αἷμα μὲν ὑπὸ τῆς συμμετρίας τῆς κατ'
αὐτήν, οἱ δ' ἄλλοι χυμοὶ διὰ τὰς ἀμετρίας γι-
γνόμενοι· καὶ τούτῳ τῷ λόγῳ πάνθ' ὁμολογεῖ τὰ
φαινόμενα. καὶ γὰρ τῶν ἐδεσμάτων ὅσα μέν ἐστι
θερμότερα φύσει, χολωδέστερα, τὰ δὲ ψυχρότερα
φλεγματικώτερα· καὶ τῶν ἡλικιῶν ὡσαύτως χο-
118 λωδέστε‖ραι μὲν αἱ θερμότεραι φύσει, φλεγματωδέστεραι δ' αἱ ψυχρότεραι· καὶ τῶν ἐπιτηδευ-
μάτων δὲ καὶ τῶν χωρῶν καὶ τῶν ὡρῶν καὶ πολὺ
δὴ πρότερον ἔτι τῶν φύσεων αὐτῶν αἱ μὲν ψυ-
χρότεραι φλεγματωδέστεραι, χολωδέστεραι δ' αἱ

---

[1] Aristotle, *Hist. Animal.*, iii. xix. ; Plato, *Timaeus*, 80 E.

in holes like corpses. Further, the question of the colour of the blood has been dealt with not only by Aristotle but also by Plato.[1] Now I, for my part, as I have already said, did not set before myself the task of stating what has been so well demonstrated by the Ancients, since I cannot surpass these men either in my views or in my method of giving them expression. Doctrines, however, which they either stated without demonstration, as being self-evident (since they never suspected that there could be sophists so degraded as to contemn the truth in these matters), or else which they actually omitted to mention at all—these I propose to discover and prove.

Now in reference to the *genesis of the humours*, I do not know that any one could add anything wiser than what has been said by Hippocrates, Aristotle, Praxagoras, Philotimus [2] and many other among the Ancients. These men demonstrated that when the nutriment becomes altered in the veins by the innate heat, blood is produced when it is in moderation, and the other humours when it is not in proper proportion. And all the observed facts [3] agree with this argument. Thus, those articles of food, which are by nature warmer are more productive of bile, while those which are colder produce more phlegm. Similarly of the periods of life, those which are naturally warmer tend more to bile, and the colder more to phlegm. Of occupations also, localities and seasons, and, above all, of natures [4] themselves, the colder are more phlegmatic, and the warmer more

[2] Philotimus succeeded Diocles and Praxagoras, who were successive leaders of the Hippocratic school. *cf.* p. 51, note 1.

[3] Lit. *phenomena.*

[4] *i.e.* living organisms ; *cf.* p. 47, note 1.

θερμότεραι· καὶ νοσημάτων τὰ μὲν ψυχρὰ τοῦ
φλέγματος ἔκγονα, τὰ δὲ θερμὰ τῆς ξανθῆς χολῆς·
καὶ ὅλως οὐδὲν ἔστιν εὑρεῖν τῶν πάντων, ὃ μὴ τούτῳ
τῷ λόγῳ μαρτυρεῖ. πῶς δ' οὐ μέλλει; διὰ γὰρ τὴν
ἐκ τῶν τεττάρων ποιὰν κρᾶσιν ἑκάστου τῶν μορίων
ὡδί πως ἐνεργοῦντος ἀνάγκη πᾶσα καὶ διὰ τὴν
βλάβην αὐτῶν ἢ διαφθείρεσθαι τελέως ἢ ἐμπο-
δίζεσθαί γε τὴν ἐνέργειαν καὶ οὕτω νοσεῖν τὸ
ζῷον ἢ ὅλον ἢ κατὰ τὰ μόρια.

Καὶ τὰ πρῶτά γε καὶ γενικώτατα νοσήματα
τέτταρα τὸν ἀριθμὸν ὑπάρχει θερμότητι καὶ
ψυχρότητι καὶ ξηρότητι καὶ ὑγρότητι διαφέροντα.
τοῦτο δὲ καὶ αὐτὸς ὁ Ἐρασίστρατος ὁμολογεῖ
καίτοι μὴ βουλόμενος. ὅταν γὰρ ἐν τοῖς πυρετοῖς
χείρους τῶν σιτίων τὰς πέψεις γίγνεσθαι λέγῃ
119 μὴ διότι τῆς ἐμφύτου ‖ θερμασίας ἡ συμμετρία
διέφθαρται, καθάπερ οἱ πρόσθεν ὑπελάμβανον,
ἀλλ' ὅτι περιστέλλεσθαι καὶ τρίβειν ἡ γαστὴρ
οὐχ ὁμοίως δύναται βεβλαμμένη τὴν ἐνέργειαν,
ἐρέσθαι δίκαιον αὐτόν, ὑπὸ τίνος ἡ τῆς γαστρὸς
ἐνέργεια βέβλαπται.

Γενομένου γάρ, εἰ τύχοι, βουβῶνος ἐπὶ προσ-
πταίσματι, πρὶν μὲν πυρέξαι τὸν ἄνθρωπον,
οὐκ ἂν χεῖρον ἡ γαστὴρ πέψειεν· οὐ γὰρ
ἱκανὸν ἦν οὐδέτερον αὐτῶν οὔθ' ὁ βουβὼν
οὔτε τὸ ἕλκος ἐμποδίσαι τι καὶ βλάψαι τὴν
ἐνέργειαν τῆς κοιλίας· εἰ δὲ πυρέξειεν, εὐθὺς μὲν
αἱ πέψεις γίγνονται χείρους, εὐθὺς δὲ καὶ τὴν
ἐνέργειαν τῆς γαστρὸς βεβλάφθαι φαμὲν ὀρθῶς
λέγοντες. ἀλλ' ὑπὸ τίνος ἐβλάβη, προσθεῖναι

---

[1] Erasistratus rejected the idea of innate heat; he held
that the heat of the body was introduced from outside.

bilious. Also cold diseases result from phlegm, and warmer ones from yellow bile. There is not a single thing to be found which does not bear witness to the truth of this account. How could it be otherwise? For, seeing that every part functions in its own special way because of the manner in which the four qualities are compounded, it is absolutely necessary that the function [activity] should be either completely destroyed, or, at least hampered, by any damage to the qualities, and that thus the animal should fall ill, either as a whole, or in certain of its parts.

Also the diseases which are primary and most generic are four in number, and differ from each other in warmth, cold, dryness and moisture. Now, Erasistratus himself confesses this, albeit unintentionally;[1] for when he says that the digestion of food becomes worse in fever, not because the innate heat has ceased to be in due proportion, as people previously supposed, but because the stomach, with its activity impaired, cannot contract and triturate as before—then, I say, one may justly ask him what it is that has impaired the activity of the stomach.

Thus, for example, when a bubo develops following an accidental wound[2] gastric digestion does not become impaired · *until after the patient has become fevered*; neither the bubo nor the sore of itself impedes in any way or damages the activity of the stomach. But if fever occurs, the digestion at once deteriorates, and we are also right in saying that the activity of the stomach at once becomes impaired. We must add, however, by what

[2] As a *bubo* is a swelling in the groin, we must suppose that the wound referred to would be in the leg or lower abdomen.

χρὴ τῷ λόγῳ. τὸ μὲν γὰρ ἕλκος οὐχ οἷόν τ᾽ ἦν
αὐτὴν βλάπτειν, ὥσπερ οὐδ᾽ ὁ βουβών· ἢ γὰρ ἂν
ἔβλαψε καὶ πρὸ τοῦ πυρετοῦ. εἰ δὲ μὴ ταῦτα,
δῆλον, ὡς ἡ τῆς θερμασίας πλεονεξία. δύο γὰρ
ταῦτα προσεγένετο τῷ βουβῶνι, ἡ τῆς κατὰ τὰς
ἀρτηρίας τε καὶ τὴν καρδίαν κινήσεως ἀλλοίωσις
καὶ ἡ τῆς κατὰ φύσιν θερμασίας πλεονεξία.
ἀλλ᾽ ἡ μὲν τῆς κινήσεως ἀλλοίωσις οὐ μόνον οὐδὲν
120 βλάψει τὴν ἐνέργειαν τῆς γα‖στρός, ἀλλὰ καὶ
προσωφελήσει κατ᾽ ἐκεῖνα τῶν ζῴων, ἐν οἷς εἰς
τὴν πέψιν ὑπέθετο πλεῖστον δύνασθαι τὸ διὰ τῶν
ἀρτηριῶν εἰς τὴν κοιλίαν ἐμπῖπτον πνεῦμα. διὰ
λοιπὴν οὖν ἔτι καὶ μόνην τὴν ἄμετρον θερμασίαν
ἡ βλάβη τῆς ἐνεργείας τῇ γαστρί. τὸ μὲν γὰρ
πνεῦμα σφοδρότερόν τε καὶ συνεχέστερον καὶ
πλέον ἐμπίπτει νῦν ἢ πρότερον. ὥστε ταύτῃ
μὲν μᾶλλον πέψει τὰ διὰ τὸ πνεῦμα καλῶς
πέττοντα ζῷα, διὰ λοιπὴν δ᾽ ἔτι τὴν παρὰ φύσιν
θερμασίαν ἀπεπτήσει. τὸ γὰρ καὶ τῷ πνεύματι
φάναι τιν᾽ ὑπάρχειν ἰδιότητα, καθ᾽ ἣν πέττει,
κἄπειτα ταύτην πυρεττόντων διαφθείρεσθαι καθ᾽
ἕτερον τρόπον ἐστὶν ὁμολογῆσαι τὸ ἄτοπον.
ἐρωτηθέντες γὰρ αὖθις, ὑπὸ τίνος ἠλλοιώθη τὸ
πνεῦμα, μόνην ἕξουσιν ἀποκρίνεσθαι τὴν παρὰ
φύσιν θερμασίαν καὶ μάλιστ᾽ ἐπὶ τοῦ κατὰ τὴν

---

[1] *i.e.* fever as a *cause* of disease.

[2] As we should say, "circulatory" changes.

[3] This is the "vital spirit" or pneuma which, according to Erasistratus and the Pneumatist school, was elaborated in the left ventricle, and thereafter carried by the arteries all over the body, there to subserve circulatory processes. It

it has been impaired. For the wound was not capable of impairing it, nor yet the bubo, for, if they had been, then they would have caused this damage before the fever as well. If it was not these that caused it, then it was the excess of heat [1] (for these two symptoms occurred besides the bubo—an alteration in the arterial and cardiac movements [2] and an excessive development of natural heat). Now the alteration of these movements will not merely not impair the function of the stomach in any way: it will actually prove an additional help among those animals in which, according to Erasistratus, the *pneuma*, which is propelled through the arteries and into the alimentary canal, is of great service in digestion; [3] there is only left, then, the disproportionate heat to account for the damage to the gastric activity. For the pneuma is driven in more vigorously and continuously, and in greater quantity now than before; thus in this case, the animal whose digestion is promoted by pneuma will digest more, whereas the remaining factor—abnormal heat—will give them indigestion. For to say, on the one hand, that the pneuma has a certain property by virtue of which it promotes digestion, and then to say that this property disappears in cases of fever, is simply to admit the absurdity. For when they are again asked what it is that has altered the pneuma, they will only be able to reply, "the abnormal heat," and particularly if it be the pneuma in the food canal which is in

---

has some analogy with oxygen, but this is also the case with the "*natural* spirit" or pneuma, whose seat was the liver and which was distributed by the *veins* through the body; it presided over the more *vegetative* processes. *cf.* p. 152, note 1; Introduction, p. xxxiv.

κοιλίαν· οὐδὲ γὰρ πλησιάζει κατ' οὐδὲν τοῦτο τῷ βουβῶνι.

Καίτοι τί τῶν ζῴων ἐκείνων, ἐν οἷς ἡ τοῦ πνεύματος ἰδιότης μέγα δύναται, μνημονεύω, παρὸν ἐπ' ἀνθρώποις, ἐν οἷς ἢ οὐδὲν ἢ παντάπασιν 121 ἀμυ‖δρόν τι καὶ μικρὸν ὠφελεῖ, ποιεῖσθαι τὸν λόγον; ἀλλ' ὅτι μὲν ἐν τοῖς πυρετοῖς οὗτοι κακῶς πέττουσιν, ὁμολογεῖ καὶ αὐτὸς καὶ τήν γ' αἰτίαν προστιθεὶς βεβλάφθαι φησὶ τῆς γαστρὸς τὴν ἐνέργειαν. οὐ μὴν ἄλλην γέ τινα πρόφασιν τῆς βλάβης εἰπεῖν ἔχει πλὴν τῆς παρὰ φύσιν θερμασίας. ἀλλ' εἰ βλάπτει τὴν ἐνέργειαν ἡ παρὰ φύσιν θερμασία μὴ κατά τι συμβεβηκός, ἀλλὰ διὰ τὴν αὑτῆς οὐσίαν τε καὶ δύναμιν, ἐκ τῶν πρώτων ἂν εἴη νοσημάτων· καὶ μὴν οὐκ ἐνδέχεται τῶν πρώτων μὲν εἶναι νοσημάτων τὴν ἀμετρίαν τῆς θερμασίας, τὴν δ' ἐνέργειαν ὑπὸ τῆς εὐκρασίας μὴ γίγνεσθαι. οὐδὲ γὰρ δι' ἄλλο τι δυνατὸν γίγνεσθαι τὴν δυσκρασίαν αἰτίαν τῶν πρώτων νοσημάτων ἀλλ' ἢ διὰ τὴν εὐκρασίαν διαφθειρομένην. τῷ γὰρ ὑπὸ ταύτης γίγνεσθαι τὰς ἐνεργείας ἀνάγκη καὶ τὰς πρώτας αὐτῶν βλάβας διαφθειρομένης γίγνεσθαι.

Ὅτι μὲν οὖν καὶ κατ' αὐτὸν τὸν Ἐρασίστρατον ἡ εὐκρασία τοῦ θερμοῦ τῶν ἐνεργειῶν αἰτία, τοῖς θεωρεῖν τὸ ἀκόλουθον δυναμένοις ἱκανῶς ἀποδεδεῖχθαι νομίζω. τούτου δ' ὑπάρχοντος 122 ἡμῖν οὐδὲν ἔτι χαλεπὸν ‖ ἐφ' ἑκάστης ἐνεργείας

---

[1] Even leaving the pneuma out of account, Galen claims that he can still prove his thesis.

[2] In other words : if *dyscrasia* is a first principle in *pathology*, then *eucrasia* must be a first principle in *physiology*.

question (since this does not come in any way near the bubo).

Yet why do I mention those animals in which the property of the pneuma plays an important part, when it is possible to base one's argument upon human beings, in whom it is either of no importance at all, or acts quite faintly and feebly?[1] But Erasistratus himself agrees that human beings digest badly in fevers, adding as the cause that the activity of the stomach has been impaired. He cannot, however, advance any other cause of this impairment than abnormal heat. But if it is not by accident that the abnormal heat impairs this activity, but by virtue of its own essence and power, then this abnormal heat must belong to the *primary diseases*. But, indeed, if *disproportion* of heat belongs to the primary diseases, it cannot but be that a *proportionate* blending [eucrasia] of the qualities produces the normal activity.[2] For a disproportionate blend [dyscrasia] can only become a cause of the primary diseases through derangement of the eucrasia. That is to say, it is because the [normal] activities arise from the eucrasia that the primary impairments of these activities necessarily arise from its derangement.

I think, then, it has been proved to the satisfaction of those people who are capable of seeing logical consequences, that, even according to Erasistratus's own argument, the cause of the normal functions is eucrasia of the Warm.[3] Now, this being so, there is nothing further to prevent us from saying

---

[3] The above is a good instance of Galen's "logical" method as applied to medical questions; an appeal to those who are capable of following "logical sequence." *cf.* p. 209, note 1.

τῇ μὲν εὐκρασίᾳ τὸ βέλτιον ἕπεσθαι λέγειν, τῇ
δὲ δυσκρασίᾳ τὰ χείρω. καὶ τοίνυν εἴπερ ταῦθ᾽
οὕτως ἔχει, τὸ μὲν αἷμα τῆς συμμέτρου θερμασίας, τὴν δὲ ξανθὴν χολὴν τῆς ἀμέτρου νομιστέον ὑπάρχειν ἔγγονον. οὕτω γὰρ καὶ ἡμῖν
ἔν τε ταῖς θερμαῖς ἡλικίαις καὶ τοῖς θερμοῖς
χωρίοις καὶ ταῖς ὥραις τοῦ ἔτους ταῖς θερμαῖς
καὶ ταῖς θερμαῖς καταστάσεσιν, ὡσαύτως δὲ καὶ
ταῖς θερμαῖς κράσεσι τῶν ἀνθρώπων καὶ τοῖς
ἐπιτηδεύμασί τε καὶ τοῖς διαιτήμασι καὶ τοῖς
νοσήμασι τοῖς θερμοῖς εὐλόγως ἡ ξανθὴ χολὴ
πλείστη φαίνεται γιγνομένη.

Τὸ δ᾽ ἀπορεῖν, εἴτ᾽ ἐν τοῖς σώμασι τῶν ἀνθρώπων ὁ χυμὸς οὗτος ἔχει τὴν γένεσιν εἴτ᾽ ἐν τοῖς
σιτίοις περιέχεται, μηδ᾽ ὅτι τοῖς ὑγιαίνουσιν
ἀμέμπτως, ὅταν ἀσιτήσωσι παρὰ τὸ ἔθος ὑπό
τινος περιστάσεως πραγμάτων ἀναγκασθέντες,
πικρὸν μὲν τὸ στόμα γίγνεται, χολώδη δὲ τὰ
οὖρα, δάκνεται δ᾽ ἡ γαστήρ, ἑωρακότος ἐστὶν
ἀλλ᾽ ὥσπερ ἐξαίφνης νῦν εἰς τὸν κόσμον ἐλη-
λυθότος καὶ μήπω τὰ κατ᾽ αὐτὸν φαινόμενα
γιγνώσκοντος. ἐπεὶ τίς οὐκ οἶδεν, ὡς ἕκαστον
τῶν ἑψομένων ἐπὶ πλέον ἁλυκώτερον μὲν τὸ
123 πρῶτον, ὕστερον ‖ δὲ πικρότερον γίγνεται; κἂν
εἰ τὸ μέλι βουληθείης αὐτὸ τὸ πάντων γλυκύτατον ἐπὶ πλεῖστον ἕψειν, ἀποδείξεις καὶ τοῦτο
πικρότατον· ὃ γὰρ τοῖς ἄλλοις, ὅσα μὴ φύσει
θερμά, παρὰ τῆς ἑψήσεως ἐγγίγνεται, τοῦτ᾽ ἐκ
φύσεως ὑπάρχει τῷ μέλιτι. διὰ τοῦτ᾽ οὖν ἑψόμενον οὐ γίγνεται γλυκύτερον· ὅσον γὰρ ἐχρῆν
εἶναι θερμότητος εἰς γένεσιν γλυκύτητος, ἀκριβῶς
αὐτῷ τοῦτο πᾶν οἴκοθεν ὑπάρχει. ὃ τοίνυν

that, in the case of each function, eucrasia is followed by the more, and dyscrasia by the less favourable alternative. And, therefore, if this be the case, we must suppose blood to be the outcome of proportionate, and yellow bile of disproportionate heat. So we naturally find yellow bile appearing in greatest quantity in ourselves at the warm periods of life, in warm countries, at warm seasons of the year, and when we are in a warm condition; similarly in people of warm temperaments, and in connection with warm occupations, modes of life, or diseases.

And to be in doubt as to whether this humour has its genesis in the human body or is contained in the food is what you would expect from one who has—I will not say failed to see that, when those who are perfectly healthy have, under the compulsion of circumstances, to fast contrary to custom, their mouths become bitter and their urine bile-coloured, while they suffer from gnawing pains in the stomach— but has, as it were, just made a sudden entrance into the world, and is not yet familiar with the phenomena which occur there. Who, in fact, does not know that anything which is overcooked grows at first salt and afterwards bitter? And if you will boil honey itself, far the sweetest of all things, you can demonstrate that even this becomes quite bitter. For what may occur as a result of boiling in the case of other articles which are not warm by nature, exists naturally in honey; for this reason it does not become sweeter on being boiled, since exactly the same quantity of heat as is needed for the production of sweetness exists from beforehand in the honey. Therefore the external heat,

ἔξωθεν τοῖς ἐλλιπῶς θερμοῖς ἦν ὠφέλιμον, τοῦτ᾿
ἐκείνῳ βλάβη τε καὶ ἀμετρία γίγνεται καὶ διὰ
τοῦτο θᾶττον τῶν ἄλλων ἑψόμενον ἀποδείκνυται
πικρόν. δι᾿ αὐτὸ δὲ τοῦτο καὶ τοῖς θερμοῖς
φύσει καὶ τοῖς ἀκμάζουσιν εἰς χολὴν ἑτοίμως
μεταβάλλεται. θερμῷ γὰρ θερμὸν πλησιάζον
εἰς ἀμετρίαν κράσεως ἑτοίμως ἐξίσταται καὶ
φθάνει χολὴ γιγνόμενον, οὐχ αἷμα. δεῖται τοίνυν ψυχρᾶς μὲν κράσεως ἀνθρώπου, ψυχρᾶς δ᾿
ἡλικίας, ἵν᾿ εἰς αἵματος ἄγηται φύσιν. οὔκουν
ἀπὸ τρόπου συνεβούλευσεν Ἱπποκράτης τοῖς
φύσει πικροχόλοις μὴ προσφέρειν τὸ μέλι, ὡς
124 ἂν θερμοτέρας ‖ δηλονότι κράσεως ὑπάρχουσιν.
οὕτω δὲ καὶ τοῖς νοσήμασι τοῖς πικροχόλοις
πολέμιον εἶναι τὸ μέλι καὶ τῇ τῶν γερόντων
ἡλικίᾳ φίλιον οὐχ Ἱπποκράτης μόνον ἀλλὰ καὶ
πάντες ἰατροὶ λέγουσιν, οἱ μὲν ἐκ τῆς φύσεως
αὐτοῦ τὴν δύναμιν ἐνδειξαμένης εὑρόντες, οἱ δ᾿
ἐκ τῆς πείρας μόνης. οὐδὲ γὰρ οὐδὲ τοῖς ἀπὸ
τῆς ἐμπειρίας ἰατροῖς ἕτερόν τι παρὰ ταῦτα
τετήρηται γιγνόμενον, ἀλλὰ χρηστὸν μὲν γέροντι,
νέῳ δ᾿ οὐ χρηστόν, καὶ τῷ μὲν φύσει πικροχόλῳ
βλαβερόν, ὠφέλιμον δὲ τῷ φλεγματώδει· καὶ
τῶν νοσημάτων ὡσαύτως τοῖς μὲν πικροχόλοις
ἐχθρόν, τοῖς δὲ φλεγματώδεσι φίλιον· ἑνὶ δὲ
λόγῳ τοῖς μὲν θερμοῖς σώμασιν ἢ διὰ φύσιν ἢ διὰ
νόσον ἢ δι᾿ ἡλικίαν ἢ δι᾿ ὥραν ἢ διὰ χώραν ἢ
δι᾿ ἐπιτήδευμα χολῆς γεννητικόν, αἵματος δὲ τοῖς
ἐναντίοις.

Καὶ μὴν οὐκ ἐνδέχεται ταὐτὸν ἔδεσμα τοῖς μὲν
χολὴν γεννᾶν, τοῖς δ᾿ αἷμα μὴ οὐκ ἐν τῷ σώματι

which would be useful for insufficiently warm substances, becomes in the honey a source of damage, in fact an excess; and it is for this reason that honey, when boiled, can be demonstrated to become bitter sooner than the others. For the same reason it is easily transmuted into bile in those people who are naturally warm, or in their prime, since warm when associated with warm becomes readily changed into a disproportionate combination and turns into bile sooner than into blood. Thus we need a cold temperament and a cold period of life if we would have honey brought to the nature of blood.[1] Therefore Hippocrates not improperly advised those who were naturally bilious not to take honey, since they were obviously of too warm a temperament. So also, not only Hippocrates, but all physicians say that honey is bad in bilious diseases but good in old age; some of them having discovered this through the indications afforded by its nature, and others simply through experiment,[2] for the Empiricist physicians too have made precisely the same observation, namely, that honey is good for an old man and not for a young one, that it is harmful for those who are naturally bilious, and serviceable for those who are phlegmatic. In a word, in bodies which are warm either through nature, disease, time of life, season of the year, locality, or occupation, honey is productive of bile, whereas in opposite circumstances it produces blood.

But surely it is impossible that the same article of diet can produce in certain persons bile and in others blood, if it be not that the genesis of these humours is

---

[1] The aim of dietetics always being the production of moderate heat—*i.e.* blood.
[2] Note contrasted methods of Rationalists and Empiricists.

τῆς γενέσεως αὐτῶν ἐπιτελουμένης. εἰ γὰρ δὴ
οἴκοθέν γε καὶ παρ᾽ ἑαυτοῦ τῶν ἐδεσμάτων
ἕκαστον ἔχον καὶ οὐκ ἐν τοῖς τῶν ζῴων σώμασι ‖
125 μεταβαλλόμενον ἐγέννα τὴν χολήν, ἐν ἅπασιν
ἂν ὁμοίως αὐτὴν τοῖς σώμασιν ἐγέννα καὶ τὸ
μὲν πικρὸν ἔξω γενομένοις ἦν ἂν οἶμαι χολῆς
ποιητικόν, εἰ δέ τι γλυκὺ καὶ χρηστόν, οὐκ ἂν
οὐδὲ τὸ βραχύτατον ἐξ αὐτοῦ χολῆς ἐγεννᾶτο. καὶ
μὴν οὐ τὸ μέλι μόνον, ἀλλὰ καὶ τῶν ἄλλων ἕκαστον
τῶν γλυκέων τοῖς προειρημένοις σώμασι τοῖς δι᾽
ὁτιοῦν τῶν εἰρημένων θερμοῖς οὖσιν εἰς χολὴν
ἑτοίμως ἐξίσταται.

Καίτοι ταῦτ᾽ οὐκ οἶδ᾽ ὅπως ἐξηνέχθην εἰπεῖν
οὐ προελόμενος ἀλλ᾽ ὑπ᾽ αὐτῆς τοῦ λόγου τῆς
ἀκολουθίας ἀναγκασθείς. εἴρηται δ᾽ ἐπὶ πλεῖστον ὑπὲρ αὐτῶν Ἀριστοτέλει τε καὶ Πραξαγόρα
τὴν Ἱπποκράτους καὶ Πλάτωνος γνώμην ὀρθῶς
ἐξηγησαμένοις.

IX

Μὴ τοίνυν ὡς ἀποδείξεις ὑφ᾽ ἡμῶν εἰρῆσθαι
νομίζειν τὰ τοιαῦτα μᾶλλον ἢ περὶ τῆς τῶν
ἄλλως γιγνωσκόντων ἀναισθησίας[1] ἐνδείξεις, οἳ
μηδὲ τὰ πρὸς ἁπάντων ὁμολογούμενα καὶ καθ᾽
ἑκάστην ἡμέραν φαινόμενα γιγνώσκουσιν· τὰς
δ᾽ ἀποδείξεις αὐτῶν τὰς κατ᾽ ἐπιστήμην ἐξ
ἐκείνων χρὴ λαμβάνειν τῶν ἀρχῶν, ὧν ἤδη
126 καὶ πρόσθεν ‖ εἴπομεν, ὡς τὸ δρᾶν καὶ πάσχειν
εἰς ἄλληλα τοῖς σώμασιν ὑπάρχει κατὰ τὸ
θερμὸν καὶ ψυχρὸν καὶ ξηρὸν καὶ ὑγρόν. καὶ

---

[1] Lit. anaesthesia. Linacre renders it indocilitas.

accomplished *in the body.* For if all articles of food contained bile from the beginning and of themselves, and did not produce it by undergoing change in the animal body, then they would produce it similarly in all bodies; the food which was bitter to the taste would, I take it, be productive of bile, while that which tasted good and sweet would not generate even the smallest quantity of bile. Moreover, not only honey but all other sweet substances are readily converted into bile in the aforesaid bodies which are warm for any of the reasons mentioned.

Well, I have somehow or other been led into this discussion,—not in accordance with my plan, but compelled by the course of the argument. This subject has been treated at great length by Aristotle and Praxagoras, who have correctly expounded the view of Hippocrates and Plato.

IX

For this reason the things that we have said are not to be looked upon as proofs but rather as indications of the dulness [1] of those who think differently, and who do not even recognise what is agreed on by everyone and is a matter of daily observation. As for the scientific proofs of all this, they are to be drawn from these principles of which I have already spoken [2]—namely, that bodies act upon and are acted upon by each other in virtue of the Warm, Cold, Moist and Dry. And if one is

[2] p. 15.

εἴτε φλέβας εἴθ᾽ ἧπαρ εἴτ᾽ ἀρτηρίας εἴτε καρδίαν
εἴτε κοιλίαν εἴτ᾽ ἄλλο τι μόριον ἐνεργεῖν τις
φήσειεν ἡντινοῦν ἐνέργειαν, ἀφύκτοις ἀνάγκαις
ἀναγκασθήσεται διὰ τὴν ἐκ τῶν τεττάρων ποιὰν
κρᾶσιν ὁμολογῆσαι τὴν ἐνέργειαν ὑπάρχειν αὐτῷ.
διὰ τί γὰρ ἡ γαστὴρ περιστέλλεται τοῖς σιτίοις,
διὰ τί δ᾽ αἱ φλέβες αἷμα γεννῶσι, παρὰ τῶν
Ἐρασιστρατείων ἐδεόμην ἀκοῦσαι. τὸ γὰρ ὅτι
περιστέλλεται μόνον αὐτὸ καθ᾽ ἑαυτὸ γιγνώσκειν
οὐδέπω χρηστόν, εἰ μὴ καὶ τὴν αἰτίαν εἰδείημεν·
οὕτω γὰρ ἂν οἶμαι καὶ τὰ σφάλματα θερα-
πεύσαιμεν. οὐ μέλει, φασίν, ἡμῖν οὐδὲ πολυ-
πραγμονοῦμεν ἔτι τὰς τοιαύτας αἰτίας· ὑπὲρ
ἰατρὸν¹ γάρ εἰσι καὶ τῷ φυσικῷ² προσήκουσι.
πότερον οὖν οὐδ᾽ ἀντερεῖτε τῷ φάσκοντι τὴν μὲν
εὐκρασίαν τὴν κατὰ φύσιν αἰτίαν εἶναι τῆς ἐνεργείας ἑκάστῳ τῶν ὀργάνων, τὴν δ᾽ αὖ δυσκρασίαν
127 νόσον τ᾽ ἤδη καλεῖσθαι καὶ πάντως ὑπ᾽ αὐ‖τῆς
βλάπτεσθαι τὴν ἐνέργειαν; ἢ πεισθήσεσθε ταῖς
τῶν παλαιῶν ἀποδείξεσιν; ἢ τρίτον τι καὶ μέσον
ἑκατέρου τούτων πράξετε μήθ᾽ ὡς ἀληθέσι τοῖς
λόγοις ἐξ ἀνάγκης πειθόμενοι μήτ᾽ ἀντιλέγοντες
ὡς ψευδέσιν, ἀλλ᾽ ἀπορητικοί τινες ἐξαίφνης καὶ
Πυρρώνειοι γενήσεσθε; καὶ μὴν εἰ τοῦτο δράσετε,
τὴν ἐμπειρίαν ἀναγκαῖον ὑμῖν προστήσασθαι. τῷ
γὰρ ἂν ἔτι τρόπῳ καὶ τῶν ἰαμάτων εὐποροίητε
τὴν οὐσίαν ἑκάστου τῶν νοσημάτων ἀγνοοῦντες;
τί οὖν οὐκ ἐξ ἀρχῆς ἐμπειρικοὺς ὑμᾶς αὐτοὺς
ἐκαλέσατε; τί δὲ πράγμαθ᾽ ἡμῖν παρέχετε φυ-

---

¹ *Iatros*: lit. "healer."
² Lit. "physicist" or "physiologist," the student of the
physis. cf. p. 70, note 2.

speaking of any activity, whether it be exercised by vein, liver, arteries, heart, alimentary canal, or any part, one will be inevitably compelled to acknowledge that this activity depends upon the way in which the four qualities are blended. Thus I should like to ask the Erasistrateans why it is that the stomach contracts upon the food, and why the veins generate blood. There is no use in recognizing the mere fact of contraction, without also knowing the *cause*; if we know this, we shall also be able to rectify the failures of function. "This is no concern of ours," they say; "we do not occupy ourselves with such causes as these; they are outside the sphere of the *practitioner*,[1] and belong to that of the *scientific investigator*." [2] Are you, then, going to oppose those who maintain that the cause of the function of every organ is a natural eucrasia,[3] that the dyscrasia is itself known as a *disease*, and that it is certainly by this that the activity becomes impaired? Or, on the other hand, will you be convinced by the proofs which the ancient writers furnished? Or will you take a midway course between these two, neither perforce accepting these arguments as true nor contradicting them as false, but suddenly becoming sceptics— Pyrrhonists, in fact? But if you do this you will have to shelter yourselves behind the Empiricist teaching. For how are you going to be successful in treatment, if you do not understand the real essence of each disease? Why, then, did you not call yourselves Empiricists from the beginning? Why do you confuse us by announcing that you are

[3] That is, a *blending* of the four principles in their natural proportion; Lat. *temperies*. Dyscrasia = *intemperies*, "distemper."

σικὰς ἐνεργείας ἐπαγγελλόμενοι ζητεῖν ἰάσεως
ἕνεκεν· .εἰ γὰρ ἀδύνατος ἡ γαστήρ ἐστί τινι
περιστέλλεσθαι καὶ τρίβειν, πῶς αὐτὴν εἰς τὸ
κατὰ φύσιν ἐπανάξομεν ἀγνοοῦντες τὴν αἰτίαν
τῆς ἀδυναμίας; ἐγὼ μέν φημι τὴν μὲν ὑπερτεθερ-
μασμένην ἐμψυκτέον ἡμῖν εἶναι, τὴν δ' ἐψυγ-
μένην θερμαντέον· οὕτω δὲ καὶ τὴν ἐξηρασμένην
ὑγραντέον, τὴν δ' ὑγρασμένην ξηραντέον. ἀλλὰ
128 καὶ ‖ κατὰ συζυγίαν, εἰ θερμοτέρα τοῦ κατὰ φύσιν
ἅμα καὶ ξηροτέρα τύχοι γεγενημένη, κεφάλαιον
εἶναι τῆς ἰάσεως ἐμψύχειν θ' ἅμα καὶ ὑγραίνειν·
εἰ δ' αὖ ψυχροτέρα τε καὶ ὑγροτέρα, θερμαίνειν
τε καὶ ξηραίνειν κἀπὶ τῶν ἄλλων ὡσαύτως· οἱ
ἀπ' Ἐρασιστράτου τί ποτε καὶ πράξουσιν οὐδ'
ὅλως ζητεῖν τῶν ἐνεργειῶν τὰς αἰτίας ὁμολογοῦν-
τες; ὁ γάρ τοι καρπὸς τῆς περὶ τῶν ἐνεργειῶν
ζητήσεως οὗτός ἐστι, τὸ τὰς αἰτίας τῶν δυσκρα-
σιῶν εἰδότα εἰς τὸ κατὰ φύσιν ἐπανάγειν αὐτάς,
ὡς αὐτό γε μόνον τὸ γνῶναι τὴν ἑκάστου τῶν
ὀργάνων ἐνέργειαν ἥτις ἐστὶν οὔπω χρηστὸν εἰς
τὰς ἰάσεις.

Ἐρασίστρατος δέ μοι δοκεῖ καὶ αὐτὸ τοῦτ'
ἀγνοεῖν, ὡς, ἥτις ἂν ἐν τῷ σώματι διάθεσις βλά-
πτῃ τὴν ἐνέργειαν μὴ κατά τι συμβεβηκὸς ἀλλὰ
πρώτως τε καὶ καθ' ἑαυτήν, αὕτη τὸ νόσημά
ἐστιν αὐτό. πῶς οὖν ἔτι διαγνωστικός τε καὶ
ἰατικὸς ἔσται τῶν νοσημάτων ἀγνοῶν ὅλως αὐτὰ
τίνα τ' ἐστὶ καὶ πόσα καὶ ποῖα; κατὰ μὲν δὴ τὴν
γαστέρα τό γε τοσοῦτον Ἐρασίστρατος ἠξίωσε

investigating natural activities with a view to treatment? If the stomach is, in a particular case, unable to exercise its peristaltic and grinding functions, how are we going to bring it back to the normal if we do not know the *cause* of its disability? What I say is[1] that we must cool the over-heated stomach and warm the chilled one; so also we must moisten the one which has become dried up, and conversely; so, too, in combinations of these conditions; if the stomach becomes at the same time warmer and drier than normally, the first principle of treatment is at once to chill and moisten it; and if it become colder and moister, it must be warmed and dried; so also in other cases. But how on earth are the followers of Erasistratus going to act, confessing as they do that they make no sort of investigation into the cause of disease? For the fruit of the enquiry into activities is that by knowing the causes of the dyscrasiae one may bring them back to the normal, since it is of no use for the purposes of treatment merely to know what the activity of each organ is.

Now, it seems to me that Erasistratus is unaware of this fact also, that the actual disease is that condition of the body which, not accidentally, but primarily and of itself, impairs the normal function. How, then, is he going to diagnose or cure diseases if he is entirely ignorant of what they are, and of what kind and number? As regards the stomach, certainly, Erasistratus held that one should at least

---

[1] This is the orthodox Hippocratic treatment, that of *opposites by opposites*. Contrast the *homoeopathic* principle which is the basis of our modern methods of *immunisation* (*similia similibus curentur*, Hahnemann).

129 ζητεῖσθαι τὸ πῶς πέττεται τὰ σιτία· ‖ τὸ δ' ἥτις
πρώτη τε καὶ ἀρχηγὸς αἰτία τούτου, πῶς οὐκ
ἐπεσκέψατο; κατὰ δὲ τὰς φλέβας καὶ τὸ αἷμα
καὶ αὐτὸ τὸ πῶς παρέλιπεν.

'Αλλ' οὔθ' Ἱπποκράτης οὔτ' ἄλλος τις ὧν
ὀλίγῳ πρόσθεν ἐμνημόνευσα φιλοσόφων ἢ ἰατρῶν
ἄξιον ᾤετ' εἶναι παραλιπεῖν· ἀλλὰ τὴν κατὰ
φύσιν ἐν ἑκάστῳ ζῴῳ θερμασίαν εὔκρατόν τε καὶ
μετρίως ὑγρὰν οὖσαν αἵματος εἶναί φασι γεννη-
τικὴν καὶ δι' αὐτό γε τοῦτο καὶ τὸ αἷμα θερμὸν
καὶ ὑγρὸν εἶναί φασι τῇ δυνάμει χυμόν, ὥσπερ
τὴν ξανθὴν χολὴν θερμὴν καὶ ξηρὰν εἶναι, εἰ
καὶ ὅτι μάλισθ' ὑγρὰ φαίνεται. διαφέρειν γὰρ
αὐτοῖς δοκεῖ τὸ κατὰ φαντασίαν ὑγρὸν τοῦ κατὰ
δύναμιν. ἢ τίς οὐκ οἶδεν, ὡς ἄλμη μὲν καὶ
θάλαττα ταριχεύει τὰ κρέα καὶ ἄσηπτα διαφυ-
λάττει, τὸ δ' ἄλλο πᾶν ὕδωρ τὸ πότιμον ἑτοίμως
διαφθείρει τε καὶ σήπει; τίς δ' οὐκ οἶδεν, ὡς
ξανθῆς χολῆς ἐν τῇ γαστρὶ περιεχομένης πολλῆς
ἀπαύστῳ δίψει συνεχόμεθα καὶ ὡς ἐμέσαντες
αὐτὴν εὐθὺς ἄδιψοι γιγνόμεθα μᾶλλον ἢ εἰ
130 πάμπολυ ποτὸν προσηράμεθα; ‖ θερμὸς οὖν
εὐλόγως ὁ χυμὸς οὗτος εἴρηται καὶ ξηρὸς κατὰ
δύναμιν, ὥσπερ γε καὶ τὸ φλέγμα ψυχρὸν
καὶ ὑγρόν. ἐναργεῖς γὰρ καὶ περὶ τούτου
πίστεις Ἱπποκράτει τε καὶ τοῖς ἄλλοις εἴρηνται
παλαιοῖς.

Πρόδικος δ' ἐν τῷ περὶ φύσεως ἀνθρώπου
γράμματι τὸ συγκεκαυμένον καὶ οἷον ὑπερωπτη-
μένον ἐν τοῖς χυμοῖς ὀνομάζων φλέγμα παρὰ τὸ
πεφλέχθαι τῇ λέξει μὲν ἑτέρως χρῆται, φυλάττει

investigate *how* it digests the food. But why was not investigation also made as to the primary originative cause of this? And, as regards the veins and the blood, he omitted even to ask the question "*how?*"

Yet neither Hippocrates nor any of the other physicians or philosophers whom I mentioned a short while ago thought it right to omit this; they say that when the heat which exists naturally in every animal is well blended and moderately moist it generates blood; for this reason they also say that the blood is a *virtually* warm and moist humour, and similarly also that yellow bile is warm and dry, even though for the most part it appears moist. (For in them the *apparently* dry would seem to differ from the *virtually* dry.) Who does not know that brine and sea-water preserve meat and keep it uncorrupted,[1] whilst all other water—the drinkable kind—readily spoils and rots it? And who does not know that when yellow bile is contained in large quantity in the stomach, we are troubled with an unquenchable thirst, and that when we vomit this up, we at once become much freer from thirst than if we had drunk very large quantities of fluid? Therefore this humour has been very properly termed warm, and also virtually dry. And, similarly, *phlegm* has been called cold and moist; for about this also clear proofs have been given by Hippocrates and the other Ancients.

Prodicus[2] also, when in his book "On the Nature of Man" he gives the name "phlegm" (from the verb πεφλέχθαι) to that element in the humours which has been burned or, as it were, over-roasted, while using

---

[1] Lit. *aseptic.*

[2] Prodicus of Ceos, a Sophist, contemporary of Socrates.

μέντοι τὸ πρᾶγμα κατὰ ταὐτὸ τοῖς ἄλλοις. τὴν
δ' ἐν τοῖς ὀνόμασι τἀνδρὸς τούτου καινοτομίαν
ἱκανῶς ἐνδείκνυται καὶ Πλάτων. ἀλλὰ τοῦτό
γε τὸ πρὸς ἁπάντων ἀνθρώπων ὀνομαζόμενον
φλέγμα τὸ λευκὸν τὴν χρόαν, ὃ βλένναν ὀνομάζει
Πρόδικος, ὁ ψυχρὸς καὶ ὑγρὸς χυμός ἐστιν οὗτος
καὶ πλεῖστος τοῖς τε γέρουσι καὶ τοῖς ὁπωσδή-
ποτε ψυγεῖσιν ἀθροίζεται καὶ οὐδεὶς οὐδὲ μαινό-
μενος ἂν ἄλλο τι ἢ ψυχρὸν καὶ ὑγρὸν εἴποι ἂν
αὐτόν.

Ἆρ' οὖν θερμὸς μέν τίς ἐστι καὶ ὑγρὸς χυμὸς
καὶ θερμὸς καὶ ξηρὸς ἕτερος καὶ ὑγρὸς καὶ
ψυχρὸς ἄλλος, οὐδεὶς δ' ἐστὶ ψυχρὸς καὶ ξηρὸς
τὴν δύναμιν, ἀλλ' ἡ τετάρτη συζυγία τῶν κρά-
131 σεων ‖ ἐν ἅπασι τοῖς ἄλλοις ὑπάρχουσα μόνοις
τοῖς χυμοῖς οὐχ ὑπάρχει; καὶ μὴν ἥ γε μέλαινα
χολὴ τοιοῦτός ἐστι χυμός, ὃν οἱ σωφρονοῦντες
ἰατροὶ καὶ φιλόσοφοι πλεονεκτεῖν ἔφασαν τῶν
μὲν ὡρῶν τοῦ ἔτους ἐν φθινοπώρῳ μάλιστα,
τῶν δ' ἡλικιῶν ἐν ταῖς μετὰ τὴν ἀκμήν. οὕτω
δὲ καὶ διαιτήματα καὶ χωρία καὶ καταστάσεις
καὶ νόσους τινὰς ψυχρὰς καὶ ξηρὰς εἶναί
φασιν· οὐ γὰρ δὴ χωλὴν ἐν ταύτῃ μόνῃ τῇ
συζυγίᾳ τὴν φύσιν εἶναι νομίζουσιν ἀλλ' ὥσπερ
τὰς ἄλλας τρεῖς οὕτω καὶ τήνδε διὰ πάντων
ἐκτετάσθαι.

Ηὐξάμην οὖν κἀνταῦθ' ἐρωτῆσαι δύνασθαι τὸν
Ἐρασίστρατον, εἰ μηδὲν ὄργανον ἡ τεχνικὴ φύσις
ἐδημιούργησε καθαρτικὸν τοῦ τοιούτου χυμοῦ,
ἀλλὰ τῶν μὲν οὔρων ἄρα τῆς διακρίσεώς ἐστιν
ὄργανα δύο καὶ τῆς ξανθῆς χολῆς ἕτερον οὐ

a different terminology, still keeps to the fact just as the others do; this man's innovations in nomenclature have also been amply done justice to by Plato.[1] Thus, the white-coloured substance which everyone else calls *phlegm,* and which Prodicus calls *blenna* [mucus],[2] is the well-known cold, moist humour which collects mostly in old people and in those who have been chilled[3] in some way, and not even a lunatic could say that this was anything else than cold and moist.

If, then, there is a warm and moist humour, and another which is warm and dry, and yet another which is moist and cold, is there none which is virtually *cold and dry?* Is the fourth combination of temperaments, which exists in all other things, non-existent in the humours alone? No; the *black bile* is such a humour. This, according to intelligent physicians and philosophers, tends to be in excess, as regards seasons, mainly in the fall of the year, and, as regards ages, mainly after the prime of life. And, similarly, also they say that there are cold and dry modes of life, regions, constitutions, and diseases. Nature, they suppose, is not defective in this single combination; like the three other combinations, it extends everywhere.

At this point, also, I would gladly have been able to ask Erasistratus whether his "artistic" Nature has not constructed any organ for *clearing away* a humour such as this. For whilst there are two organs for the excretion of urine, and another of considerable size for that of yellow bile, does the

[1] Plato, *Timaeus,* 83–86, *passim.*

[2] *cf.* the term *blennorrhoea,* which is still used.

[3] *cf.* the Scotch term "colded" for "affected with a cold"; Germ. *erkältet.*

σμικρόν, ὁ δὲ τούτων κακοηθέστερος χυμὸς ἀλᾶται διὰ παντὸς ἐν ταῖς φλεψὶν ἀναμεμιγμένος τῷ αἵματι. καίτοι " Δυσεντερίη," φησί που Ἱπποκράτης, " ἣν ἀπὸ χολῆς μελαίνης ἄρξηται, θανάσιμον," οὐ μὴν ἥ γ᾽ ἀπὸ τῆς ξαν‖θῆς χολῆς ἀρχομένη πάντως ὀλέθριος, ἀλλ᾽ οἱ πλείους ἐξ αὐτῆς διασῴζονται. τοσούτῳ κακοηθεστέρα τε καὶ δριμυτέρα τὴν δύναμιν ἡ μέλαινα χολὴ τῆς ξανθῆς ἐστιν. ἆρ᾽ οὖν οὔτε τῶν ἄλλων ἀνέγνω τι τῶν τοῦ Ἱπποκράτους γραμμάτων ὁ Ἐρασίστρατος οὐδὲν οὔτε τὸ περὶ φύσεως ἀνθρώπου βιβλίον, ἵν᾽ οὕτως ἀργῶς παρέλθοι τὴν περὶ τῶν χυμῶν ἐπίσκεψιν, ἢ γιγνώσκει μέν, ἑκὼν δὲ παραλείπει καλλίστην τῆς τέχνης θεωρίαν; ἐχρῆν οὖν αὐτὸν μηδὲ περὶ τοῦ σπληνὸς εἰρηκέναι τι μηδ᾽ ἀσχημονεῖν ὑπὸ τῆς τεχνικῆς φύσεως ὄργανον τηλικοῦτον μάτην ἡγούμενον κατεσκευάσθαι. καὶ μὴν οὐχ Ἱπποκράτης μόνον ἢ Πλάτων, οὐδέν τι χείρους Ἐρασιστράτου περὶ φύσιν ἄνδρες, ἔν τι τῶν καθαιρόντων τὸ αἷμα καὶ τοῦτ᾽ εἶναί φασι τὸ σπλάγχνον, ἀλλὰ καὶ μυρίοι σὺν αὐτοῖς ἄλλοι τῶν παλαιῶν ἰατρῶν τε καὶ φιλοσόφων, ὧν ἁπάντων προσποιησάμενος ὑπερφρονεῖν ὁ γενναῖος Ἐρασίστρατος οὔτ᾽ ἀντεῖπεν οὔθ᾽ ὅλως τῆς δόξης αὐτῶν ἐμνημόνευσε. καὶ μὴν ὅσοις γε τὸ σῶμα θάλλει, τούτοις ὁ σπλὴν φθίνει, φησὶν Ἱπποκράτης, καὶ οἱ ἀπὸ τῆς ‖ ἐμπειρίας ὁρμώμενοι πάντες ὁμολογοῦσιν ἰατροί. καὶ ὅσοις γ᾽ αὖ μέγας καὶ ὕπουλος

humour which is more pernicious than these wander about persistently in the veins mingled with the blood? Yet Hippocrates says, " Dysentery is a fatal condition if it proceeds from black bile " ; while that proceeding from yellow bile is by no means deadly, and most people recover from it ; this proves how much more pernicious and acrid in its potentialities is black than yellow bile. Has Erasistratus, then, not read the book, " On the Nature of Man," any more than any of the rest of Hippocrates's writings, that he so carelessly passes over the consideration of the humours? Or, does he know it, and yet voluntarily neglect one of the finest studies [1] in medicine? Thus he ought not to have said anything about the *spleen*,[2] nor have stultified himself by holding that an artistic Nature would have prepared so large an organ for no purpose. As a matter of fact, not only Hippocrates and Plato—who are no less authorities on Nature than is Erasistratus—say that this viscus also is one of those which cleanse the blood, but there are thousands of the ancient physicians and philosophers as well who are in agreement with them. Now, all of these the high and mighty Erasistratus affected to despise, and he neither contradicted them nor even so much as mentioned their opinion. Hippocrates, indeed, says that the spleen wastes in those people in whom the body is in good condition, and all those physicians also who base themselves on experience [3] agree with this. Again, in those cases in which the spleen is large and is increasing from

[1] The word *theôria* used here is not the same as our *theory*. It is rather a " contemplation," the process by which a theory is arrived at. *cf.* p. 226, note 2.
[2] Erasistratus on the uselessness of the spleen. *cf.* p. 143.
[3] The Empirical school. *cf.* p. 193.

αὐξάνεται, τούτοις καταφθείρει τε καὶ κακόχυμα
τὰ σώματα τίθησιν, ὡς καὶ τοῦτο πάλιν οὐχ
Ἱπποκράτης μόνον ἀλλὰ καὶ Πλάτων ἄλλοι τε
πολλοὶ καὶ οἱ ἀπὸ τῆς ἐμπειρίας ὁμολογοῦσιν
ἰατροί. καὶ οἱ ἀπὸ σπληνὸς δὲ κακοπραγοῦντος
ἴκτεροι μελάντεροι καὶ τῶν ἑλκῶν αἱ οὐλαὶ
μέλαιναι. καθόλου γάρ, ὅταν ἐνδεέστερον ἢ
προσῆκεν εἰς ἑαυτὸν ἕλκῃ τὸν μελαγχολικὸν
χυμόν, ἀκάθαρτον μὲν τὸ αἷμα, κακόχρουν δὲ
τὸ πᾶν γίγνεται σῶμα. πότε δ' ἐνδεέστερον
ἕλκει; ἢ δῆλον ὅτι κακῶς διακείμενος; ὥσπερ
οὖν τοῖς νεφροῖς ἐνεργείας οὔσης ἕλκειν τὰ
οὖρα κακῶς ἕλκειν ὑπάρχει κακοπραγοῦσιν,
οὕτω καὶ τῷ σπληνὶ ποιότητος μελαγχολικῆς
ἑλκτικὴν ἐν ἑαυτῷ δύναμιν ἔχοντι σύμφυτον
ἀρρωστήσαντί ποτε ταύτην ἀναγκαῖον ἕλκειν
κακῶς κἂν τῷδε παχύτερον ἤδη καὶ μελάντερον
γίγνεσθαι τὸ αἷμα.

Ταῦτ' οὖν ἅπαντα πρός τε τὰς διαγνώσεις
τῶν νοσημάτων καὶ τὰς ἰάσεις μεγίστην παρεχό-
134 μενα χρείαν ‖ ὑπερεπήδησε τελέως ὁ Ἐρασίστρατος καὶ καταφρονεῖν προσεποιήσατο τηλι-
κούτων ἀνδρῶν ὁ μηδὲ τῶν τυχόντων καταφρονῶν
ἀλλ' ἀεὶ φιλοτίμως ἀντιλέγων ταῖς ἠλιθιωτάταις
δόξαις· ᾧ καὶ δῆλον, ὡς οὐδὲν ἔχων οὔτ'
ἀντειπεῖν τοῖς πρεσβυτέροις ὑπὲρ ὧν ἀπεφήναντο
περὶ σπληνὸς ἐνεργείας τε καὶ χρείας οὔτ' αὐτὸς
ἐξευρίσκων τι καινὸν εἰς τὸ μηδὲν ὅλως εἰπεῖν
ἀφίκετο. ἀλλ' ἡμεῖς γε πρῶτον μὲν ἐκ τῶν
αἰτίων, οἷς ἅπαντα διοικεῖται τὰ κατὰ τὰς

---

[1] Enlargement and suppuration (?) of spleen associated
with toxaemia or "cacochymy."　　[2] Lit. "melancholic."

internal suppuration, it destroys the body and fills it with evil humours;[1] this again is agreed on, not only by Hippocrates, but also by Plato and many others, including the Empiric physicians. And the jaundice which occurs when the spleen is out of order is darker in colour, and the cicatrices of ulcers are dark. For, generally speaking, when the spleen is drawing the atrabiliary[2] humour into itself to a less degree than is proper, the blood is unpurified, and the whole body takes on a bad colour. And when does it draw this in to a less degree than proper? Obviously, when it [the spleen] is in a bad condition. Thus, just as the kidneys, whose function it is to attract the urine, do this badly when they are out of order, so also the spleen, which has in itself a native power of attracting an atrabiliary quality,[3] if it ever happens to be weak, must necessarily exercise this attraction badly, with the result that the blood becomes thicker and darker.

Now all these points, affording as they do the greatest help in the diagnosis and in the cure of disease were entirely passed over by Erasistratus, and he pretended to despise these great men—he who does not despise ordinary people, but always jealously attacks the most absurd doctrines. Hence, it was clearly because he had nothing to say against the statements made by the ancients regarding the function and utility of the spleen, and also because he could discover nothing new himself, that he ended by saying nothing at all. I, however, for my part, have demonstrated, firstly from the *causes* by which everything throughout nature is governed (by

[3] *i.e.* the combination of sensible qualities which we call black bile. *cf.* p. 8, note 3.

GALEN

φύσεις, τοῦ θερμοῦ λέγω καὶ ψυχροῦ καὶ ξηροῦ
καὶ ὑγροῦ, δεύτερον δ' ἐξ αὐτῶν τῶν ἐναργῶς
φαινομένων κατὰ τὸ σῶμα ψυχρὸν καὶ ξηρὸν
εἶναί τινα χρῆναι χυμὸν ἀπεδείξαμεν. ἑξῆς δ',
ὅτι καὶ μελαγχολικὸς οὗτος ὑπάρχει καὶ τὸ
καθαῖρον αὐτὸν σπλάγχνον ὁ σπλήν ἐστιν, διὰ
βραχέων ὡς ἔνι μάλιστα τῶν τοῖς παλαιοῖς ἀπο-
δεδειγμένων ἀναμνήσαντες ἐπὶ τὸ λεῖπον ἔτι τοῖς
παροῦσι λόγοις ἀφιξόμεθα.

Τί δ' ἂν εἴη λεῖπον ἄλλο γ' ἢ ἐξηγήσασθαι
135 σαφῶς, οἷόν τι βούλονταί τε ‖ καὶ ἀποδεικνύουσι
περὶ τὴν τῶν χυμῶν γένεσιν οἱ παλαιοὶ συμβαίνειν. ἐναργέστερον δ' ἂν γνωσθείη διὰ παρα-
δείγματος. οἶνον δή μοι νόει γλεύκινον οὐ πρὸ
πολλοῦ τῶν σταφυλῶν ἐκτεθλιμμένον ζέοντά τε
καὶ ἀλλοιούμενον ὑπὸ τῆς ἐν αὐτῷ θερμασίας·
ἔπειτα κατὰ τὴν αὐτοῦ μεταβολὴν δύο γεννώμενα
περιττώματα τὸ μὲν κουφότερόν τε καὶ ἀερω-
δέστερον, τὸ δὲ βαρύτερόν τε καὶ γεωδέστερον,
ὧν τὸ μὲν ἄνθος, οἶμαι, τὸ δὲ τρύγα καλοῦσι.
τούτων τῷ μὲν ἑτέρῳ τὴν ξανθὴν χολήν, τῷ δ'
ἑτέρῳ τὴν μέλαιναν εἰκάζων οὐκ ἂν ἁμάρτοις, οὐ
τὴν αὐτὴν ἐχόντων ἰδέαν τῶν χυμῶν τούτων ἐν
τῷ κατὰ φύσιν διοικεῖσθαι τὸ ζῷον, οἵαν καὶ
παρὰ φύσιν ἔχοντος ἐπιφαίνονται πολλάκις.
ἡ μὲν γὰρ ξανθὴ λεκιθώδης γίγνεται· καὶ γὰρ
ὀνομάζουσιν οὕτως αὐτήν, ὅτι ταῖς τῶν ᾠῶν
λεκίθοις ὁμοιοῦται κατά τε χρόαν καὶ πάχος.
ἡ δ' αὖ μέλαινα κακοηθέστερα μὲν πολὺ καὶ

the causes I mean the Warm, Cold, Dry and Moist)
and secondly, from obvious bodily phenomena, that
there must needs be a cold and dry humour.[1] And
having in the next place drawn attention to the fact
that this humour is black bile [atrabiliary] and that
the viscus which clears it away is the spleen—having
pointed this out by help of as few as possible of the
proofs given by ancient writers, I shall now proceed
to what remains of the subject in hand.

What else, then, remains but to explain clearly
what it is that happens in the generation of the
humours, according to the belief and demonstration
of the Ancients? This will be more clearly understood from a comparison. Imagine, then, some new
wine which has been not long ago pressed from the
grape, and which is fermenting and undergoing
alteration through the agency of its contained heat.[2]
Imagine next two residual substances produced
during this process of alteration, the one tending to
be light and air-like and the other to be heavy and
more of the nature of earth; of these the one, as I
understand, they call the *flower* and the other the
lees. Now you may correctly compare yellow bile to
the first of these, and black bile to the latter,
although these humours have not the same appearance when the animal is in normal health as that
which they often show when it is not so; for then
the yellow bile becomes *vitelline*,[3] being so termed
because it becomes like the yolk of an egg, both in
colour and density; and again, even the black bile
itself becomes much more malignant than when in

[1] Thus Galen has demonstrated the functions of the spleen
both deductively and inductively. For another example of
the combined method *cf.* Book III., chaps. i. and ii. ; *cf.* also
Introd. p. xxxii.   [2] *i.e.* its innate heat.   [3] Lit. *lecithoid.*

αὕτη τῆς κατὰ φύσιν· ὄνομα δ᾽ οὐδὲν ἴδιον κεῖται
τῷ τοιούτῳ χυμῷ, πλὴν εἴ πού τινες ἢ ξυστικὸν
ἢ ὀξώδη κεκλήκασιν αὐτόν, ὅτι καὶ δριμὺς ὁμοίως
136 ὄξει γίγνεται καὶ ‖ ξύει γε τὸ σῶμα τοῦ ζῴου
καὶ τὴν γῆν, εἰ κατ᾽ αὐτῆς ἐκχυθείη, καί τινα
μετὰ πομφολύγων οἷον ζύμωσίν τε καὶ ζέσιν
ἐργάζεται, σηπεδόνος ἐπικτήτου προσελθούσης
ἐκείνῳ τῷ κατὰ φύσιν ἔχοντι χυμῷ τῷ μέλανι.
καί μοι δοκοῦσιν οἱ πλεῖστοι τῶν παλαιῶν
ἰατρῶν αὐτὸ μὲν τὸ κατὰ φύσιν ἔχον τοῦ τοιούτου
χυμοῦ καὶ διαχωροῦν κάτω καὶ πολλάκις ἐπι-
πολάζον ἄνω μέλανα καλεῖν χυμόν, οὐ μέλαιναν
χολήν, τὸ δ᾽ ἐκ συγκαύσεώς τινος καὶ σηπεδόνος
εἰς τὴν ὀξεῖαν μεθιστάμενον ποιότητα μέλαιναν
ὀνομάζειν χολήν. ἀλλὰ περὶ μὲν τῶν ὀνομάτων οὐ χρὴ διαφέρεσθαι, τὸ δ᾽ ἀληθὲς ὧδ᾽ ἔχον
εἰδέναι.

Κατὰ τὴν τοῦ αἵματος γένεσιν ὅσον ἂν ἱκανῶς
παχὺ καὶ γεῶδες ἐκ τῆς τῶν σιτίων φύσεως
ἐμφερόμενον τῇ τροφῇ μὴ δέξηται καλῶς τὴν
ἐκ τῆς ἐμφύτου θερμασίας ἀλλοίωσιν, ὁ σπλὴν
εἰς ἑαυτὸν ἕλκει τοῦτο. τὸ δ᾽ ὀπτηθέν, ὡς ἄν
τις εἴποι, καὶ συγκαυθὲν τῆς τροφῆς, εἴη δ᾽ ἂν
τοῦτο τὸ θερμότατον ἐν αὐτῇ καὶ γλυκύτατον,
οἷον τό τε μέλι καὶ ἡ πιμελή, ξανθὴ γενόμενον
χολὴ διὰ τῶν χοληδόχων ὀνομαζομένων ἀγγείων
137 ἐκκαθαίρεται. ‖ λεπτὸν δ᾽ ἐστὶ τοῦτο καὶ ὑγρὸν
καὶ ῥυτὸν οὐχ ὥσπερ ὅταν ὀπτηθὲν ἐσχάτως
ξανθὸν καὶ πυρῶδες καὶ παχὺ γένηται ταῖς τῶν

---

[1] Note that there can be "normal" black bile.
[2] The term *food* here means the food as introduced into
the stomach; the term *nutriment* (*trophé*) means the same

its normal condition,[1] but no particular name has been given to [such a condition of] the humour, except that some people have called it *corrosive* or *acetose*, because it also becomes sharp like vinegar and corrodes the animal's body—as also the earth, if it be poured out upon it—and it produces a kind of fermentation and seething, accompanied by bubbles— an abnormal putrefaction having become added to the natural condition of the black humour. It seems to me also that most of the ancient physicians give the name *black humour* and not *black bile* to the normal portion of this humour, which is discharged from the bowel and which also frequently rises to the top [of the stomach-contents]; and they call *black bile* that part which, through a kind of combustion and putrefaction, has had its quality changed to acid. There is no need, however, to dispute about names, but we must realise the facts, which are as follow :—

In the genesis of blood, everything in the nutri- ment[2] which belongs naturally to the thick and earth-like part of the food,[2] and which does not take on well the alteration produced by the innate heat—all this the spleen draws into itself. On the other hand, that part of the nutriment which is roasted, so to speak, or burnt (this will be the warmest and sweetest part of it, like honey and fat), becomes *yellow bile*, and is cleared away through the so-called biliary[3] vessels; now, this is thin, moist, and fluid, not like what it is when, having been roasted to an *excessive* degree, it becomes yellow, fiery, and thick, like the yolk of

food in the digested condition, as it is conveyed to the tissues. *cf.* pp. 41–43. Note idea of imperfectly oxidized material being absorbed by the spleen. *cf.* p. 214, note 1.

[3] Lit. *choledochous*, bile-receiving.

ὠῶν ὅμοιον λεκίθοις. τοῦτο μὲν γὰρ ἤδη παρὰ
φύσιν· θάτερον δὲ τὸ πρότερον εἰρημένον κατὰ
φύσιν ἐστίν· ὥσπερ γε καὶ τοῦ μέλανος χυμοῦ
τὸ μὲν μήπω τὴν οἷον ζέσιν τε καὶ ζύμωσιν τῆς
γῆς ἐργαζόμενον κατὰ φύσιν ἐστί, τὸ δ' εἰς
τοιαύτην μεθιστάμενον ἰδέαν τε καὶ δύναμιν ἤδη
παρὰ φύσιν, ὡς ἂν τὴν ἐκ τῆς συγκαύσεως τοῦ
παρὰ φύσιν θερμοῦ προσειληφὸς δριμύτητα καὶ
οἷον τέφρα τις ἤδη γεγονός. ὡδέ πως καὶ ἡ
κεκαυμένη τρὺξ τῆς ἀκαύστου διήνεγκε. θερμὸν
γάρ τι χρῆμα αὕτη γ' ἱκανῶς ἐστιν, ὥστε καίειν
τε καὶ τήκειν καὶ διαφθείρειν τὴν σάρκα. τῇ δ'
ἑτέρᾳ τῇ μήπω κεκαυμένῃ τοὺς ἰατροὺς ἔστιν
εὑρεῖν χρωμένους εἰς ὅσαπερ καὶ τῇ γῇ τῇ καλου-
μένῃ κεραμίτιδι καὶ τοῖς ἄλλοις, ὅσα ξηραίνειν θ'
ἅμα καὶ ψύχειν πέφυκεν.

Εἰς τὴν τῆς οὕτω συγκαυθείσης μελαίνης
χολῆς ἰδέαν καὶ ἡ λεκιθώδης ἐκείνη μεθίσταται
πολλάκις, ὅταν καὶ αὐτή ποθ' οἷον ὀπτηθεῖσα
138 τύχῃ πυρώδει θερμασίᾳ. τὰ δ' ἄλλα ‖ τῶν χολῶν
εἴδη σύμπαντα τὰ μὲν ἐκ τῆς τῶν εἰρημένων
κράσεως γίγνεται, τὰ δ' οἷον ὁδοί τινές εἰσι τῆς
τούτων γενέσεώς τε καὶ εἰς ἄλληλα μεταβολῆς.
διαφέρουσι δὲ τῷ τὰς μὲν ἀκράτους εἶναι καὶ
μόνας, τὰ δ' οἷον ὀρροῖς τισιν ἐξυγρασμένας. ἀλλ'
οἱ μὲν ὀρροὶ τῶν χυμῶν ἅπαντες περιττώματα
καὶ καθαρὸν αὐτῶν εἶναι δεῖται τοῦ ζῴου τὸ σῶμα.
τῶν δ' εἰρημένων χυμῶν ἐστί τις χρεία τῇ φύσει
καὶ τοῦ παχέος καὶ τοῦ λεπτοῦ καὶ καθαίρεται
πρός τε τοῦ σπληνὸς καὶ τῆς ἐπὶ τῷ ἥπατι
κύστεως τὸ αἷμα καὶ ἀποτίθεται τοσοῦτόν τε καὶ
τοιοῦτον ἑκατέρου μέρος, ὅσον καὶ οἷον, εἴπεο εἰς

eggs ; for this latter is already abnormal, while the previously mentioned state is natural. Similarly with the black humour: that which does not yet produce, as I say, this seething and fermentation on the ground, is natural, while that which has taken over this character and faculty is unnatural; it has assumed an acridity owing to the combustion caused by abnormal heat, and has practically become transformed into ashes.[1] In somewhat the same way burned lees differ from unburned. The former is a warm substance, able to burn, dissolve, and destroy the flesh. The other kind, which has not yet undergone combustion, one may find the physicians employing for the same purposes that one uses the so-called *potter's earth* and other substances which have naturally a combined drying and chilling action.

Now the vitelline bile also may take on the appearance of this combusted black bile, if ever it chance to be roasted, so to say, by fiery heat. And all the other forms of bile are produced, some from a blending of those mentioned, others being, as it were, transition-stages in the genesis of these or in their conversion into one another. And they differ in that those first mentioned are unmixed and unique, while the latter forms are diluted with various kinds of *serum*. And all the serums in the humours are waste substances, and the animal body needs to be purified from them. There is, however, a natural use for the humours first mentioned, both thick and thin; the blood is purified both by the spleen and by the bladder beside the liver, and a part of each of the two humours is put away, of such quantity and

---

[1] Thus *over-roasting*—shall we say excessive *oxidation?*—produces the abnormal forms of both black and yellow bile.

GALEN

ὅλον ἠνέχθη τοῦ ζῴου τὸ σῶμα, βλάβην ἄν τιν'
εἰργάσατο. τὸ γὰρ ἱκανῶς παχὺ καὶ γεῶδες καὶ
τελέως διαπεφευγὸς τὴν ἐν τῷ ἥπατι μεταβολὴν
ὁ σπλὴν εἰς ἑαυτὸν ἕλκει· τὸ δ' ἄλλο τὸ μετρίως
παχὺ σὺν τῷ κατειργάσθαι πάντη φέρεται. δεῖται
γὰρ ἐν πολλοῖς τοῦ ζῴου μορίοις παχύτητός τινος
139 τὸ αἷμα καθάπερ οἶμαι καὶ τῶν ‖ ἐμφερομένων
ἰνῶν. καὶ εἴρηται μὲν καὶ Πλάτωνι περὶ τῆς
χρείας αὐτῶν, εἰρήσεται δὲ καὶ ἡμῖν ἐν ἐκείνοις
τοῖς γράμμασιν, ἐν οἷς ἂν τὰς χρείας τῶν μορίων
διερχώμεθα· δεῖται δ' οὐχ ἥκιστα καὶ τοῦ ξανθοῦ
χυμοῦ τοῦ μήπω πυρώδους ἐσχάτως γεγενημένου
τὸ αἷμα καὶ τίς αὐτῷ καὶ ἡ παρὰ τοῦδε χρεία,
δι' ἐκείνων εἰρήσεται.

Φλέγματος δ' οὐδὲν ἐποίησεν ἡ φύσις ὄργανον
καθαρτικόν, ὅτι ψυχρὸν καὶ ὑγρόν ἐστι καὶ οἷον
ἡμίπεπτός τις τροφή. δεῖται τοίνυν οὐ κενοῦσθαι
τὸ τοιοῦτον ἀλλ' ἐν τῷ σώματι μένον ἀλλοιοῦσθαι.
τὸ δ' ἐξ ἐγκεφάλου καταρρέον περίττωμα τάχα
μὲν ἂν οὐδὲ φλέγμα τις ὀρθῶς ἀλλὰ βλένναν τε
καὶ κόρυζαν, ὥσπερ οὖν καὶ ὀνομάζεται, καλοίη.
εἰ δὲ μή, ἀλλ' ὅτι γε τῆς τούτου κενώσεως ὀρθῶς
ἡ φύσις προὐνοήσατο, καὶ τοῦτ' ἐν τοῖς περὶ
χρείας μορίων εἰρήσεται. καὶ γὰρ οὖν καὶ τὸ
κατά τε τὴν γαστέρα καὶ τὰ ἔντερα συνιστάμενον
φλέγμα ὅπως ἂν ἐκκενωθῇ καὶ αὐτὸ τάχιστά τε
καὶ κάλλιστα, τὸ παρεσκευασμένον τῇ φύσει
μηχάνημα δι' ἐκείνων εἰρήσεται καὶ αὐτὸ τῶν

---

[1] cf. p. 277, note 2.
[2] Timaeus, 82 c–d.
[3] cf. p. 90, note 1. The term "catarrh" refers to this
"running down," which was supposed to take place through

214

quality that, if it were carried all over the body, it would do a certain amount of harm. For that which is decidedly thick and earthy in nature, and has entirely escaped alteration in the liver, is drawn by the spleen into itself[1]; the other part which is only moderately thick, after being elaborated [in the liver], is carried all over the body. For the blood in many parts of the body has need of a certain amount of thickening, as also, I take it, of the *fibres* which it contains. And the use of these has been discussed by Plato,[2] and it will also be discussed by me in such of my treatises as may deal with the use of parts. And the blood also needs, not least, the yellow humour, which has as yet not reached the extreme stage of combustion; in the treatises mentioned it will be pointed out what purpose is subserved by this.

Now Nature has made no organ for clearing away *phlegm*, this being cold and moist, and, as it were, half-digested nutriment; such a substance, therefore, does not need to be evacuated, but remains in the body and undergoes *alteration* there. And perhaps one cannot properly give the name of *phlegm* to the surplus-substance which runs down from the brain,[3] but one should call it *mucus* [blenna] or *coryza*—as, in fact, it is actually termed; in any case it will be pointed out, in the treatise "On the Use of Parts," how Nature has provided for the evacuation of this substance. Further, the device provided by Nature which ensures that the phlegm which forms in the stomach and intestines may be evacuated in the most rapid and effective way possible—this also will be described in that com-

the pores of the cribriform plate of the ethmoid into the nose.

140 ὑπομνη‖μάτων. ὅσον οὖν ἐμφέρεται ταῖς φλεψὶ
φλέγμα χρήσιμον ὑπάρχον τοῖς ζῴοις, οὐδεμιᾶς
δεῖται κενώσεως. προσέχειν δὲ χρὴ κἀνταῦθα
τὸν νοῦν καὶ γιγνώσκειν, ὥσπερ τῶν χολῶν
ἑκατέρας τὸ μέν τι χρήσιμόν ἐστι καὶ κατὰ φύσιν
τοῖς ζῴοις, τὸ δ᾽ ἄχρηστόν τε καὶ παρὰ φύσιν,
οὕτω καὶ τοῦ φλέγματος, ὅσον μὲν ἂν ᾖ γλυκύ,
χρηστὸν εἶναι τοῦτο τῷ ζῴῳ καὶ κατὰ φύσιν, ὅσον
δ᾽ ὀξὺ καὶ ἁλμυρὸν ἐγένετο, τὸ μὲν ὀξὺ τελέως
ἠπεπτῆσθαι, τὸ δ᾽ ἁλμυρὸν διασεσῆφθαι. τελείαν
δ᾽ ἀπεψίαν φλέγματος ἀκούειν χρὴ τὴν τῆς
δευτέρας πέψεως δηλονότι τῆς ἐν φλεψίν· οὐ
γὰρ δὴ τῆς γε πρώτης τῆς κατὰ τὴν κοιλίαν·
ἢ οὐδ᾽ ἂν ἐγεγένητο τὴν ἀρχὴν χυμός, εἰ καὶ
ταύτην διεπεφεύγει.

Ταῦτ᾽ ἀρκεῖν μοι δοκεῖ περὶ γενέσεώς τε καὶ
διαφθορᾶς χυμῶν ὑπομνήματ᾽ εἶναι τῶν Ἱπποκράτει τε καὶ Πλάτωνι καὶ Ἀριστοτέλει καὶ
Πραξαγόρᾳ καὶ Διοκλεῖ καὶ πολλοῖς ἄλλοις τῶν
παλαιῶν εἰρημένων· οὐ γὰρ ἐδικαίωσα πάντα
μεταφέρειν εἰς τόνδε τὸν λόγον τὰ τελέως ἐκείνοις
γεγραμμένα. τοσοῦτον δὲ μόνον ὑπὲρ ἑκάστου
141 εἶπον, ὅσον ἐξορμήσει τε τοὺς ‖ ἐντυγχάνοντας,
εἰ μὴ παντάπασιν εἶεν σκαιοί, τοῖς τῶν παλαιῶν
ὁμιλῆσαι γράμμασι καὶ τὴν εἰς τὸ ῥᾷον αὐτοῖς
συνεῖναι βοήθειαν παρέξει. γέγραπται δέ που
καὶ δι᾽ ἑτέρου λόγου περὶ τῶν κατὰ Πραξαγόραν
τὸν Νικάρχου χυμῶν. εἰ γὰρ καὶ ὅτι μάλιστα

mentary. As to that portion of the phlegm which is carried in the veins, seeing that this is of service to the animal it requires no evacuation. Here too, then, we must pay attention and recognise that, just as in the case of each of the two kinds of bile, there is one part which is useful to the animal and in accordance with its nature, while the other part is useless and contrary to nature, so also is it with the phlegm ; such of it as is sweet is useful to the animal and according to nature, while, as to such of it as has become bitter or salt, that part which is bitter is completely undigested, while that part which is salt has undergone putrefaction. And the term "*complete indigestion*" refers of course to the second digestion—that which takes place in the veins ; it is not a failure of the first digestion—that in the alimentary canal—for it would not have become a humour at the outset if it had escaped this digestion also.

It seems to me that I have made enough reference to what has been said regarding the genesis and destruction of humours by Hippocrates, Plato, Aristotle, Praxagoras, and Diocles, and many others among the Ancients ; I did not deem it right to transport the whole of their final pronouncements into this treatise. I have said only so much regarding each of the humours as will stir up the reader, unless he be absolutely inept, to make himself familiar with the writings of the Ancients, and will help him to gain more easy access to them. In another treatise [1] I have written on the humours according to Praxagoras, son of Nicarchus ; although this authority makes as many as ten humours, not

[1] Now lost.

δέκα ποιεῖ χωρὶς τοῦ αἵματος, ἐνδέκατος γὰρ ἂν
εἴη χυμὸς αὐτὸ τὸ αἷμα, τῆς Ἱπποκράτους οὐκ
ἀποχωρεῖ διδασκαλίας. ἀλλ' εἰς εἴδη τινὰ καὶ
διαφορὰς τέμνει τοὺς ὑπ' ἐκείνου πρώτου πάντων
ἅμα ταῖς οἰκείαις ἀποδείξεσιν εἰρημένους χυμούς.

Ἐπαινεῖν μὲν οὖν χρὴ τούς τ' ἐξηγησαμένους
τὰ καλῶς εἰρημένα καὶ τοὺς εἴ τι παραλέλειπται
προστιθέντας· οὐ γὰρ οἷόν τε τὸν αὐτὸν ἄρξασθαί
τε καὶ τελειῶσαι· μέμφεσθαι δὲ τοὺς οὕτως
ἀταλαιπώρους, ὡς μηδὲν ὑπομένειν μαθεῖν τῶν
ὀρθῶς εἰρημένων, καὶ τοὺς εἰς τοσοῦτον φιλοτί-
μους, ὥστ' ἐπιθυμίᾳ νεωτέρων δογμάτων ἀεὶ
πανουργεῖν τι καὶ σοφίζεσθαι, τὰ μὲν ἑκόντας
παραλιπόντας, ὥσπερ Ἐρασίστρατος ἐπὶ τῶν
142 χυμῶν ἐποίησε, τὰ δὲ πα‖νούργως ἀντιλέγοντας,
ὥσπερ αὐτός θ' οὗτος καὶ ἄλλοι πολλοὶ τῶν
νεωτέρων.

Ἀλλ' οὗτος μὲν ὁ λόγος ἐνταυθοῖ τελευτάτω,
τὸ δ' ὑπόλοιπον ἅπαν ἐν τῷ τρίτῳ προσθήσω.

including the blood (the blood itself being an eleventh), this is not a departure from the teaching of Hippocrates; for Praxagoras divides into species and varieties the humours which Hippocrates first mentioned, with the demonstration proper to each.

Those, then, are to be praised who explain the points which have been duly mentioned, as also those who add what has been left out; for it is not possible for the same man to make both a beginning and an end. Those, on the other hand, deserve censure who are so impatient that they will not wait to learn any of the things which have been duly mentioned, as do also those who are so ambitious that, in their lust after novel doctrines, they are always attempting some fraudulent sophistry, either purposely neglecting certain subjects, as Erasistratus does in the case of the humours, or unscrupulously attacking other people, as does this same writer, as well as many of the more recent authorities.

But let this discussion come to an end here, and I shall add in the third book all that remains.

BOOK III

Γ

I

143 Ὅτι μὲν οὖν ἡ θρέψις ἀλλοιουμένου τε καὶ
ὁμοιουμένου γίγνεται τοῦ τρέφοντος τῷ τρεφομένῳ καὶ ὡς ἐν ἑκάστῳ τῶν τοῦ ζῴου μορίων
ἐστί τις δύναμις, ἣν ἀπὸ τῆς ἐνεργείας ἀλλοιω-
τικὴν μὲν κατὰ γένος, ὁμοιωτικὴν δὲ καὶ θρεπτικὴν κατ᾽ εἶδος ὀνομάζομεν, ἐν τῷ πρόσθεν
δεδήλωται λόγῳ. τὴν δ᾽ εὐπορίαν τῆς ὕλης, ἣν
τροφὴν ἑαυτῷ ποιεῖται τὸ τρεφόμενον, ἐξ ἑτέρας
τινὸς ἔχειν ἐδείκνυτο δυνάμεως ἐπισπᾶσθαι
πεφυκυίας τὸν οἰκεῖον χυμόν, εἶναι δ᾽ οἰκεῖον
144 ἑκάστῳ τῶν μορίων χυμόν, ὃς ἂν ‖ ἐπιτήδειος
εἰς τὴν ἐξομοίωσιν ᾖ, καὶ τὴν ἕλκουσαν αὐτὸν
δύναμιν ἀπὸ τῆς ἐνεργείας ἑλκτικήν τέ τινα καὶ
ἐπισπαστικὴν ὀνομάζεσθαι. δέδεικται δὲ καί, ὡς
πρὸ μὲν τῆς ὁμοιώσεως ἡ πρόσφυσίς ἐστιν,
ἐκείνης δ᾽ ἔμπροσθεν ἡ πρόσθεσις γίγνεται, τέλος,
ὡς ἂν εἴποι τις, οὖσα τῆς κατὰ τὴν ἐπισπαστικὴν
δύναμιν ἐνεργείας. αὐτὸ μὲν γὰρ τὸ παράγεσθαι
τὴν τροφὴν ἐκ τῶν φλεβῶν εἰς ἕκαστον τῶν
μορίων τῆς ἑλκτικῆς ἐνεργούσης γίγνεται δυνά-

---

[1] "Of food to feeder," *i.e.* of the environment to the
organism. *cf.* p. 39, chap. xi.

[2] "Drawing"; *cf.* p. 116, note 2.

BOOK III

I

It has been made clear in the preceding discussion that nutrition occurs by an *alteration* or *assimilation* of that which nourishes to that which receives nourishment,[1] and that there exists in every part of the animal a faculty which in view of its activity we call, in general terms, *alterative,* or, more specifically, *assimilative* and *nutritive.* It was also shown that a sufficient supply of the matter which the part being nourished makes into nutriment for itself is ensured by virtue of another faculty which naturally attracts its *proper juice* [humour] that that juice is proper to each part which is adapted for assimilation, and that the faculty which attracts the juice is called, by reason of its activity, *attractive* or *epispastic.*[2] It has also been shown that assimilation is preceded by *adhesion,* and this, again, by *presentation,*[3] the latter stage being, as one might say, the end or goal of the attractive activity corresponding to the attractive faculty. For the actual bringing up of nutriment from the veins into each of the parts takes place through the activation of the attractive faculty,[4] whilst to

[3] For these terms (*prosthesis* and *prosphysis* in Greek) *cf.* p. 39, notes 5 and 6.
[4] Lit. "through the *energizing* (or *functioning*) of the attractive faculty"; the faculty (δύναμις) *in operation* is an activity (ἐνέργεια). *cf.* p. 3, note 2.

μεως, τὸ δ' ἤδη παρῆχθαί τε καὶ προστίθεσθαι
τῷ μορίῳ τὸ τέλος ἐστὶν αὐτό, δι' ὃ καὶ τῆς τοιαύτης ἐνεργείας ἐδεήθημεν· ἵνα γὰρ προστεθῇ, διὰ
τοῦθ' ἕλκεται. χρόνου δ' ἐντεῦθεν ἤδη πλείονος
εἰς τὴν θρέψιν τοῦ ζῴου δεῖ· ἑλχθῆναι μὲν γὰρ
καὶ διὰ ταχέων τι δύναται, προσφῦναι δὲ καὶ
ἀλλοιωθῆναι καὶ τελέως ὁμοιωθῆναι τῷ τρεφομένῳ καὶ μέρος αὐτοῦ γενέσθαι παραχρῆμα μὲν
οὐχ οἷόν τε, χρόνῳ δ' ἂν πλείονι συμβαίνοι
καλῶς. ἀλλ' εἰ μὴ μένοι κατὰ τὸ μέρος ὁ προσ-
τεθεὶς οὗτος χυμός, εἰς ἕτερον δέ τι μεθίσταιτο
καὶ παραρρέοι διὰ παντὸς ἀμείβων τε καὶ ὑπαλ-
145 λάττων τὰ χωρία, κατ' οὐδὲν αὐτῶν ‖ οὔτε πρόσφυσις οὔτ' ἐξομοίωσις ἔσται. δεῖ δὲ κἀνταῦθά
τινος τῇ φύσει δυνάμεως ἑτέρας εἰς πολυχρόνιον
μονὴν τοῦ προστεθέντος τῷ μορίῳ χυμοῦ καὶ
ταύτης οὐκ ἔξωθέν ποθεν ἐπιρρεούσης ἀλλ' ἐν
αὐτῷ τῷ θρεψομένῳ κατῳκισμένης, ἣν ἀπὸ τῆς
ἐνεργείας πάλιν οἱ πρὸ ἡμῶν ἠναγκάσθησαν ὀνο-
μάσαι καθεκτικήν.

Ὁ μὲν δὴ λόγος ἤδη σαφῶς ἐνεδείξατο τὴν
ἀνάγκην τῆς γενέσεως τῆς τοιαύτης δυνάμεως καὶ
ὅστις ἀκολουθίας σύνεσιν ἔχει, πέπεισται βε-
βαίως ἐξ ὧν εἴπομεν, ὡς ὑποκειμένου τε καὶ
προαποδεδειγμένου τοῦ τεχνικὴν εἶναι τὴν φύσιν
καὶ τοῦ ζῴου κηδεμονικὴν ἀναγκαῖον ὑπάρχειν
αὐτῇ καὶ τὴν τοιαύτην δύναμιν.

have been finally brought up and presented to the part is the actual end for which we desired such an activity; it is attracted in order that it may be presented. After this, considerable time is needed for the nutrition of the animal; whilst a thing may be even rapidly attracted, on the other hand to become adherent, altered, and entirely assimilated to the part which is being nourished and to become a part of it, cannot take place suddenly, but requires a considerable amount of time. But if the nutritive juice, so presented, does not remain in the part, but withdraws to another one, and keeps flowing away, and constantly changing and shifting its position, neither adhesion nor complete assimilation will take place in any of them. Here too, then, the [animal's] nature has need of some other faculty for ensuring a prolonged stay of the presented juice at the part, and this not a faculty which comes in from somewhere outside but one which is resident in the part which is to be nourished. This faculty, again, in view of its activity our predecessors were obliged to call *retentive*.

Thus our argument has clearly shown[1] the necessity for the genesis of such a faculty, and whoever has an appreciation of logical sequence must be firmly persuaded from what we have said that, if it be laid down and proved by previous demonstration that Nature is artistic and solicitous for the animal's welfare, it necessarily follows that she must also possess a faculty of this kind.

[1] This chapter is an excellent example of Galen's method of reasoning *a priori*. The complementary inductive method, however, is employed in the next chapter. *cf.* p. 209, note 1.

II

Ἀλλ' ἡμεῖς οὐ τούτῳ μόνῳ τῷ γένει τῆς ἀποδείξεως εἰθισμένοι χρῆσθαι, προστιθέντες δ' αὐτῷ καὶ τὰς ἐκ τῶν ἐναργῶς φαινομένων ἀναγκαζούσας τε καὶ βιαζομένας πίστεις ἐπὶ τὰς τοιαύτας καὶ νῦν ἀφιξόμεθα καὶ δείξομεν ἐπὶ μέν τινων μορίων τοῦ σώματος οὕτως ἐναργῆ τὴν καθεκτικὴν δύ-
146 ναμιν, ὡς αὐταῖς ταῖς αἰσθήσεσι ‖ διαγιγνώσκεσθαι τὴν ἐνέργειαν αὐτῆς, ἐπὶ δέ τινων ἧττον μὲν ἐναργῶς ταῖς αἰσθήσεσι, λόγῳ δὲ κἀνταῦθα φωραθῆναι δυναμένην.

Ἀρξώμεθ' οὖν τῆς διδασκαλίας ἀπ' αὐτοῦ τοῦ τέως πρῶτον μεθόδῳ τινὶ προχειρίσασθαι μόρι' ἄττα τοῦ σώματος, ἐφ' ὧν ἀκριβῶς ἔστι βασανίσαι τε καὶ ζητῆσαι τὴν καθεκτικὴν δύναμιν ὁποία ποτ' ἐστίν.

Ἀρ' οὖν ἄμεινον ἄν τις ἑτέρωθεν ἢ ἀπὸ τῶν μεγίστων τε καὶ κοιλοτάτων ὀργάνων ὑπάρξαιτο τῆς ζητήσεως; ἐμοὶ μὲν οὖν οὐκ ἂν δοκεῖ βέλτιον. ἐναργεῖς γοῦν εἰκὸς ἐπὶ τούτων φανῆναι τὰς ἐνεργείας διὰ τὸ μέγεθος· ὡς τά γε σμικρὰ τάχ' ἄν, εἰ καὶ σφοδρὰν ἔχει τὴν τοιαύτην δύναμιν, ἀλλ' οὐκ αἰσθήσει γ' ἑτοίμην διαγιγνώσκεσθαι τὴν ἐνέργειαν αὐτῆς.

Ἀλλ' ἔστιν ἐν τοῖς μάλιστα κοιλότατα καὶ μέγιστα τῶν τοῦ ζῴου μορίων ἥ τε γαστὴρ καὶ <αἱ> μῆτραί τε καὶ ὑστέραι καλούμεναι. τί οὖν κωλύει ταῦτα πρῶτα προχειρισαμένους ἐπισκέψασθαι τὰς ἐνεργείας αὐτῶν, ὅσαι μὲν καὶ πρὸ τῆς ἀνατομῆς

---

[1] The deductive.
[2] The *logos* is the argument or " theory " arrived at by the

II

SINCE, however, it is not our habit to employ this kind of demonstration[1] alone, but to add thereto cogent and compelling proofs drawn from obvious facts, we will also proceed to the latter kind in the present instance : we will demonstrate that in certain parts of the body *the retentive faculty* is so obvious that its operation can be actually recognised by the *senses*, whilst in other parts it is less obvious to the senses, but is capable even here of being detected by the *argument.*[2]

Let us begin our exposition, then, by first dealing systematically for a while with certain definite parts of the body, in reference to which we may accurately test and enquire what sort of thing the retentive faculty is.

Now, could one begin the enquiry in any better way than with the largest and hollowest organs ? Personally I do not think one could. It is to be expected that in these, owing to their size, the activities will show quite clearly, whereas with respect to the small organs, even if they possess a strong faculty of this kind, its activation will not at once be recognisable to sense.

Now those parts of the animal which are especially hollow and large are the stomach and the organ which is called the womb or uterus.[3] What prevents us, then, from taking up these first and considering their activities, conducting the enquiry on our own

process of λογικὴ θεωρία or "theorizing"; *cf.* p. 151, note 3; p. 205, note 1.

[3] The Greek words for the uterus (*mêtrae* and *hysterae*) probably owe their plural form to the belief that the organ was bicornuate in the human, as it is in some of the lower species. **227**

δῆλαι, τὴν ἐξέτασιν ἐφ᾽ ἡμῶν αὐτῶν ποιουμένους, ὅσαι δ᾽ ἀμυδρότεραι, τὰ παραπλήσια διαιροῦντας 147 ἀνθρώπῳ ζῷα, ‖ οὐχ ὡς οὐκ ἂν ἱκανῶς τό γε καθόλου περὶ τῆς ζητουμένης δυνάμεως καὶ τῶν ἀνομοίων ἐνδειξομένων, ἀλλ᾽ ὡς ἵν᾽ ἅμα τῷ κοινῷ καὶ τὸ ἴδιον ἐφ᾽ ἡμῶν αὐτῶν ἐγνωκότες εἴς τε τὰς διαγνώσεις τῶν νοσημάτων καὶ τὰς ἰάσεις εὐπορώτεροι γιγνώμεθα.

Περὶ μὲν οὖν ἀμφοτέρων τῶν ὀργάνων ἅμα λέγειν ἀδύνατον, ἐν μέρει δ᾽ ὑπὲρ ἑκατέρου ποιησόμεθα τὸν λόγον ἀπὸ τοῦ σαφέστερον ἐνδείξασθαι δυναμένου τὴν καθεκτικὴν δύναμιν ἀρξάμενοι. κατέχει μὲν γὰρ καὶ ἡ γαστὴρ τὰ σιτία, μέχρι περ ἂν ἐκπέψῃ, κατέχουσι δὲ καὶ αἱ μῆτραι τὸ ἔμβρυον, ἔστ᾽ ἂν τελειώσωσιν· ἀλλὰ πολλαπλάσιός ἐστιν ὁ τῆς τῶν ἐμβρύων τελειώσεως χρόνος τῆς τῶν σιτίων πέψεως.

III

Εἰκὸς οὖν καὶ τὴν δύναμιν ἐναργέστερον ἐν ταῖς μήτραις φωράσειν ἡμᾶς τὴν καθεκτικήν, ὅσῳ καὶ πολυχρονιωτέραν τῆς γαστρὸς τὴν ἐνέργειαν κέκτηται. μησὶ γὰρ ἐννέα που ταῖς πλείσταις τῶν γυναικῶν ἐν αὐταῖς τελειοῦται τὰ κυήματα, μεμυκυίαις μὲν ἅπαντι τῷ αὐχένι, περιεχούσαις δὲ πανταχόθεν αὐτὰ σὺν τῷ χορίῳ. ‖ 148 καὶ πέρας γε τῆς τοῦ στόματος μύσεως καὶ τῆς τοῦ κυουμένου κατὰ τὰς μήτρας μονῆς ἡ χρεία τῆς ἐνεργείας ἐστίν· οὐ γὰρ ὡς ἔτυχεν οὐδ᾽ ἀλόγως ἱκανὰς περιστέλλεσθαι καὶ κατέχειν τὸ

persons in regard to those activities which are obvious
without dissection, and, in the case of those which
are more obscure, dissecting animals which are near
to man ; [1] not that even animals unlike him will
not show, in a general way, the faculty in question,
but because in this manner we may find out at once
what is common to all and what is peculiar to ourselves, and so may become more resourceful in the
diagnosis and treatment of disease.

Now it is impossible to speak of both organs at
once, so we shall deal with each in turn, beginning
with the one which is capable of demonstrating the
retentive faculty most plainly. For the stomach
retains the food until it has quite digested it, and
the uterus retains the embryo until it brings it to
completion, but the time taken for the completion of
the embryo is many times more than that for the
digestion of food.

III

WE may expect, then, to detect the retentive
faculty in the uterus more clearly in proportion to
the longer duration of its activity as compared with
that of the stomach. For, as we know, it takes nine
months in most women for the foetus to attain
maturity in the womb, this organ having its neck
quite closed, and entirely surrounding the embryo
together with the *chorion.* Further, it is the utility
of the function which determines the closure of the os
and the stay of the foetus in the uterus. For it is
not casually nor without reason that Nature has made

[1] Note this expression. For Galen's views on the origin
of species, *cf.* Introduction, p. xxxi., footnote.

ἔμβρυον ἡ φύσις ἀπείργασατο τὰς ὑστέρας, ἀλλ᾽
ἵν᾽ εἰς τὸ πρέπον ἀφίκηται μέγεθος τὸ κυούμενον.
ὅταν οὖν, οὗ χάριν ἐνήργουν τῇ καθεκτικῇ δυνάμει, συμπεπληρωμένον ᾖ, ταύτην μὲν ἀνέπαυσάν
τε καὶ εἰς ἠρεμίαν ἐπανήγαγον, ἀντ᾽ αὐτῆς δ᾽
ἑτέρᾳ χρῶνται τῇ τέως ἡσυχαζούσῃ, τῇ προωσ-
τικῇ. ἦν δ᾽ ἄρα καὶ τῆς ἐκείνης ἡσυχίας ὅρος ἡ
χρεία καὶ τῆς γ᾽ ἐνεργείας ὡσαύτως ἡ χρεία·
καλούσης μὲν γὰρ αὐτῆς ἐνεργεῖ, μὴ καλούσης δ᾽
ἡσυχάζει.

Καὶ χρὴ πάλιν κἀνταῦθα καταμαθεῖν τῆς
φύσεως τὴν τέχνην, ὡς οὐ μόνον ἐνεργειῶν χρη-
σίμων δυνάμεις ἐνέθηκεν ἑκάστῳ τῶν ὀργάνων,
ἀλλὰ καὶ τοῦ τῶν ἡσυχιῶν τε καὶ κινήσεων
καιροῦ προὐνοήσατο. καλῶς μὲν γὰρ ἁπάντων
γιγνομένων τῶν κατὰ τὴν κύησιν ἡ ἀποκριτικὴ
δύναμις ἡσυχάζει τελέως ὥσπερ οὐκ οὖσα, κα-
κοπραγίας δέ τινος γενομένης ἢ περὶ τὸ χορίον ἢ
149 περί τινα τῶν ἄλλων ‖ ὑμένων ἢ περὶ τὸ κυούμενον αὐτὸ καὶ τῆς τελειώσεως αὐτοῦ παντάπασιν
ἀπογνωσθείσης οὐκέτ᾽ ἀναμένουσι τὸν ἐννεάμηνον
αἱ μῆτραι χρόνον, ἀλλ᾽ ἡ μὲν καθεκτικὴ δύναμις
αὐτίκα δὴ πέπαυταί τε καὶ παραχωρεῖ κινεῖσθαι τῇ
πρότερον ἀργούσῃ, πράττει δ᾽ ἤδη τι καὶ πραγμα-
τεύεται χρηστὸν ἡ ἀποκριτική τε καὶ προωστική·
καὶ γὰρ οὖν καὶ ταύτην οὕτως ἐκάλεσαν ἀπὸ τῶν
ἐνεργειῶν αὐτῇ τὰ ὀνόματα θέμενοι καθάπερ καὶ
ταῖς ἄλλαις.

Καί πως ὁ λόγος ἔοικεν ὑπὲρ ἀμφοτέρων
ἀποδείξειν ἅμα· καὶ γάρ τοι καὶ διαδεχομένας
αὐτὰς ἀλλήλας καὶ παραχωροῦσαν ἀεὶ τὴν
ἑτέραν τῇ λοιπῇ, καθότι ἂν ἡ χρεία κελεύῃ, καὶ

the uterus capable of contracting upon, and of retaining the embryo, but in order that the latter may
arrive at a proper size. When, therefore, the object
for which the uterus brought its retentive faculty
into play has been fulfilled, it then stops this faculty
and brings it back to a state of rest, and employs
instead of it another faculty hitherto quiescent—the
propulsive faculty. In this case again the quiescent
and active states are both determined by utility;
when this calls, there is activity; when it does
not, there is rest.

Here, then, once more, we must observe well
the Art [artistic tendency] of Nature—how she has
not merely placed in each organ the capabilities
of useful activities, but has also fore-ordained the
times both of rest and movement. For when everything connected with the pregnancy proceeds properly, the *eliminative* faculty remains quiescent as
though it did not exist, but if anything goes wrong
in connection either with the chorion or any of
the other membranes or with the foetus itself, and
its completion is entirely despaired of, then the
uterus no longer awaits the nine-months period, but
the retentive faculty forthwith ceases and allows the
heretofore inoperative faculty to come into action.
Now it is that something is done—in fact, useful
work effected—by the *eliminative or propulsive faculty*
(for so it, too, has been called, receiving, like the
rest, its names from the corresponding activities).

Further, our theory can, I think, demonstrate both
together; for seeing that they succeed each other,
and that the one keeps giving place to the other
according as utility demands, it seems not unreason-

τὴν διδασκαλίαν κοινὴν οὐκ ἀπεικός ἐστι δέχε-
σθαι. τῆς μὲν οὖν καθεκτικῆς δυνάμεως ἔργον
περιστεῖλαι τὰς μήτρας τῷ κυουμένῳ πανταχόθεν, ὥστ' εὐλόγως ἁπτομέναις μὲν ταῖς μαιευ-
τρίαις τὸ στόμα μεμυκὸς αὐτῶν φαίνεται, ταῖς
κυούσαις δ' αὐταῖς κατὰ τὰς πρώτας ἡμέρας καὶ
μάλιστα κατ' αὐτὴν ἐκείνην, ἐν ᾗπερ ἂν ἡ τῆς
γονῆς σύλληψις γένηται, κινουμένων τε καὶ συν-
150 τρεχουσῶν εἰς ἑαυτὰς τῶν ὑστερῶν αἴσθη‖σις γίγνεται καὶ ἢν ἄμφω ταῦτα συμβῇ, μῦσαι μὲν τὸ
στόμα χωρὶς φλεγμονῆς ἤ τινος ἄλλου παθήματος, αἴσθησιν δὲ τῆς κατὰ τὰς μήτρας κινήσεως ἀκολουθῆσαι, πρὸς αὐτὰς ἤδη τὸ σπέρμα
τὸ παρὰ τἀνδρὸς εἰληφέναι τε καὶ κατέχειν αἱ
γυναῖκες νομίζουσι.

Ταῦτα δ' οὐχ ἡμεῖς νῦν ἀναπλάττομεν ἡμῖν
αὐτοῖς, ἀλλ' ἐκ μακρᾶς πείρας δοκιμασθέντα
πᾶσι γέγραπται σχεδόν τι τοῖς περὶ τούτων
πραγματευσαμένοις. Ἡρόφιλος μέν γε καὶ ὡς
οὐδὲ πυρῆνα μήλης ἂν δέχοιτο τῶν μητρῶν τὸ
στόμα, πρὶν ἀποκυεῖν τὴν γυναῖκα, καὶ ὡς οὐδὲ
τοὐλάχιστον ἔτι διέστηκεν, ἢν ὑπάρξηται κύειν,
καὶ ὡς ἐπὶ πλέον ἀναστομοῦνται κατὰ τὰς τῶν
ἐπιμηνίων φοράς, οὐκ ὤκνησε γράφειν· συνομο-
λογοῦσι δ' αὐτῷ καὶ οἱ ἄλλοι πάντες οἱ περὶ
τούτων πραγματευσάμενοι καὶ πρῶτός γ' ἁπάντων
ἰατρῶν τε καὶ φιλοσόφων Ἱπποκράτης ἀπεφήνατο
μύειν τὸ στόμα τῶν ὑστερῶν ἔν τε ταῖς κυήσεσι
καὶ ταῖς φλεγμοναῖς, ἀλλ' ἐν μὲν ταῖς κυήσεσιν
οὐκ ἐξιστάμενον τῆς φύσεως, ἐν δὲ ταῖς φλεγμοναῖς
σκληρὸν γιγνόμενον.

able to accept a common demonstration also for both. Thus it is the work of the retentive faculty to make the uterus contract upon the foetus at every point, so that, naturally enough, when the midwives palpate it, the os is found to be closed, whilst the pregnant women themselves, during the first days—and particularly on that on which conception takes place—experience a sensation as if the uterus were moving and contracting upon itself. Now, if both of these things occur—if the os closes apart from inflammation or any other disease, and if this is accompanied by a feeling of movement in the uterus—then the women believe that they have received the semen which comes from the male, and that they are retaining it.

Now we are not inventing this for ourselves : one may say the statement is based on prolonged experience of those who occupy themselves with such matters. Thus Herophilus[1] does not hesitate to state in his writings that up to the time of labour the os uteri will not admit so much as the tip of a probe, that it no longer opens to the slightest degree if pregnancy has begun—that, in fact, it dilates more widely at the times of the menstrual flow. With him are in agreement all the others who have applied themselves to this subject ; and particularly Hippocrates, who was the first of all physicians and philosophers to declare that the os uteri closes during pregnancy and inflammation, albeit in pregnancy it does not depart from its own nature, whilst in inflammation it becomes hard.

---

[1] Herophilus of Chalcedon (*circa* 300 B.C.) was, like Erasistratus, a representative of the anatomical school of Alexandria. His book on Midwifery was known for centuries. *cf.* Introduction, p. xii.

GALEN

Ἐπὶ δέ γε τῆς ἐναντίας τῆς ἐκκριτικῆς ἀνοίγνυται μὲν τὸ στόμα, προέρχεται δ' ὁ πυθμὴν ‖
151 ἅπας ὅσον οἷόν τ' ἐγγυτάτω τοῦ στόματος
ἀπωθούμενος ἔξω τὸ ἔμβρυον, ἅμα δ' αὐτῷ καὶ
τὰ συνεχῆ μέρη τὰ οἷον πλευρὰ τοῦ παντὸς
ὀργάνου συνεπιλαμβανόμενα τοῦ ἔργου θλίβει
τε καὶ προωθεῖ πᾶν ἔξω τὸ ἔμβρυον. καὶ πολλαῖς
τῶν γυναικῶν ὠδῖνες βίαιοι τὰς μήτρας ὅλας
ἐκπεσεῖν ἠνάγκασαν ἀμέτρως χρησαμέναις τῇ
τοιαύτῃ δυνάμει, παραπλησίου τινὸς γιγνομένου
τῷ πολλάκις ἐν πάλαις τισὶ καὶ φιλονεικίαις
συμβαίνοντι, ὅταν ἀνατρέψαι τε καὶ καταβαλεῖν
ἑτέρους σπεύδοντες αὐτοὶ συγκαταπέσωμεν.
οὕτω γὰρ καὶ αἱ μῆτραι τὸ ἔμβρυον ὠθοῦσαι
συνεξέπεσον ἐνίοτε καὶ μάλισθ', ὅταν οἱ πρὸς τὴν
ῥάχιν αὐτῶν σύνδεσμοι χαλαροὶ φύσει τυγχάνωσιν
ὄντες.

Ἔστι δὲ καὶ τοῦτο θαυμαστόν τι τῆς φύσεως
σόφισμα, τὸ ζῶντος μὲν τοῦ κυήματος ἀκριβῶς
πάνυ μεμυκέναι τὸ στόμα τῶν μητρῶν, ἀπο-
θανόντος δὲ παραχρῆμα διανοίγεσθαι τοσοῦτον,
ὅσον εἰς τὴν ἔξοδον αὐτοῦ διαφέρει. καὶ μέντοι
καὶ αἱ μαῖαι τὰς τικτούσας οὐκ εὐθὺς ἀνιστᾶσιν
οὐδ' ἐπὶ τὸν δίφρον καθίζουσιν, ἀλλ' ἅπτονται
152 πρότερον ἀνοιγομένου τοῦ στόματος ‖ κατὰ βραχὺ
καὶ πρῶτον μέν, ὥστε τὸν μικρὸν δάκτυλον
καθιέναι, διεστηκέναι φασίν, ἔπειτ' ἤδη καὶ
μεῖζον καὶ κατὰ βραχὺ δὴ πυνθανομένοις ἡμῖν
ἀποκρίνονται τὸ μέγεθος τῆς διαστάσεως ἐπαυ-
ξανόμενον. ὅταν δ' ἱκανὸν ᾖ πρὸς τὴν τοῦ
κυουμένου δίοδον, ἀνιστᾶσιν αὐτὰς καὶ καθίζουσι

In the case of the opposite (the eliminative) faculty, the os opens, whilst the whole fundus approaches as near as possible to the os, expelling the embryo as it does so; and along with the fundus the contiguous parts—which form as it were a girdle round the whole organ—co-operate in the work; they squeeze upon the embryo and propel it bodily outwards. And, in many women who exercise such a faculty immoderately, violent pains cause forcible prolapse of the whole womb; here almost the same thing happens as frequently occurs in wrestling-bouts and struggles, when in our eagerness to overturn and throw others we are ourselves upset along with them; for similarly when the uterus is forcing the embryo forward it sometimes becomes entirely prolapsed, and particularly when the ligaments connecting it with the spine happen to be naturally lax.[1]

A wonderful device of Nature's also is this—that, when the foetus is alive, the os uteri is closed with perfect accuracy, but if it dies, the os at once opens up to the extent which is necessary for the foetus to make its exit. The midwife, however, does not make the parturient woman get up at once and sit down on the [obstetric] chair, but she begins by palpating the os as it gradually dilates, and the first thing she says is that it has dilated " enough to admit the little finger," then that " it is bigger now," and as we make enquiries from time to time, she answers that the size of the dilatation is increasing. And when it is sufficient to allow of the transit of the foetus,[2] she then makes the patient get up from her bed and

---

[1] Relaxation of utero-sacral ligaments as an important predisposing cause of prolapsus uteri.
[2] That is, at the end of the first stage of labour.

καὶ προθυμεῖσθαι κελεύουσιν ἀπώσασθαι τὸ
παιδίον. ἔστι δ᾽ ἤδη τοῦτο τὸ ἔργον, ὃ παρ᾽
ἑαυτῶν αἱ κύουσαι προστιθέασιν, οὐκέτι τῶν
ὑστερῶν, ἀλλὰ τῶν κατ᾽ ἐπιγάστριον μυῶν, οἳ
πρὸς τὴν ἀποπάτησίν τε καὶ τὴν οὔρησιν ἡμῖν
συνεργοῦσιν.

IV

Οὕτω μὲν ἐπὶ τῶν μητρῶν ἐναργῶς αἱ δύο
φαίνονται δυνάμεις, ἐπὶ δὲ τῆς γαστρὸς ὧδε.
πρῶτον μὲν τοῖς κλύδωσιν, οἳ δὴ καὶ πεπίστευνται
τοῖς ἰατροῖς ἀρρώστου κοιλίας εἶναι συμπτώματα
καὶ κατὰ λόγον πεπίστευνται· ἐνίοτε μὲν γὰρ
ἐλάχιστα προσενηνεγμένων οὐ γίγνονται περι-
στελλομένης ἀκριβῶς αὐτοῖς τῆς γαστρὸς καὶ
σφιγγούσης πανταχόθεν, ἐνίοτε δὲ μεστὴ μὲν ἡ
153 γαστήρ ἐστιν, οἱ κλύ‖δωνες δ᾽ ὡς ἐπὶ κενῆς
ἐξακούονται. κατὰ φύσιν μὲν γὰρ ἔχουσα καὶ
χρωμένη καλῶς τῇ περισταλτικῇ δυνάμει, κἂν
ὀλίγον ᾖ τὸ περιεχόμενον, ἅπαν αὐτὸ περι-
λαμβάνουσα χώραν οὐδεμίαν ἀπολείπει κενήν,
ἀρρωστοῦσα δέ, καθότι ἂν ἀδυνατήσῃ περιλαβεῖν
ἀκριβῶς, ἐνταῦθ᾽ εὐρυχωρίαν τιν᾽ ἐργαζομένη
συγχωρεῖ τοῖς περιεχομένοις ὑγροῖς κατὰ τὰς
τῶν σχημάτων μεταλλαγὰς ἄλλοτ᾽ ἀλλαχόσε
μεταρρέουσι κλύδωνας ἀποτελεῖν.

Εὐλόγως οὖν, ὅτι μηδὲ πέψουσιν ἱκανῶς, οἱ ἐν
τῷδε τῷ συμπτώματι γενόμενοι προσδοκῶσιν· οὐ
γὰρ ἐνδέχεται πέψαι καλῶς ἄρρωστον γαστέρα.
τοῖς τοιούτοις δὲ καὶ μέχρι πλείονος ἐν αὐτῇ

sit on the chair, and bids her make every effort to expel the child. Now, this additional work which the patient does of herself is no longer the work of the uterus but of the epigastric* muscles, which also help us in defaecation and micturition.

IV

THUS the two faculties are clearly to be seen in the case of the uterus; in the case of the *stomach* they appear as follows:—Firstly in the condition of *gurgling*, which physicians are persuaded, and with reason, to be a symptom of weakness of the stomach; for sometimes when the very smallest quantity of food has been ingested this does not occur, owing to the fact that the stomach is contracting accurately upon the food and constricting it at every point; sometimes when the stomach is full the gurglings yet make themselves heard as though it were empty. For if it be in a natural condition, employing its contractile faculty in the ordinary way, then, even if its contents be very small, it grasps the whole of them and does not leave any empty space. When it is weak, however, being unable to lay hold of its contents accurately, it produces a certain amount of vacant space, and allows the liquid contents to flow about in different directions in accordance with its changes of shape, and so to produce gurglings.

Thus those who are troubled with this symptom expect, with good reason, that they will also be unable to digest adequately; proper digestion cannot take place in a weak stomach. In such people also, the mass of food may be plainly seen to remain

GALEN

φαίνεται παραμένον τὸ βάρος, ὡς ἂν καὶ βραδύ-
τερον πέττουσι. καὶ μὴν θαυμάσειεν ἄν τις ἐπ'
αὐτῶν τούτων μάλιστα τὸ πολυχρόνιον τῆς ἐν τῇ
γαστρὶ διατριβῆς οὐ τῶν σιτίων μόνον ἀλλὰ καὶ
τοῦ πόματος· οὐ γάρ, ὅπερ ἂν οἰηθείη τις, ὡς τὸ
τῆς γαστρὸς στόμα τὸ κάτω στενὸν ἱκανῶς
ὑπάρχον οὐδὲν παρίησι πρὶν ἀκριβῶς λειωθῆναι,
τοῦτ' αἴτιον ὄντως ἐστί. πολλὰ γοῦν πολλάκις
154 ὀπωρῶν ὀστᾶ μέγιστα καταπίνουσι ‖ πάμπολλοι
καί τις δακτύλιον χρυσοῦν ἐν τῷ στόματι φυ-
λάττων ἄκων κατέπιε καὶ ἄλλος τις νόμισμα καὶ
ἄλλος ἄλλο τι σκληρὸν καὶ δυσκατέργαστον,
ἀλλ' ὅμως ἅπαντες οὗτοι ῥᾳδίως ἀπεπάτησαν, ἃ
κατέπιον, οὐδενὸς αὐτοῖς ἀκολουθήσαντος συμ-
πτώματος. εἰ δέ γ' ἡ στενότης τοῦ πόρου τῆς
γαστρὸς αἰτία τοῦ μένειν ἐπὶ πλέον ἦν τοῖς
ἀτρίπτοις σιτίοις, οὐδὲν ἂν τούτων ποτὲ διεχώ-
ρησεν. ἀλλὰ καὶ τὸ τὰ πόματ' αὐτοῖς ἐν τῇ
γαστρὶ παραμένειν ἐπὶ πλεῖστον ἱκανὸν ἀπάγειν
τὴν ὑπόνοιαν τοῦ πόρου τῆς στενότητος· ὅλως
γάρ, εἴπερ ἦν ἐν τῷ κεχυλῶσθαι τὸ θᾶττον
ὑπιέναι, τά τε ῥοφήματ' ἂν οὕτω καὶ τὸ γάλα καὶ
ὁ τῆς πτισάνης χυλὸς αὐτίκα διεξῄει πᾶσιν.
ἀλλ' οὐχ ὧδ' ἔχει· τοῖς μὲν γὰρ ἀσθενέσιν ἐπὶ
πλεῖστον ἐμπλεῖ ταῦτα καὶ κλύδωνας ἐργάζεται
παραμένοντα καὶ θλίβει καὶ βαρύνει τὴν γαστέρα,
τοῖς δ' ἰσχυροῖς οὐ μόνον τούτων οὐδὲν συμβαίνει,
ἀλλὰ καὶ πολὺ πλῆθος ἄρτων καὶ κρεῶν ὑπο-
χωρεῖ ταχέως.

---

[1] The pylorus.
[2] "Chylosis," chylification. cf. p. 240, note 1.

an abnormally long time in the stomach, as would be natural if their digestion were slow. Indeed, the chief way in which these people will surprise one is in the length of time that not food alone but even fluids will remain in their stomachs. Now, the actual cause of this is not, as one would imagine, that the lower outlet of the stomach,[1] being fairly narrow, will allow nothing to pass before being reduced to a fine state of division. There are a great many people who frequently swallow large quantities of big fruit-stones; one person, who was holding a gold ring in his mouth, inadvertently swallowed it; another swallowed a coin, and various people have swallowed various hard and indigestible objects; yet all these people easily passed by the bowel what they had swallowed, without there being any subsequent symptoms. Now surely if narrowness of the gastric outlet were the cause of untriturated food remaining for an abnormally long time, none of these articles I have mentioned would ever have escaped. Furthermore, the fact that it is liquids which remain longest in these people's stomachs is sufficient to put the idea of narrowness of the outlet out of court. For, supposing a rapid descent were dependent upon emulsification,[2] then soups, milk, and barley-emulsion[3] would at once pass along in every case. But as a matter of fact this is not so. For in people who are extremely asthenic it is just these fluids which remain undigested, which accumulate and produce gurglings, and which oppress and overload the stomach, whereas in strong persons not merely do none of these things happen, but even a large quantity of bread or meat passes rapidly down.

[3] Lit. barley-"chyle," *i.e.* barley-water.

Οὐ μόνον δ' ἐκ τοῦ περιτετάσθαι τὴν γαστέρα
155 καὶ βαρύνεσθαι ‖ καὶ μεταρρεῖν ἄλλοτ' εἰς ἄλλα
μέρη μετὰ κλύδωνος τὸ παραμένειν ἐπὶ πλέον ἐν
αὐτῇ πάντως τοῖς οὕτως ἔχουσι τεκμήραιτ' ἄν
τις ἀλλὰ κἀκ τῶν ἐμέτων· ἔνιοι γὰρ οὐ μετὰ
τρεῖς ὥρας ἢ τέτταρας ἀλλὰ νυκτῶν ἤδη μέσων
παμπόλλου μεταξὺ χρόνου διελθόντος ἐπὶ ταῖς
προσφοραῖς ἀνήμεσαν ἀκριβῶς ἅπαντα τὰ ἐδη-
δεσμένα.

Καὶ μὲν δὴ καὶ ζῷον ὁτιοῦν ἐμπλήσας ὑγρᾶς
τροφῆς, ὥσπερ ἡμεῖς πολλάκις ἐπὶ συῶν ἐπειρά-
θημεν ἐξ ἀλεύρων μεθ' ὕδατος οἷον κυκεῶνά τινα
δόντες αὐτοῖς, ἔπειτα μετὰ τρεῖς που καὶ τέτταρας
ὥρας ἀνατεμόντες, εἰ οὕτω καὶ σὺ πράξειας,
εὑρήσεις ἔτι κατὰ τὴν γαστέρα τὰ ἐδηδεσμένα·
πέρας γὰρ αὐτοῖς ἐστι τῆς ἐνταῦθα μονῆς οὐχ ἡ
χύλωσις, ἣν καὶ ἐκτὸς ἔτι ὄντων μηχανήσασθαι
δυνατόν ἐστιν, ἀλλ' ἡ πέψις, ἕτερόν τι τῆς χυλώσεως οὖσα, καθάπερ αἱμάτωσίς τε καὶ θρέψις.
ὡς γὰρ κἀκεῖνα δέδεικται ποιοτήτων μεταβολῇ
γιγνόμενα, τὸν αὐτὸν τρόπον καὶ ἡ ἐν τῇ γαστρὶ
πέψις τῶν σιτίων εἰς τὴν οἰκείαν ἐστὶ τῷ τρεφο-
156 μένῳ ποιότητα ‖ μεταβολὴ καὶ ὅταν γε πεφθῇ
τελέως, ἀνοίγνυται μὲν τηνικαῦτα τὸ κάτω στόμα,
διεκπίπτει δ' αὐτοῦ τὰ σιτία ῥᾳδίως, εἰ καὶ
πλῆθός τι μεθ' ἑαυτῶν ἔχοντα τύχοι λίθων ἢ
ὀστῶν ἢ γιγάρτων ἤ τινος ἄλλου χυλωθῆναι
μὴ δυναμένου. καί σοι τοῦτ' ἔνεστιν ἐπὶ ζῴου

And it is not only because the stomach is distended and loaded and because the fluid runs from one part of it to another accompanied by gurglings—it is not only for these reasons that one would judge that there was an unduly long continuance of the food in it, in those people who are so disposed, but also from the *vomiting*. Thus, there are some who vomit up every particle of what they have eaten, not after three or four hours, but actually in the middle of the night, a lengthy period having elapsed since their meal.

Suppose you fill any animal whatsoever with liquid food—an experiment I have often carried out in pigs, to whom I give a sort of mess of wheaten flour and water, thereafter cutting them open after three or four hours ; if you will do this yourself, you will find the food still in the stomach. For it is not *chylification*[1] which determines the length of its stay here—since this can also be effected outside the stomach ; the determining factor is *digestion*[2] which is a different thing from chylification, as are blood-production and nutrition. For, just as it has been shown[3] that these two processes depend upon a *change of qualities*, similarly also the digestion of food in the stomach involves a transmutation of it into the quality proper to that which is receiving nourishment.[4] Then, when it is completely digested, the lower outlet opens and the food is quickly ejected through it, even if there should be amongst it abundance of stones, bones, grape-pips, or other things which cannot be reduced to chyle. And you may observe this

---

[1] *i.e.* not the mere mechanical breaking down of food, but a distinctively vital action of " alteration."
[2] *Pepsis.* [3] Book I., chaps. x., xi. [4] *cf.* p. 222, note 1.

θεάσασθαι στοχασαμένῳ τὸν καιρὸν τῆς κάτω διεξόδου. καὶ μέν γε καὶ εἰ σφαλείης ποτὲ τοῦ καιροῦ καὶ μηδὲν μήπω κάτω παρέρχοιτο πεττομένων ἔτι κατὰ τὴν γαστέρα τῶν σιτίων, οὐδ' οὕτως ἄκαρπος ἡ ἀνατομή σοι γενήσεται· θεάσῃ γὰρ ἐπ' αὐτῶν, ὅπερ ὀλίγῳ πρόσθεν ἐλέγομεν, ἀκριβῶς μὲν μεμυκότα τὸν πυλωρόν, ἅπασαν δὲ τὴν γαστέρα περιεσταλμένην τοῖς σιτίοις τρόπον ὁμοιότατον, οἱόνπερ καὶ αἱ μῆτραι τοῖς κυουμένοις. οὐ γὰρ ἔστιν οὐδέποτε κενὴν εὑρεῖν χώραν οὔτε κατὰ τὰς ὑστέρας οὔτε κατὰ τὴν κοιλίαν οὔτε κατὰ τὰς κύστεις ἀμφοτέρας οὔτε κατὰ τὴν χοληδόχον[1] ὀνομαζομένην οὔτε τὴν ἑτέραν· ἀλλ' εἴτ' ὀλίγον εἴη τὸ περιεχόμενον ἐν αὐταῖς εἴτε πολύ, μεσταὶ καὶ πλήρεις αὐτῶν αἱ κοιλίαι φαίνονται περιστελλομένων[2] ἀεὶ τῶν χιτώνων τοῖς περιεχομένοις, ὅταν γε κατὰ φύσιν ἔχῃ τὸ ζῷον. ‖

157 Ἐρασίστρατος δ' οὐκ οἶδ' ὅπως τὴν περιστολὴν τῆς γαστρὸς ἁπάντων αἰτίαν ἀποφαίνει καὶ τῆς λειώσεως τῶν σιτίων καὶ τῆς τῶν περιττωμάτων ὑποχωρήσεως καὶ τῆς τῶν κεχυλωμένων ἀναδόσεως.

Ἐγὼ μὲν γὰρ μυριάκις ἐπὶ ζῶντος ἔτι τοῦ ζῴου διελὼν τὸ περιτόναιον εὗρον ἀεὶ τὰ μὲν ἔντερα πάντα περιστελλόμενα τοῖς ἐνυπάρχουσι, τὴν κοιλίαν δ' οὐχ ἁπλῶς, ἀλλ' ἐπὶ μὲν ταῖς ἐδωδαῖς ἄνωθέν τε καὶ κάτωθεν αὐτὰ καὶ πανταχόθεν ἀκρι-

---

[1] *Choledochous.* [2] More exactly *peristolé*; *cf.* p. 97, note 1.
[3] Neuburger says of Erasistratus that "dissection had taught him to think in terms of anatomy." It was chiefly

yourself in an animal, if you will try to hit upon the time at which the descent of food from the stomach takes place. But even if you should fail to discover the time, and nothing was yet passing down, and the food was still undergoing digestion in the stomach, still even then you would find dissection not without its uses. You will observe, as we have just said, that the pylorus is accurately closed, and that the whole stomach is in a state of contraction upon the food very much as the womb contracts upon the foetus. For it is never possible to find a vacant space in the uterus, the stomach, or in either of the two bladders—that is, either in that called bile-receiving[1] or in the other ; whether their contents be abundant or scanty, their cavities are seen to be replete and full, owing to the fact that their coats contract constantly upon the contents—so long, at least, as the animal is in a natural condition.

Now Erasistratus for some reason declares that it is the contractions[2] of the stomach which are the cause of everything—that is to say, of the softening of the food,[3] the removal of waste matter, and the absorption of the food when chylified [emulsified].

Now I have personally, on countless occasions, divided the peritoneum of a still living animal and have always found all *the intestines* contracting peristaltically[4] upon their contents. The condition of *the stomach*, however, is found less simple ; as regards the substances freshly swallowed, it had grasped these accurately both above and below, in fact at every point, and was as devoid of movement

the gross movements or structure of organs with which he concerned himself. Where an organ had no obvious function, he dubbed it " useless " ; *e.g.* the spleen (*cf.* p. 143).

[4] *i.e.* contracting and dilating ; no longitudinal movements involved ; *cf.* p. 263, note 2.

βῶς περιειληφυῖαν ἀκίνητον, ὡς δοκεῖν ἡνῶσθαι καὶ περιπεφυκέναι τοῖς σιτίοις· ἐν δὲ τούτῳ καὶ τὸν πυλωρὸν εὕρισκον ἀεὶ μεμυκότα καὶ κεκλεισμένον ἀκριβῶς ὥσπερ τὸ τῶν ὑστερῶν στόμα ταῖς ἐγκύμοσιν.

Ἐπὶ μέντοι ταῖς πέψεσι συμπεπληρωμέναις ἀνέῳκτο μὲν ὁ πυλωρός, ἡ γαστὴρ δὲ περισταλτικῶς ἐκινεῖτο παραπλησίως τοῖς ἐντέροις.

V

Ἅπαντ᾽ οὖν ἀλλήλοις ὁμολογεῖ ταῦτα καὶ τῇ γαστρὶ καὶ ταῖς ὑστέραις καὶ ταῖς κύστεσιν εἶναί τινας ἐμφύτους δυνάμεις καθεκτικὰς μὲν τῶν
158 οἰκείων ποιοτήτων, ‖ ἀποκριτικὰς δὲ τῶν ἀλλοτρίων. ὅτι μὲν γὰρ ἕλκει τὴν χολὴν εἰς ἑαυτὴν ἡ ἐπὶ τῷ ἥπατι κύστις, ἔμπροσθεν δέδεικται, ὅτι δὲ καὶ ἀποκρίνει καθ᾽ ἑκάστην ἡμέραν εἰς τὴν γαστέρα, καὶ τοῦτ᾽ ἐναργῶς φαίνεται. καὶ μὴν εἰ διεδέχετο τὴν ἑλκτικὴν δύναμιν ἡ ἐκκριτικὴ καὶ μὴ μέση τις ἀμφοῖν ἦν ἡ καθεκτική, διὰ παντὸς ἐχρῆν ἀνατεμνομένων τῶν ζῴων ἴσον πλῆθος χολῆς εὑρίσκεσθαι κατὰ τὴν κύστιν· οὐ μὴν εὑρίσκεταί γε. ποτὲ μὲν γὰρ πληρεστάτη, ποτὲ δὲ κενοτάτη, ποτὲ δὲ τὰς ἐν τῷ μεταξὺ διαφορὰς ἔχουσα θεωρεῖται, καθάπερ καὶ ἡ ἑτέρα κύστις ἡ τὸ οὖρον ὑποδεχομένη. ταύτης μέν γε καὶ πρὸ τῆς ἀνατομῆς αἰσθανόμεθα, πρὶν ἀνιαθῆναι τῷ πλήθει βαρυνθεῖσαν ἢ τῇ δριμύτητι δηχθεῖσαν,

---

[1] cf. p. 282, note 1. [2] Book II., chaps. ii. and viii.

as though it had grown round and become united with the food.[1] At the same time I found the pylorus persistently closed and accurately shut, like the os uteri on the foetus.

In the cases, however, where digestion had been completed the pylorus had opened, and the stomach was undergoing peristaltic movements, similar to those of the intestines.

V

THUS all these facts agree that the stomach, uterus, and bladders possess certain inborn faculties which are retentive of their own proper qualities and eliminative of those that are foreign. For it has been already shown[2] that the bladder by the liver draws bile into itself, while it is also quite obvious that it eliminates this daily into the stomach. Now, of course, if the eliminative were to succeed the attractive faculty and there were not a *retentive* faculty between the two, there would be found, on every occasion that animals were dissected, an equal quantity of bile in the gall-bladder. This however, we do not find. For the bladder is sometimes observed to be very full, sometimes quite empty, while at other times you find in it various intermediate degrees of fulness, just as is the case with the other bladder—that which receives the urine; for even without resorting to anatomy we may observe that the urinary bladder continues to collect urine up to the time that it becomes uncomfortable through the increasing quantity of urine or the irritation caused by its acidity—the presumption

ἀθροιζούσης ἔτι τὸ οὖρον, ὡς οὔσης τινὸς κἀνταῦθα δυνάμεως καθεκτικῆς.

Οὕτω δὲ καὶ ἡ γαστὴρ ὑπὸ δριμύτητος πολλάκις δηχθεῖσα πρωιαίτερον τοῦ δέοντος ἄπεπτον ἔτι τὴν τροφὴν ἀποτρίβεται. αὖθις δ' ἄν ποτε τῷ πλήθει βαρυνθεῖσα ἢ καὶ κατ' ἄμφω συνελθόντα κακῶς διατεθεῖσα διαρροίαις ἑάλω. καὶ μέν γε καὶ οἱ ἔμετοι, τῷ πλήθει βαρυνθείσης ∥
159 αὐτῆς ἢ τὴν ποιότητα τῶν ἐν αὐτῇ σιτίων τε καὶ περιττωμάτων μὴ φερούσης, ἀνάλογόν τι ταῖς διαρροίαις πάθημα τῆς ἄνω γαστρός ἐστιν. ὅταν μὲν γὰρ ἐν τοῖς κάτω μέρεσιν αὐτῆς ἡ τοιαύτη γένηται διάθεσις, ἐρρωμένων τῶν κατὰ τὸν στόμαχον, εἰς διαρροίας ἐτελεύτησεν, ὅταν δ' ἐν τοῖς κατὰ τὸ στόμα, τῶν ἄλλων εὐρωστούντων, εἰς ἐμέτους.

VI

Ἔνεστι δὲ καὶ τοῦτο πολλάκις ἐναργῶς ἰδεῖν ἐπὶ τῶν ἀποσίτων· ἀναγκαζόμενοι γὰρ ἐσθίειν οὔτε καταπίνειν εὐσθενοῦσιν οὔτ', εἰ καὶ βιάσαιντο, κατέχουσιν, ἀλλ' εὐθὺς ἀνεμοῦσι. καὶ οἱ ἄλλως δὲ τῶν ἐδεσμάτων πρὸς ὁτιοῦν δυσχεραίνοντες βιασθέντες ἐνίοτε προσάρασθαι ταχέως ἐξεμοῦσιν, ἢ εἰ κατάσχοιεν βιασάμενοι, ναυτιώδεις τ' εἰσὶ καὶ τῆς γαστρὸς ὑπτίας αἰσθάνονται καὶ σπευδούσης ἀποθέσθαι τὸ λυποῦν.

Οὕτως ἐξ ἁπάντων τῶν φαινομένων, ὅπερ ἐξ ἀρχῆς ἐρρέθη, μαρτυρεῖται τὸ δεῖν ὑπάρχειν τοῖς τοῦ ζῴου μορίοις σχεδὸν ἅπασιν ἔφεσιν μέν τινα

thus being that here, too, there is a retentive faculty.

Similarly, too, the stomach, when, as often happens, it is irritated by acidity, gets rid of the food, although still undigested, earlier than proper ; or again, when oppressed by the quantity of its contents, or disordered from the co-existence of both conditions, it is seized with *diarrhoea*. *Vomiting* also is an affection of the upper [part of the] stomach analogous to diarrhoea, and it occurs when the stomach is overloaded or is unable to stand the quality of the food or surplus substances which it contains. Thus, when such a condition develops in the lower parts of the stomach, while the parts about the inlet are normal, it ends in diarrhoea, whereas if this condition is in the upper stomach, the lower parts being normal, it ends in vomiting.

VI

THIS may often be clearly observed in those who are disinclined for food ; when obliged to eat, they have not the strength to swallow, and, even if they force themselves to do so, they cannot retain the food, but at once vomit it up. And those especially who have a dislike to some particular kind of food, sometimes take it under compulsion, and then promptly bring it up ; or, if they force themselves to keep it down, they are nauseated and feel their stomach turned up, and endeavouring to relieve itself of its discomfort.

Thus, as was said at the beginning, all the observed facts testify that there must exist in almost all parts of the animal a certain inclination towards, or, so to

καὶ οἷον ὄρεξιν τῆς οἰκείας ποιότητος, ἀποστροφὴν
160 δέ τινα ‖ καὶ οἷον μῖσός τι τῆς ἀλλοτρίας. ἀλλ᾽
ἐφιέμενα μὲν ἕλκειν εὔλογον, ἀποστρεφόμενα δ᾽
ἐκκρίνειν.

Κἀκ τούτων πάλιν ἥ θ᾽ ἑλκτικὴ δύναμις
ἀποδείκνυται καθ᾽ ἅπαν ὑπάρχουσα καὶ ἡ προ-
ωστική.

᾽Αλλ᾽ εἴπερ ἔφεσίς τέ τίς ἐστι καὶ ἕλξις, εἴη ἄν
τις καὶ ἀπόλαυσις· οὐδὲν γὰρ τῶν ὄντων ἕλκει τι
δι᾽ αὐτὸ τὸ ἕλκειν, ἀλλ᾽ ἵν᾽ ἀπολαύσῃ τοῦ διὰ
τῆς ὁλκῆς εὐπορηθέντος. καὶ μὴν ἀπολαύειν οὐ
δύναται μὴ κατασχόν. κἂν τούτῳ πάλιν ἡ
καθεκτικὴ δύναμις ἀποδείκνυται τὴν γένεσιν
ἀναγκαίαν ἔχουσα· σαφῶς γὰρ ἐφίεται μὲν τῶν
οἰκείων ποιοτήτων ἡ γαστήρ, ἀποστρέφεται δὲ
τὰς ἀλλοτρίας.

᾽Αλλ᾽ εἴπερ ἐφίεταί τε καὶ ἕλκει καὶ ἀπολαύει
κατέχουσα καὶ περιστελλομένη, εἴη ἄν τι καὶ
πέρας αὐτῇ τῆς ἀπολαύσεως κἀπὶ τῷδ᾽ ὁ καιρὸς
ἤδη τῆς ἐκκριτικῆς δυνάμεως ἐνεργούσης.

VII

᾽Αλλ᾽ εἰ καὶ κατέχει καὶ ἀπολαύει, κατα-
χρῆται πρὸς ὃ πέφυκε. πέφυκε δὲ τοῦ προσ-
161 ήκοντος ἑαυτῇ ‖ κατὰ ποιότητα καὶ οἰκείου

---

[1] Note use of psychological terms in biology. cf. also
p. 133, note 3.
[2] "In everything." cf. p. 66, note 3.

speak, an appetite for their own special quality, and an aversion to, or, as it were, a hatred [1] of the foreign quality. And it is natural that when they feel an inclination they should attract, and that when they feel aversion they should expel.

From these facts, then, again, both the attractive and the propulsive faculties have been demonstrated to exist in everything.[2]

But if there be an inclination or attraction, there will also be some benefit derived; for no existing thing attracts anything else for the mere sake of attracting, but in order to benefit by what is acquired by the attraction. And of course it cannot benefit by it if it cannot retain it. Herein, then, again, the retentive faculty is shown to have its necessary origin: for the stomach obviously inclines towards its own proper qualities and turns away from those that are foreign to it.[3]

But if it aims at and attracts its food and benefits by it while retaining and contracting upon it, we may also expect that there will be some *termination* to the benefit received, and that thereafter will come the time for the exercise of the eliminative faculty.

VII

BUT if the stomach both retains and benefits by its food, then it employs it for the end for which it [the stomach] naturally exists. And it exists to partake of that which is of a quality befitting and proper to

---

[3] Galen confuses the nutrition of organs with that of the ultimate living elements or cells; the stomach does not, of course, feed itself in the way a cell does. *cf.* Introduction, p. xxxii.

μεταλαμβάνειν· ὥσθ' ἕλκει τῶν σιτίων ὅσον
χρηστότατον ἀτμωδῶς τε καὶ κατὰ βραχὺ καὶ
τοῦτο τοῖς ἑαυτῆς χιτῶσιν ἐναποτίθεταί τε καὶ
προστίθησιν. ὅταν δ' ἱκανῶς ἐμπλησθῇ, καθάπερ ἄχθος τι τὴν λοιπὴν ἀποτίθεται τροφὴν
ἐσχηκυῖάν τι χρηστὸν ἤδη καὶ αὐτὴν ἐκ τῆς πρὸς
τὴν γαστέρα κοινωνίας· οὐδὲ γὰρ ἐνδέχεται δύο
σώματα δρᾶν καὶ πάσχειν ἐπιτήδεια συνελθόντα
μὴ οὐκ ἤτοι πάσχειν θ' ἅμα καὶ δρᾶν ἢ θάτερον
μὲν δρᾶν, θάτερον δὲ πάσχειν. ἐὰν μὲν γὰρ
ἰσάζῃ ταῖς δυνάμεσιν, ἐξ ἴσου δράσει τε καὶ
πείσεται, ἂν δ' ὑπερέχῃ πολὺ καὶ κρατῇ θάτερον,
ἐνεργήσει περὶ τὸ πάσχον· ὥστε δράσει μέγα
μέν τι καὶ αἰσθητόν, αὐτὸ δ' ἤτοι σμικρόν τι καὶ
οὐκ αἰσθητὸν ἢ παντάπασιν οὐδὲν πείσεται. ἀλλ'
ἐν τούτῳ δὴ καὶ μάλιστα διήνεγκε φαρμάκου
δηλητηρίου τροφή· τὸ μὲν γὰρ κρατεῖ τῆς ἐν τῷ
σώματι δυνάμεως, ἡ δὲ κρατεῖται.

Οὔκουν ἐνδέχεται τροφὴν μὲν εἶναί τι τῷ ζῴῳ
προσήκουσαν, οὐ μὴν καὶ κρατεῖσθαί γ' ὁμοίως
162 πρὸς τῶν || ἐν τῷ ζῴῳ ποιοτήτων· τὸ κρατεῖσθαι
δ' ἦν ἀλλοιοῦσθαι. ἀλλ' ἐπεὶ τὰ μὲν ἰσχυρότερα
ταῖς δυνάμεσίν ἐστι μόρια, τὰ δ' ἀσθενέστερα,
κρατήσει μὲν πάντα τῆς οἰκείας τῷ ζῴῳ τροφῆς,
οὐχ ὁμοίως δὲ πάντα· κρατήσει δ' ἄρα καὶ ἡ
γαστὴρ καὶ ἀλλοιώσει μὲν τὴν τροφήν, οὐ μὴν
ὁμοίως ἥπατι καὶ φλεψὶ καὶ ἀρτηρίαις καὶ
καρδίᾳ.

Πόσον οὖν ἐστιν, ὃ ἀλλοιοῖ, καὶ δὴ θεασώμεθα·
πλέον μὲν ἢ κατὰ τὸ στόμα, μεῖον δ' ἢ κατὰ τὸ

---

[1] cf. Asclepiades's theory regarding the urine, p. 51.
[2] The process of *application* or *prosthesis*. cf. p. 223, note 3.

it. Thus it attracts all the most useful parts of the food in a vaporous [1] and finely divided condition, storing this up in its own coats, and applying [2] it to them. And when it is sufficiently full it' puts away from it, as one might something troublesome, the rest of the food, this having itself meanwhile obtained some profit from its association with the stomach. For it is impossible for two bodies which are adapted for acting and being acted upon to come together without either both acting or being acted upon, or else one acting and the other being acted upon. For if their forces are equal they will act and be acted upon equally, and if the one be much superior in strength, it will exert its activity upon its passive neighbour; thus, while producing a great and appreciable effect, it will itself be acted upon either little or not at all. But it is herein also that the main difference lies between nourishing food and a deleterious drug; the latter masters the forces of the body, whereas the former is mastered by them. [3]

There cannot, then, be food which is suited for the animal which is not also correspondingly subdued by the qualities existing in the animal. And to be subdued means to undergo *alteration*. [4] Now, some parts are stronger in power and others weaker; therefore, while all will subdue the nutriment which is proper to the animal, they will not all do so equally. Thus the stomach will subdue and alter its food, but not to the same extent as will the liver, veins, arteries, and heart.

We must therefore observe to what extent it does alter it. The alteration is more than that which

[3] Mutual influence of organism and environment.
[4] Qualitative change. *cf.* Book I., chap. ii.

251

ἧπάρ τε καὶ τὰς φλέβας. αὕτη μὲν γὰρ ἡ
ἀλλοίωσις εἰς αἵματος οὐσίαν ἄγει τὴν τροφήν,
ἡ δ' ἐν τῷ στόματι μεθίστησι μὲν αὐτὴν ἐναργῶς
εἰς ἕτερον εἶδος, οὐ μὴν εἰς τέλος γε μετακοσμεῖ.
μάθοις δ' ἂν ἐπὶ τῶν ἐγκαταλειφθέντων ταῖς
διαστάσεσι τῶν ὀδόντων σιτίων καὶ καταμεινάν-
των δι' ὅλης νυκτός· οὔτε γὰρ ἄρτος ἀκριβῶς ὁ
ἄρτος οὔτε κρέας ἐστὶ τὸ κρέας, ἀλλ' ὄζει μὲν
τοιοῦτον, οἷόνπερ καὶ τοῦ ζῴου τὸ στόμα, δια-
λέλυται δὲ καὶ διατέτηκε καὶ τὰς ἐν τῷ ζῴῳ τῆς
σαρκὸς ἀπομέμακται ποιότητας. ἔνεστι δέ σοι
163 θεάσασθαι τὸ μέγεθος τῆς ἐν τῷ στόματι ‖ τῶν
σιτίων ἀλλοιώσεως, εἰ πυροὺς μασησάμενος
ἐπιθείης ἀπέπτοις δοθιῆσιν· ὄψει γὰρ αὐτοὺς
τάχιστα μεταβάλλοντάς τε καὶ συμπέττοντας,
οὐδὲν τοιοῦτον, ὅταν ὕδατι φυραθῶσιν, ἐργάσα-
σθαι δυναμένους. καὶ μὴ θαυμάσῃς· τὸ γάρ τοι
φλέγμα τουτὶ τὸ κατὰ τὸ στόμα καὶ λειχῆνων
ἐστὶν ἄκος καὶ σκορπίους ἀναιρεῖ παραχρῆμα καὶ
πολλὰ τῶν ἰοβόλων θηρίων τὰ μὲν εὐθέως
ἀποκτείνει, τὰ δ' ἐς ὕστερον· ἅπαντα γοῦν
βλάπτει μεγάλως. ἀλλὰ τὰ μεμασημένα σιτία
πρῶτον μὲν τούτῳ τῷ φλέγματι βέβρεκταί τε
καὶ πεφύραται, δεύτερον δὲ καὶ τῷ χρωτὶ τοῦ
στόματος ἅπαντα πεπλησίακεν, ὥστε πλείονα
μεταβολὴν εἴληφε τῶν ἐν ταῖς κεναῖς χώραις τῶν
ὀδόντων ἐσφηνωμένων.

Ἀλλ' ὅσον τὰ μεμασημένα τούτων ἐπὶ πλέον
ἠλλοίωται, τοσοῦτον ἐκείνων τὰ καταποθέντα

occurs in the mouth, but less than that in the liver
and veins. For the latter alteration changes the
nutriment into the *substance* of blood, whereas that in
the mouth obviously changes it into a new *form*, but
certainly does not completely transmute it. This
you may discover in the food which is left in the
intervals between the teeth, and which remains there
all night; the bread is not exactly bread, nor the
meat meat, for they have a smell similar to that of
the animal's mouth, and have been disintegrated and
dissolved, and have had the qualities of the animal's
flesh impressed upon them. And you may observe
the extent of the alteration which occurs to food in
the mouth if you will chew some corn and then
apply it to an unripe [undigested] boil: you will see
it rapidly transmuting—in fact entirely digesting—
the boil, though it cannot do anything of the kind if
you mix it with water. And do not let this surprise
you; this phlegm [saliva] in the mouth is also a cure
for *lichens* [1]; it even rapidly destroys scorpions;
while, as regards the animals which emit venom,
some it kills at once, and others after an interval;
to all of them in any case it does great damage.
Now, the masticated food is all, firstly, soaked in and
mixed up with this phlegm; and secondly, it is
brought into contact with the actual skin of the
mouth; thus it undergoes more change than the
food which is wedged into the vacant spaces between
the teeth.

But just as masticated food is more altered than
the latter kind, so is food which has been swallowed
more altered than that which has been merely

---

[1] Apparently skin-diseases in which a superficial crust
(resembling the lichen on a tree-trunk) forms—*e.g.* psoriasis.

μὴ γὰρ οὐδὲ παραβλητὸν ᾖ τὸ τῆς ὑπερβολῆς,
εἰ τὸ κατὰ τὴν κοιλίαν ἐννοήσαιμεν φλέγμα καὶ
χολὴν καὶ πνεῦμα καὶ θερμασίαν καὶ ὅλην τὴν
οὐσίαν τῆς γαστρός. εἰ δὲ καὶ συνεπινοήσαις
164 αὐτῇ τὰ παρακείμενα ‖ σπλάγχνα καθάπερ τινὶ
λέβητι μεγάλῳ πυρὸς ἑστίας πολλάς, ἐκ δεξιῶν
μὲν τὸ ἧπαρ, ἐξ ἀριστερῶν δὲ τὸν σπλῆνα, τὴν
καρδίαν δ᾽ ἐκ τῶν ἄνω, σὺν αὐτῇ δὲ καὶ τὰς
φρένας αἰωρουμένας τε καὶ διὰ παντὸς κινουμένας,
ἐφ᾽ ἅπασι δὲ τούτοις σκέπον τὸ ἐπίπλοον, ἐξαί-
σιόν τινα πεισθήσῃ τὴν ἀλλοίωσιν γίγνεσθαι τῶν
εἰς τὴν γαστέρα καταποθέντων σιτίων.

Πῶς δ᾽ ἂν ἠδύνατο ῥᾳδίως αἱματοῦσθαι μὴ
προπαρασκευασθέντα τῇ τοιαύτῃ μεταβολῇ; δέδεικται γὰρ οὖν καὶ πρόσθεν, ὡς οὐδὲν εἰς τὴν
ἐναντίαν ἀθρόως μεθίσταται ποιότητα. πῶς οὖν
ὁ ἄρτος αἷμα γίγνεται, πῶς δὲ τὸ τεῦτλον ἢ ὁ
κύαμος ἤ τι τῶν ἄλλων, εἰ μὴ πρότερόν τιν᾽ ἑτέραν
ἀλλοίωσιν ἐδέξατο; πῶς δ᾽ ἡ κόπρος ἐν τοῖς
λεπτοῖς ἐντέροις ἀθρόως γεννηθήσεται; τί γὰρ ἐν
τούτοις σφοδρότερον εἰς ἀλλοίωσίν ἐστι τῶν κατὰ
τὴν γαστέρα; πότερα τῶν χιτώνων τὸ πλῆθος ἢ
τῶν γειτνιώντων σπλάγχνων ἡ περίθεσις ἢ τῆς
μονῆς ὁ χρόνος ἢ σύμφυτός τις ἐν τοῖς ὀργάνοις
θερμασία; καὶ μὴν κατ᾽ οὐδὲν τούτων πλεονεκτεῖ
τὰ ἔντερα τῆς γαστρός. τί ποτ᾽ οὖν ἐν μὲν τῇ
165 γαστρὶ νυκτὸς ‖ ὅλης πολλάκις μείναντα τὸν
ἄρτον ἔτι φυλάττεσθαι βούλονται τὰς ἀρχαίας
διασῴζοντα ποιότητας, ἐπειδὰν δ᾽ ἅπαξ ἐμπέσῃ

---

[1] Note especially pneuma and innate heat, which practically stand for oxygen and the heat generated in oxidation.
cf. p. 41, note 3.   [2] Book I., chap. x.

masticated. Indeed, there is no comparison between these two processes ; we have only to consider what the stomach contains—phlegm, bile, pneuma, [innate] heat,[1] and, indeed the whole substance of the stomach. And if one considers along with this the adjacent viscera, like a lot of burning hearths around a great cauldron—to the right the liver, to the left the spleen, the heart above, and along with it the diaphragm (suspended and in a state of constant movement), and the omentum sheltering them all— you may believe what an extraordinary alteration it is which occurs in the food taken into the stomach.

How could it easily become blood if it were not previously prepared by means of a change of this kind ? It has already been shown[2] that nothing is altered all at once from one quality to its opposite. How then could bread, beef, beans, or any other food turn into blood if they had not previously undergone some other alteration ? And how could the faeces be generated right away in the small intestine ?[3] For what is there in this organ more potent in producing alteration than the factors in the stomach ? Is it the number of the coats, or the way it is surrounded by neighbouring viscera, or the time that the food remains in it, or some kind of innate heat which it contains ? Most assuredly the intestines have the advantage of the stomach in none of these respects. For what possible reason, then, will objectors have it that bread may often remain a whole night in the stomach and still preserve its original qualities, whereas when once it is projected into the

[3] That is to say, faeces are obviously altered food. This alteration cannot have taken place entirely in the small intestine : therefore alteration of food must take place in the stomach.

τοῖς ἐντέροις, εὐθὺς γίγνεσθαι κόπρον; εἰ μὲν
γὰρ ὁ τοσοῦτος χρόνος ἀδύνατος ἀλλοιοῦν, οὐδ' ὁ
βραχὺς ἱκανός· εἰ δ' οὗτος αὐτάρκης, πῶς οὐ
πολὺ μᾶλλον ὁ μακρός; ἆρ' οὖν ἀλλοιοῦται μὲν
ἡ τροφὴ κατὰ τὴν κοιλίαν, ἄλλην δέ τιν' ἀλλοίωσιν καὶ οὐχ οἵαν ἐκ τῆς φύσεως ἴσχει τοῦ μετα-
βάλλοντος ὀργάνου; ἢ ταύτην μέν, οὐ μὴν τήν
γ' οἰκείαν τῷ τοῦ ζῴου σώματι; μακρῷ τοῦτ'
ἀδυνατώτερόν ἐστι. καὶ μὴν οὐκ ἄλλο γ' ἦν ἡ
πέψις ἢ ἀλλοίωσις εἰς τὴν οἰκείαν τοῦ τρεφομένου
ποιότητα. εἴπερ οὖν ἡ πέψις τοῦτ' ἔστι καὶ ἡ
τροφὴ κατὰ τὴν γαστέρα δέδεικται δεχομένη
ποιότητα τῷ μέλλοντι πρὸς αὐτῆς θρέψεσθαι ζῴω
προσήκουσαν, ἱκανῶς ἀποδέδεικται τὸ πέττεσθαι
κατὰ τὴν γαστέρα τὴν τροφήν.

Καὶ γελοῖος μὲν Ἀσκληπιάδης οὔτ' ἐν ταῖς
ἐρυγαῖς λέγων ἐμφαίνεσθαί ποτε τὴν ποιότητα
τῶν πεφθέντων σιτίων οὔτ' ἐν τοῖς ἐμέτοις οὔτ'
166 ἐν ταῖς ἀνα‖τομαῖς· αὐτὸ γὰρ δὴ τὸ τοῦ σώματος
ἐξόζειν αὐτὰ τῆς κοιλίας ἐστὶ τὸ πεπέφθαι. ὁ δ'
οὕτως ἐστὶν εὐήθης, ὥστ', ἐπειδὴ τῶν παλαιῶν
ἀκούει λεγόντων ἐπὶ τὸ χρηστὸν ἐν τῇ γαστρὶ
μεταβάλλειν τὰ σιτία, δοκιμάζει ζητεῖν οὐ τὸ
κατὰ δύναμιν ἀλλὰ τὸ κατὰ γεῦσιν χρηστόν,
ὥσπερ ἢ τοῦ μήλου μηλωδεστέρου—χρὴ γὰρ
οὕτως αὐτῷ διαλέγεσθαι—γιγνομένου κατὰ τὴν
κοιλίαν ἢ τοῦ μέλιτος μελιτωδεστέρου.

---

[1] cf. p. 39.
[2] Asclepiades held that there was no such thing as real

intestines, it straightway becomes ordure? For, if such a long period of time is incapable of altering it, neither will the short period be sufficient, or, if the latter is enough, surely the longer time will be much more so! Well, then, can it be that, while the nutriment does undergo an alteration in the stomach, this is a different kind of alteration and one which is not dependent on the nature of the organ which alters it? Or if it be an alteration of this latter kind, yet one perhaps which is not proper to the body of the animal? This is still more impossible. Digestion was shown to be nothing else than an alteration to the quality proper to that which is receiving nourishment.[1] Since, then, this is what digestion means and since the nutriment has been shown to take on in the stomach a quality appropriate to the animal which is about to be nourished by it, it has been demonstrated adequately that nutriment does undergo digestion in the stomach.

And Asclepiades is absurd when he states that the quality of the digested food never shows itself either in eructations or in the vomited matter, or on dissection.[2] For of course the mere fact that the food smells of the body shows that it has undergone gastric digestion. But this man is so foolish that, when he hears the Ancients saying that the food is converted in the stomach into something "good," he thinks it proper to look out not for what is good in its possible effects, but for what is good to the taste: this is like saying that apples (for so one has to argue with him) become more apple-like [in flavour] in the stomach, or honey more honey-like!

qualitative change; the food was merely broken up into its constituent molecules, and absorbed unaltered. cf. p. 49, note 5.

Πολὺ δ' εὐηθέστερός ἐστι καὶ γελοιότερος ὁ
Ἐρασίστρατος ἢ μὴ νοῶν, ὅπως εἴρηται πρὸς τῶν
παλαιῶν ἡ πέψις ἑψήσει παραπλήσιος ὑπάρχειν,
ἢ ἑκὼν σοφιζόμενος ἑαυτόν. ἑψήσει μὲν οὖν,
φησίν, οὕτως ἐλαφρὰν ἔχουσαν θερμασίαν οὐκ
εἰκὸς εἶναι παραπλησίαν τὴν πέψιν, ὥσπερ ἢ τὴν
Αἴτνην δέον ὑποθεῖναι τῇ γαστρὶ ἢ ἄλλως αὐτῆς
ἀλλοιῶσαι τὰ σιτία μὴ δυναμένης ἢ δυναμένης
μὲν ἀλλοιοῦν, οὐ κατὰ τὴν ἔμφυτον δὲ θερμασίαν,
ὑγρὰν οὖσαν δηλονότι καὶ διὰ τοῦθ' ἕψειν οὐκ
ὀπτᾶν εἰρημένην.

Ἐχρῆν δ' αὐτόν, εἴπερ περὶ πραγμάτων ἀντιλέγειν ἐβούλετο, πειραθῆναι δεῖξαι μάλιστα μὲν
167 καὶ ‖ πρῶτον, ὡς οὐδὲ μεταβάλλει τὴν ἀρχὴν οὐδ'
ἀλλοιοῦται κατὰ ποιότητα πρὸς τῆς γαστρὸς τὰ
σιτία, δεύτερον δ', εἴπερ μὴ οἷός τ' ἦν τοῦτο
πιστώσασθαι, τὸ τὴν ἀλλοίωσιν αὐτῶν ἄχρηστον
εἶναι τῷ ζῴῳ· εἰ δὲ μηδὲ τοῦτ' εἶχε διαβάλλειν,
ἐξελέγξαι τὴν περὶ τὰς δραστικὰς ἀρχὰς ὑπόληψιν καὶ δεῖξαι τὰς ἐνεργείας ἐν τοῖς μορίοις οὐ
διὰ τὴν ἐκ θερμοῦ καὶ ψυχροῦ καὶ ξηροῦ καὶ
ὑγροῦ ποιὰν κρᾶσιν ὑπάρχειν ἀλλὰ δι' ἄλλο τι·
εἰ δὲ μηδὲ τοῦτ' ἐτόλμα διαβάλλειν, ἀλλ' ὅτι γε
μὴ τὸ θερμόν ἐστιν ἐν τοῖς ὑπὸ φύσεως διοικουμένοις τὸ τῶν ἄλλων δραστικώτατον. ἢ εἰ μήτε
τοῦτο μήτε τῶν ἄλλων τι τῶν ἔμπροσθεν εἶχεν
ἀποδεικνύναι, μὴ ληρεῖν ὀνόματι προσπαλαίοντα

---

[1] i.e. denial of forethought in the Physis.

Erasistratus, however, is still more foolish and absurd, either through not perceiving in what sense the Ancients said that digestion is similar to the process of *boiling*, or because he purposely confused himself with sophistries. It is, he says, inconceivable that digestion, involving as it does such trifling warmth, should be related to the boiling process. This is as if we were to suppose that it was necessary to put the fires of Etna under the stomach before it could manage to alter the food ; or else that, while it was capable of altering the food, it did not do this by virtue of its innate heat, which of course was moist, so that the word *boil* was used instead of *bake*.

What he ought to have done, if it was facts that he wished to dispute about, was to have tried to show, first and foremost, that the food is not transmuted or altered in quality by the stomach at all, and secondly, if he could not be confident of this, he ought to have tried to show that this alteration was not of any advantage to the animal.[1] If, again, he were unable even to make this misrepresentation, he ought to have attempted to confute the postulate concerning *the active principles*—to show, in fact, that the functions taking place in the various parts do not depend on the way in which the Warm, Cold, Dry, and Moist are mixed, but on some other factor. And if he had not the audacity to misrepresent facts even so far as this, still he should have tried at least to show that the Warm is not the most active of all the principles which play a part in things governed by Nature. But if he was unable to demonstrate this any more than any of the previous propositions, then he ought not to have made himself ridiculous by quarrelling uselessly

GALEN

μάτην, ὥσπερ οὐ σαφῶς Ἀριστοτέλους ἔν τ'
ἄλλοις πολλοῖς κἀν τῷ τετάρτῳ τῶν μετεωρολογι-
κῶν ὅπως ἡ πέψις ἑψήσει παραπλήσιος εἶναι
λέγεται, καὶ ὅτι μὴ πρώτως μηδὲ κυρίως ὀνομα-
ζόντων, εἰρηκότος.

Ἀλλ', ὡς ἤδη λέλεκται πολλάκις, ἀρχὴ τούτων
ἁπάντων ἐστὶ μία τὸ περὶ θερμοῦ καὶ ψυχροῦ καὶ
ξηροῦ καὶ ὑγροῦ διασκέψασθαι, καθάπερ Ἀριστοτέλης ἐποίησεν ἐν τῷ δευτέρῳ περὶ γενέσεως καὶ
168 φθορᾶς, ἀπο‖δείξας ἁπάσας τὰς κατὰ τὰ σώματα
μεταβολὰς καὶ ἀλλοιώσεις ὑπὸ τούτων γίγνεσθαι.
ἀλλ' Ἐρασίστρατος οὔτε τούτοις οὔτ' ἄλλῳ τινὶ
τῶν προειρημένων ἀντειπὼν ἐπὶ τοὔνομα μόνον
ἐτράπετο τῆς ἑψήσεως.

VIII

Ἐπὶ μὲν οὖν τῆς πέψεως, εἰ καὶ τἄλλα πάντα
παρέλιπε, τὸ γοῦν ὅτι διαφέρει τῆς ἐκτὸς ἑψήσεως
ἡ ἐν τοῖς ζῴοις πέψις, ἐπειράθη δεικνύναι, περὶ
δὲ τῆς καταπόσεως οὐδ' ἄχρι τοσούτου. τί γάρ
φησιν;

"Ὁλκὴ μὲν οὖν τῆς κοιλίας οὐδεμία φαίνεται
εἶναι."

Καὶ μὴν δύο χιτῶνας ἡ γαστὴρ ἔχει πάντως
ἔνεκά του γεγονότας καὶ διήκουσιν οὗτοι μέχρι
τοῦ στόματος, ὁ μὲν ἔνδον, οἷός ἐστι κατὰ τὴν
γαστέρα, τοιοῦτος διαμένων, ὁ δ' ἕτερος ἐπὶ τὸ

with a mere name—as though Aristotle had not clearly stated in the fourth book of his " Meteorology," as well as in many other passages, in what way digestion can be said to be allied to boiling, and also that the latter expression is not used in its primitive or strict sense.

But, as has been frequently said already,[1] the one starting-point of all this is a thoroughgoing enquiry into the question of the Warm, Cold, Dry and Moist ; this Aristotle carried out in the second of his books "On Genesis and Destruction," where he shows that all the transmutations and alterations throughout the body take place as a result of these principles. Erasistratus, however, advanced nothing against these or anything else that has been said above, but occupied himself merely with the word " boiling."

VIII

THUS, as regards *digestion*, even though he neglected everything else, he did at least attempt to prove his point—namely, that digestion in animals differs from boiling carried on outside ; in regard to the question of *deglutition*, however, he did not go even so far as this. What are his words?

" The stomach does not appear to exercise any traction." [2]

Now the fact is that the stomach possesses two coats, which certainly exist for some purpose ; they extend as far as the mouth, the internal one remaining throughout similar to what it is in the stomach, and the other one tending to become of a more fleshy

[1] *v.* p. 9, *et passim.*   [2] *cf.* p. 97.

σαρκωδέστερον ἐν τῷ στομάχῳ τρεπόμενος. ὅτι
μὲν οὖν ἐναντίας ἀλλήλαις τὰς ἐπιβολὰς τῶν
ἰνῶν ἔχουσιν οἱ χιτῶνες οὗτοι, τὸ φαινόμενον
αὐτὸ μαρτυρεῖ. τίνος δ᾽ ἕνεκα τοιοῦτοι γεγόνασιν, Ἐρασίστρατος μὲν οὐδ᾽ ἐπεχείρησεν εἰπεῖν,
ἡμεῖς δ᾽ ἐροῦμεν.

Ὁ μὲν ἔνδον εὐθείας ἔχει τὰς ἶνας, ὁλκῆς γὰρ
169 ἕνεκα γέ‖γονεν· ὁ δ᾽ ἔξωθεν ἐγκαρσίας ὑπὲρ τοῦ
κατὰ κύκλον περιστέλλεσθαι· ἑκάστῳ γὰρ τῶν
κινουμένων ὀργάνων ἐν τοῖς σώμασι κατὰ τὰς
τῶν ἰνῶν θέσεις αἱ κινήσεις εἰσίν. ἐπ᾽ αὐτῶν δὲ
πρῶτον τῶν μυῶν, εἰ βούλει, βασάνισον τὸν
λόγον, ἐφ᾽ ὧν καὶ αἱ ἶνες ἐναργέσταται καὶ αἱ
κινήσεις αὐτῶν ὁρῶνται διὰ σφοδρότητα. μετὰ
δὲ τοὺς μῦς ἐπὶ τὰ φυσικὰ τῶν ὀργάνων ἴθι καὶ
πάντ᾽ ὄψει κατὰ τὰς ἶνας κινούμενα καὶ διὰ τοῦθ᾽
ἑκάστῳ μὲν τῶν ἐντέρων στρογγύλαι καθ᾽ ἑκάτερον τῶν χιτώνων αἱ ἶνές εἰσι· περιστέλλονται
γὰρ μόνον, ἕλκουσι δ᾽ οὐδέν. ἡ γαστὴρ δὲ τῶν
ἰνῶν τὰς μὲν εὐθείας ἔχει χάριν ὁλκῆς, τὰς δ᾽
ἐγκαρσίας ἕνεκα περιστολῆς· ὥσπερ γὰρ ἐν τοῖς
μυσὶν ἑκάστης τῶν ἰνῶν τεινομένης τε καὶ πρὸς
τὴν ἀρχὴν ἑλκομένης αἱ κινήσεις γίγνονται, κατὰ
τὸν αὐτὸν λόγον κἂν τῇ γαστρί· τῶν μὲν οὖν
ἐγκαρσίων ἰνῶν τεινομένων ἔλαττον ἀνάγκη γί-

---

[1] It appears to me, from comparison between this and other
passages in Galen's writings (notably *Use of Parts,* iv., 8),
that he means by the "two coats" simply the mucous and
the muscular coats. In this case the "straight" or "longitudinal" fibres of the inner coat would be the *rugae* ; the
"circular" fibres of the inner intestinal coat would be the
valvulae conniventes.

nature in the gullet. Now simple observation will testify that these coats have their fibres inserted in contrary directions.[1] And, although Erasistratus did not attempt to say for what reason they are like this, I am going to do so.

The inner coat has its fibres straight, since it exists for the purpose of traction. The outer coat has its fibres transverse, for the purpose of peristalsis.[2] In fact, the movements of each of the *mobile* organs of the body depend on the setting of the fibres. Now please test this assertion first in the muscles themselves; in these the fibres are most distinct, and their movements visible owing to their vigour. And after the muscles, pass to the *physical* organs,[3] and you will see that they all move in correspondence with their fibres. This is why the fibres throughout the intestines are circular in both coats—they only contract peristaltically, they do not exercise traction. The stomach, again, has some of its fibres longitudinal for the purpose of traction and the others transverse for the purpose of peristalsis.[2] For just as the movements in the muscles[4] take place when each of the fibres becomes tightened and drawn towards its origin, such also is what happens in the stomach ; when the transverse fibres tighten, the breadth of

[2] The term here rendered *peristalsis* is *peristolé* in Greek ; it is applied only to the intermittent movements of muscles placed circularly round a lumen or cavity, and comprehends *systolé* or contraction and *diastolé* or dilatation. In its modern significance, *peristalsis*, however, also includes the movements of *longitudinal* fibres. *cf.* p. 97, note 1.

[3] *i.e.* those containing non-striped or "involuntary" muscle fibres ; organs governed by the "natural" pneuma ; *cf.* p. 186, note 3.

[4] By this term is meant only what we should call the "voluntary" muscles.

γνεσθαι τὸ εὖρος τῆς περιεχομένης ὑπ' αὐτῶν
κοιλότητος, τῶν δ' εὐθειῶν ἑλκομένων τε καὶ εἰς
ἑαυτὰς συναγομένων οὐκ ἐνδέχεται μὴ οὐ συναι-
170 ρεῖσθαι τὸ μῆκος. ἀλλὰ μὴν || ἐναργῶς γε φαίνεται
καταπινόντων συναιρούμενον καὶ τοσοῦτον ὁ
λάρυγξ ἀνατρέχων, ὅσον ὁ στόμαχος κατασπᾶ-
ται, καὶ ὅταν γε συμπληρωθείσης τῆς ἐν τῷ
καταπίνειν ἐνεργείας ἀφεθῇ τῆς τάσεως ὁ στόμαχος, ἐναργῶς πάλιν φαίνεται καταφερόμενος ὁ
λάρυγξ· ὁ γὰρ ἔνδον χιτὼν τῆς γαστρὸς ὁ τὰς
εὐθείας ἶνας ἔχων ὁ καὶ τὸν στόμαχον ὑπαλείφων
καὶ τὸ στόμα τοῖς ἐντὸς μέρεσιν ἐπεκτείνεται τοῦ
λάρυγγος, ὥστ' οὐκ ἐνδέχεται κατασπώμενον
αὐτὸν ὑπὸ τῆς κοιλίας μὴ οὐ συνεπισπᾶσθαι καὶ
τὸν λάρυγγα.

Ὅτι δ' αἱ περιφερεῖς ἶνες, αἷς περιστέλλεται
τά τ' ἄλλα μόρια καὶ ἡ γαστήρ, οὐ συναιροῦσι
τὸ μῆκος, ἀλλὰ συστέλλουσι καὶ στενοῦσι τὴν
εὐρύτητα, καὶ παρ' αὐτοῦ λαβεῖν ἔστιν ὁμολογούμενον Ἐρασιστράτου· περιστέλλεσθαι γάρ φησι
τοῖς σιτίοις τὴν γαστέρα κατὰ τὸν τῆς πέψεως
ἅπαντα χρόνον. ἀλλ' εἰ περιστέλλεται μέν,
οὐδὲν δὲ τοῦ μήκους ἀφαιρεῖται τῆς κοιλίας, οὐκ
ἔστι τῆς περισταλτικῆς κινήσεως ἴδιον τὸ κατα-
σπᾶν κάτω τὸν στόμαχον. ὅπερ γὰρ αὐτὸς ὁ
Ἐρασίστρατος εἶπε, τοῦτο μόνον αὐτὸ συμ-
171 βήσεται τὸ τῶν ἄνω συστελ||λομένων διαστέλ-
λεσθαι τὰ κάτω. τοῦτο δ' ὅτι, κἂν εἰς νεκροῦ τὸν
στόμαχον ὕδατος ἐγχέῃς, φαίνεται γιγνόμενον,
οὐδεὶς ἀγνοεῖ. ταῖς γὰρ τῶν ὑλῶν διὰ στενοῦ

---

[1] cf. p. 97.

the cavity contained by them becomes less; and when the longitudinal fibres contract and draw in upon themselves, the length must necessarily be curtailed. This curtailment of length, indeed, is well seen in the act of swallowing : the larynx is seen to rise upwards to exactly the same degree that the gullet is drawn downwards; while, after the process of swallowing has been completed and the gullet is released from tension, the larynx can be clearly seen to sink down again. This is because the inner coat of the stomach, which has the longitudinal fibres and which also lines the gullet and the mouth, extends to the interior of the larynx, and it is thus impossible for it to be drawn down by the stomach without the larynx being involved in the traction.

Further, it will be found acknowledged in Erasistratus's own writings that the circular fibres (by which the stomach as well as other parts performs its contractions) do not curtail its length, but contract and lesson its breadth. For he says that the stomach contracts peristaltically round the food during the whole period of digestion. But if it contracts, without in any way being diminished in length, this is because downward traction of the gullet is not a property of the movement of circular peristalsis. For what alone happens, as Erasistratus himself said, is that when the upper parts contract the lower ones dilate.[1] And everyone knows that this can be plainly seen happening even in a dead man, if water be poured down his throat; this symptom [2] results from the passage of matter through a narrow

---

[2] For " symptom," *cf.* p. 13, and p. 12, note 3. " Transitum namque materiae per angustum corpus id accidens consequitur " (Linacre). Less a " result " or " consequence " than an " accompaniment."

σώματος ὁδοιπορίαις ἀκόλουθόν ἐστι τὸ σύμ-
πτωμα· θαυμαστὸν γάρ, εἰ διερχομένου τινὸς
αὐτὸν ὄγκου μὴ διασταλήσεται. οὐκοῦν τὸ μὲν
τῶν ἄνω συστελλομένων διαστέλλεσθαι τὰ κάτω
κοινόν ἐστι καὶ τοῖς νεκροῖς σώμασι, ' δι' ὧν
ὁπωσοῦν τι διεξέρχεται, καὶ τοῖς ζῶσιν, εἴτε
περιστέλλοιτο τοῖς διερχομένοις εἴθ' ἕλκοιτο.

Τὸ δὲ τῆς τοῦ μήκους συναιρέσεως ἴδιον τῶν
τὰς εὐθείας ἶνας ἐχόντων ὀργάνων, ἵν' ἐπισπά-
σωνταί τι. ἀλλὰ μὴν ἐδείχθη κατασπώμενος ὁ
στόμαχος, οὐ γὰρ ἂν εἷλκε τὸν λάρυγγα· δῆλον
οὖν, ὡς ἡ γαστὴρ ἕλκει τὰ σιτία διὰ τοῦ
στομάχου.

Καὶ ἡ κατὰ τὸν ἔμετον δὲ τῶν ἐμουμένων ἄχρι
τοῦ στόματος φορὰ πάντως μέν που καὶ αὐτὴ τὰ
μὲν ὑπὸ τῶν ἀναφερομένων διατεινόμενα μέρη
τοῦ στομάχου διεστῶτα κέκτηται, τῶν πρόσω δ'
ὅ τι ἂν ἑκάστοτ' ἐπιλαμβάνηται, τοῦτ' ἀρχόμενοι
172 διαστέλλεται, τὸ δ' ‖ ὄπισθεν καταλείπει δηλονότι
συστελλόμενον, ὥσθ' ὁμοίαν εἶναι πάντῃ τὴν
διάθεσιν τοῦ στομάχου κατά γε τοῦτο τῇ τῶν
καταπινόντων· ἀλλὰ τῆς ὁλκῆς μὴ παρούσης τὸ
μῆκος ὅλον ἴσον ἐν τοῖς τοιούτοις συμπτώμασι
διαφυλάττεται.

Διὰ τοῦτο δὲ καὶ καταπίνειν ῥᾷόν ἐστιν ἢ ἐμεῖν,
ὅτι καταπίνεται μὲν ἀμφοῖν τῆς γαστρὸς τῶν
χιτώνων ἐνεργούντων, τοῦ μὲν ἐντὸς ἕλκοντος, τοῦ
δ' ἐκτὸς περιστελλομένου τε καὶ συνεπωθοῦντος,
ἐμεῖται δὲ θατέρου μόνου τοῦ ἔξωθεν ἐνεργοῦντος,

---

[1] i.e. this is a purely mechanical process.

channel; it would be extraordinary if the channel
did not dilate when a mass was passing through it.[1]
Obviously then the dilatation of the lower parts
along with the contraction of the upper is common
both to dead bodies, when anything whatsoever is
passing through them, and to living ones, whether
they contract peristaltically round their contents or
attract them.[2]

Curtailment of length, on the other hand, is
peculiar to organs which possess longitudinal fibres
for the purpose of attraction. But the gullet was
shown to be pulled down; for otherwise it would
not have drawn upon the larynx. It is therefore
clear that the stomach attracts food by the gullet.

Further, in *vomiting*, the mere passive conveyance
of rejected matter up to the mouth will certainly
itself suffice to keep open those parts of the oesophagus which are distended by the returned food;
as it occupies each part in front [above], it first
dilates this, and of course leaves the part behind
[below] contracted. Thus, in this respect at least, the
condition of the gullet is precisely similar to what it
is in the act of swallowing.[3] But there being no
traction, the whole length remains equal in such
cases.

And for this reason it is easier to swallow than
to vomit, for deglutition results from *both* coats
of the stomach being brought into action, the inner
one exerting a pull and the outer one helping by
peristalsis and propulsion, whereas emesis occurs
from the outer coat alone functioning, without there

[2] *i.e.* this phenomenon is a proof neither of *peristolé* nor
of attraction. *cf.* p. 97, note 2.
[3] Contraction and dilatation of course being reversed.

267

οὐδενὸς ἕλκοντος εἰς τὸ στόμα. οὐ γὰρ δὴ ὥσπερ
ἡ τῆς γαστρὸς ὄρεξις προηγεῖτο τοῦ καταπίνειν
τὰ σιτία, τὸν αὐτὸν τρόπον κἂν τοῖς ἐμέτοις
ἐπιθυμεῖ τι τῶν κατὰ τὸ στόμα μορίων τοῦ γιγνομένου παθήματος, ἀλλ' ἄμφω τῆς γαστρὸς αὐτῆς
εἰσιν ἐναντίαι διαθέσεις, ὀρεγομένης μὲν καὶ
προσιεμένης τὰ χρήσιμά τε καὶ οἰκεῖα, δυσχεραι-
νούσης δὲ καὶ ἀποτριβομένης τὰ ἀλλότρια. διὸ
καὶ τὸ καταπίνειν αὐτὸ τοῖς μὲν ἱκανῶς ὀρεγο-
μένοις τῶν οἰκείων ἐδεσμάτων τῇ γαστρὶ τάχιστα
γίγνεται, σαφῶς ἑλκούσης αὐτὰ καὶ κατασπώσης
πρὶν ἢ μασηθῆναι, τοῖς δ' ἤτοι φάρμακόν τι κατ'
173 ἀνάγ‖κην πίνουσιν ἢ σιτίον ἐν χώρᾳ φαρμάκου
προσφερομένοις ἀνιαρὰ καὶ μόγις ἡ κατάποσις
αὐτῶν ἐπιτελεῖται.

Δῆλος οὖν ἐστιν ἐκ τῶν εἰρημένων ὁ μὲν ἔνδον
χιτὼν τῆς γαστρὸς ὁ τὰς εὐθείας ἔχων ἶνας τῆς
ἐκ τοῦ στόματος εἰς αὐτὴν ὁλκῆς ἕνεκα γεγονὼς
καὶ διὰ τοῦτ' ἐν ταῖς καταπόσεσι μόναις ἐνεργῶν,
ὁ δ' ἔξωθεν ὁ τὰς ἐγκαρσίας ἔχων ἕνεκα μὲν τοῦ
περιστέλλεσθαι τοῖς ἐνυπάρχουσι καὶ προωθεῖν
αὐτὰ τοιοῦτος ἀποτελεσθείς, ἐνεργῶν δ' οὐδὲν
ἧττον ἐν τοῖς ἐμέτοις ἢ ταῖς καταπόσεσιν. ἐναρ-
γέστατα δὲ μαρτυρεῖ τῷ λεγομένῳ καὶ τὸ κατὰ
τὰς χάννας τε καὶ τοὺς συνόδοντας γιγνόμενον·
εὑρίσκεται γὰρ ἐνίοτε τούτων ἡ γαστὴρ ἐν τῷ
στόματι καθάπερ καὶ ὁ Ἀριστοτέλης ἐν ταῖς περὶ

---

[1] The *channa* is a kind of sea-perch; "a species of *Ser-
ranus*, either *S. scriba* or *S. cabrilla*" (D'Arcy W. Thompson).
cf. Aristotle's *Nat. Hist.* (D'Arcy Thompson's edition, Oxford, 1910), IV., xi., 538 A, 20. The *synodont* "is not to be
identified with certainty, but is supposed to be *Dentex vul-*

being any kind of pull towards the mouth. For, although the swallowing of food is ordinarily preceded by a feeling of desire on the part of the stomach, there is in the case of vomiting no corresponding desire from the mouth-parts for the experience; the two are opposite dispositions of the stomach itself; it yearns after and tends towards what is advantageous and proper to it, it loathes and rids itself of what is foreign. Thus the actual process of swallowing occurs very quickly in those who have a good appetite for such foods as are proper to the stomach; this organ obviously draws them in and down before they are masticated; whereas in the case of those who are forced to take a medicinal draught or who take food as medicine, the swallowing of these articles is accomplished with distress and difficulty.

From what has been said, then, it is clear that the inner coat of the stomach (that containing longitudinal fibres) exists for the purpose of exerting a pull from mouth to stomach, and that it is only in deglutition that it is active, whereas the external coat, which contains transverse fibres, has been so constituted in order that it may contract upon its contents and propel them forward; this coat furthermore, functions in vomiting no less than in swallowing. The truth of my statement is also borne out by what happens in the case of the *channae* and *synodonts* [1]; the stomachs of these animals are sometimes found in their mouths, as also Aristotle writes in his *History*

garis," that is, an edible Mediterranean perch. "It is not the stomach," adds Prof. Thompson, "but the air-bladder that gets everted and hangs out of the mouth in fishes, especially when they are hauled in from a considerable depth." *cf.* *H. A.*, VIII., ii., 591 B, 5.

ζῴων ἔγραψεν ἱστορίαις καὶ προστίθησί γε τὴν
αἰτίαν ὑπὸ λαιμαργίας αὐτοῖς τοῦτο συμβαίνειν
φάσκων.

Ἔχει γὰρ ὧδε· κατὰ τὰς σφοδροτέρας ὀρέξεις
ἄνω προστρέχει πᾶσι τοῖς ζῴοις ἡ γαστήρ, ὥστε
τινὲς τοῦ πάθους αἴσθησιν ἐναργῆ σχόντες
ἐξέρπειν αὐτοῖς φασι τὴν κοιλίαν, ἐνίων δὲ μασω-
174 μένων ἔτι καὶ μήπω ‖ καλῶς ἐν τῷ στόματι
τὰ σιτία κατεργασαμένων ἐξαρπάζει φανερῶς
ἀκόντων. ἐφ᾽ ὧν οὖν ζῴων φύσει λαιμάργων
ὑπαρχόντων ἥ τ᾽ εὐρυχωρία τοῦ στόματός ἐστι
δαψιλὴς ἥ τε τῆς γαστρὸς θέσις ἐγγύς, ὡς ἐπὶ
συνόδοντός τε καὶ χάννης, οὐδὲν θαυμαστόν, ὅταν
ἱκανῶς πεινάσαντα διώκῃ τι τῶν μικροτέρων
ζῴων, εἶτ᾽ ἤδη πλησίον ᾖ τοῦ συλλαβεῖν, ἀνα-
τρέχειν ἐπειγούσης τῆς ἐπιθυμίας εἰς τὸ στόμα
τὴν γαστέρα. γενέσθαι δ᾽ ἄλλως ἀμήχανον τοῦτο
μὴ οὐχ ὥσπερ διὰ χειρὸς τοῦ στομάχου τῆς
γαστρὸς ἐπισπωμένης εἰς ἑαυτὴν τὰ σιτία. καθάπερ γὰρ καὶ ἡμεῖς ὑπὸ προθυμίας ἐνίοτε τῇ χειρὶ
συνεπεκτείνομεν ὅλους ἡμᾶς αὐτοὺς ἕνεκα τοῦ
θᾶττον ἐπιδράξασθαι τοῦ προκειμένου σώματος,
οὕτω καὶ ἡ γαστὴρ οἷον χειρὶ τῷ στομάχῳ
συνεπεκτείνεται. καὶ διὰ τοῦτ᾽ ἐφ᾽ ὧν ζῴων ἅμα
τὰ τρία ταυτὶ συνέπεσεν, ἔφεσίς τε σφοδρὰ τῆς
τροφῆς ὅ τε στόμαχος μικρὸς ἥ τ᾽ εὐρυχωρία τοῦ
στόματος δαψιλής, ἐπὶ τούτων ὀλίγη ῥοπὴ τῆς
ἐπεκτάσεως εἰς τὸ στόμα τὴν κοιλίαν ὅλην ἀνα-
φέρει.

Ἥρκει μὲν οὖν ἴσως ἀνδρὶ φυσικῷ παρ᾽ αὐτῆς
175 μόνης τῆς κατασκευῆς τῶν ὀργά‖νων τὴν ἔνδειξιν
τῆς ἐνεργείας λαμβάνειν. οὐ γὰρ δὴ μάτην γ᾽

of Animals; he also adds the cause of this : he says that it is owing to their voracity.

The facts are as follows. In all animals, when the appetite is very intense, the stomach rises up, so that some people who have a clear perception of this condition say that their stomach "creeps out" of them ; in others, who are still masticating their food and have not yet worked it up properly in the mouth, the stomach obviously snatches away the food from them against their will. In those animals, therefore, which are naturally voracious, in whom the mouth cavity is of generous proportions, and the stomach situated close to it (as in the case of the synodont and channa), it is in no way surprising that, when they are sufficiently hungry and are pursuing one of the smaller animals, and are just on the point of catching it, the stomach should, under the impulse of desire, spring into the mouth. And this cannot possibly take place in any other way than by the stomach drawing the food to itself by means of the gullet, as though by a hand. In fact, just as we ourselves, in our eagerness to grasp more quickly something lying before us, sometimes stretch out our whole bodies along with our hands, so also the stomach stretches itself forward along with the gullet, which is, as it were, its hand. And thus, in these animals in whom those three factors co-exist—an excessive propensity for food, a small gullet, and ample mouth proportions—in these, any slight tendency to movement forwards brings the whole stomach into the mouth.

Now the constitution of the organs might itself suffice to give a naturalist an indication of their functions. For Nature would never have purpose-

ἂν ἡ φύσις ἐκ δυοῖν χιτώνων ἐναντίως ἀλλήλοις
ἐχόντων ἀπειργάσατο τὸν οἰσοφάγον, εἰ μὴ καὶ
διαφόρως ἑκάτερος αὐτῶν ἐνεργεῖν ἔμελλεν. ἀλλ'
ἐπεὶ πάντα μᾶλλον ἢ τὰ τῆς φύσεως ἔργα δια-
γιγνώσκειν οἱ περὶ τὸν Ἐρασίστρατόν εἰσιν
ἱκανοί, φέρε κἀκ τῆς τῶν ζώων ἀνατομῆς ἐπι-
δείξωμεν αὐτοῖς, ὡς ἑκάτερος τῶν χιτώνων ἐνεργεῖ
τὴν εἰρημένην ἐνέργειαν. εἰ δή τι λαβὼν ζῷον,
εἶτα γυμνώσας αὐτοῦ τὰ περικείμενα τῷ στομάχῳ
σώματα χωρὶς τοῦ διατεμεῖν τινα τῶν νεύρων ἢ
τῶν ἀρτηριῶν ἢ τῶν φλεβῶν τῶν αὐτόθι τεταγμένων ἐθέλοις ἀπὸ τῆς γένυος ἕως τοῦ θώρακος
εὐθείαις τομαῖς διελεῖν τὸν ἔξω χιτῶνα τὸν τὰς
ἐγκαρσίας ἶνας ἔχοντα κἄπειτα τῷ ζῴῳ τροφὴν
προσενέγκοις, ὄψει καταπίνον αὐτὸ καίτοι τῆς
περισταλτικῆς ἐνεργείας ἀπολωλυίας. εἰ δ' αὖ
πάλιν ἐφ' ἑτέρου ζῴου διατέμοις ἀμφοτέρους τοὺς
χιτῶνας τομαῖς ἐγκαρσίαις, θεάσῃ καὶ τοῦτο
καταπίνον οὐκέτ' ἐνεργούντως τοῦ ἐντός. ᾧ δῆλον,
ὅτι καὶ διὰ θατέρου μὲν αὐτῶν καταπίνειν οἷόν
176 τ' ἐστίν, ‖ ἀλλὰ χεῖρον ἢ δι' ἀμφοτέρων. πρὸς
γὰρ αὖ τοῖς ἄλλοις καὶ τοῦτ' ἔστι θεάσασθαι
σαφῶς ἐπὶ τῆς εἰρημένης ἀνατομῆς, ὡς ἐν τῷ
καταπίνειν ὑποπίμπλαται πνεύματος ὁ στόμαχος
τοῦ συγκαταπινομένου τοῖς σιτίοις, ὃ περιστελλομένου μὲν τοῦ ἔξωθεν χιτῶνος ὠθεῖται ῥᾳδίως
εἰς τὴν γαστέρα σὺν τοῖς ἐδέσμασι, μόνου δὲ τοῦ
ἔνδον ὑπάρχοντος ἐμποδὼν ἵσταται τῇ φορᾷ τῶν

---

¹ Under the term "neura," tendons were often included
as well as nerves. Similarly in modern Dutch the word
zenuw ("sinew") means both a tendon and a nerve; zenuw-
achtig = "nervous."

lessly constructed the oesophagus of two coats with
contrary dispositions; they must also have each been
meant to have a different action. The Erasistratean
school, however, are capable of anything rather than
of recognizing the effects of Nature. Come, therefore,
let us demonstrate to them by animal dissection as
well that each of the two coats does exercise the
activity which I have stated. Take an animal, then;
lay bare the structures surrounding the gullet, without
severing any of the nerves,[1] arteries, or veins which
are there situated; next divide with vertical incisions,
from the lower jaw to the thorax, the outer coat of the
oesophagus (that containing transverse fibres); then
give the animal food and you will see that it still swallows although the peristaltic function has been abol-
ished. If, again, in another animal, you cut through
both coats [2] with transverse incisions, you will observe
that this animal also swallows although the inner coat
is no longer functioning. From this it is clear that
the animal can also swallow by either of the two
coats, although not so well as by both. For the
following also, in addition to other points, may be
distinctly observed in the dissection which I have
described—that during deglutition the gullet becomes slightly filled with air which is swallowed
along with the food, and that, when the outer coat is
contracting, this air is easily forced with the food
into the stomach, but that, when there only exists
an inner coat, the air impedes the conveyance of

[2] Rather than the alternative reading, τὸν ἔσωθεν χιτῶνα.
Galen apparently supposes that the outer coat will not be
damaged, as the cuts will pass *between* its fibres. These cuts
would be, presumably, short ones, at various levels, no single
one of them involving the whole circumference of the gullet.

σιτίων διατεῖνόν τ᾽ αὐτὸν καὶ τὴν ἐνέργειαν
ἐμποδίζον.

Ἀλλ᾽ οὔτε τούτων οὐδὲν Ἐρασίστρατος εἶπεν
οὔθ᾽ ὡς ἡ σκολιὰ θέσις τοῦ στομάχου διαβάλλει
σαφῶς τὸ δόγμα τῶν νομιζόντων ὑπὸ τῆς ἄνωθεν
βολῆς μόνης ποδηγούμενα μέχρι τῆς γαστρὸς
ἰέναι τὰ καταπινόμενα. μόνον δ᾽ ὅτι πολλὰ τῶν
μακροτραχήλων ζῴων ἐπικεκυφότα καταπίνει,
καλῶς εἶπεν. ᾧ δῆλον, ὅτι τὸ φαινόμενον οὐ
τὸ πῶς καταπίνομεν ἀποδείκνυσιν, ἀλλὰ τὸ πῶς
οὐ καταπίνομεν· ὅτι γὰρ μὴ διὰ μόνης τῆς ἄνωθεν
βολῆς, ἐκ τούτου δῆλον· οὐ μὴν εἴθ᾽ ἑλκούσης
τῆς κοιλίας εἴτε παράγοντος αὐτὰ τοῦ στομάχου,
177 δῆλον ἤδη πω. ἀλλ᾽ ἡμεῖς γε ‖ πάντας τοὺς
λογισμοὺς εἰπόντες τούς τ᾽ ἐκ τῆς κατασκευῆς
τῶν ὀργάνων ὁρμωμένους καὶ τοὺς ἀπὸ τῶν ἄλλων
συμπτωμάτων τῶν τε πρὸ τοῦ γυμνωθῆναι τὸν
στόμαχον καὶ γυμνωθέντος, ὡς ὀλίγῳ πρόσθεν
ἐλέγομεν, ἱκανῶς ἐνεδειξάμεθα τοῦ μὲν ἕλκειν
ἕνεκα τὸν ἐντὸς χιτῶνα, τοῦ δ᾽ ἀπωθεῖν τὸν ἐκτὸς
γεγονέναι.

Προυθέμεθα μὲν οὖν ἀποδεῖξαι τὴν καθεκτικὴν
δύναμιν ἐν ἑκάστῳ τῶν ὀργάνων οὖσαν, ὥσπερ
ἐν τῷ πρόσθεν λόγῳ τὴν ἑλκτικήν τε καὶ προσέτι
τὴν ἀλλοιωτικήν. ὑπὸ δὲ τῆς ἀκολουθίας τοῦ
λόγου τὰς τέτταρας ἀπεδείξαμεν ὑπαρχούσας τῇ
γαστρί, τὴν ἑλκτικὴν μὲν ἐν τῷ καταπίνειν, τὴν
καθεκτικὴν δ᾽ ἐν τῷ πέττειν, τὴν ἀπωστικὴν δ᾽ ἐν
τοῖς ἐμέτοις καὶ ταῖς τῶν πεπεμμένων σιτίων εἰς
τὸ λεπτὸν ἔντερον ὑποχωρήσεσιν, αὐτὴν δὲ τὴν
πέψιν ἀλλοίωσιν ὑπάρχειν.

food, by distending this coat and hindering its action.

But Erasistratus said nothing about this, nor did he point out that the oblique situation of the gullet clearly confutes the teaching of those who hold that it is simply by virtue of the impulse from above that food which is swallowed reaches the stomach. The only correct thing he said was that many of the long-necked animals bend down to swallow. Hence, clearly, the observed fact does not show how we swallow but how we do not swallow. For from this observation it is clear that swallowing is not due merely to the impulse from above; it is yet, however, not clear whether it results from the food being attracted by the stomach, or conducted by the gullet. For our part, however, having enumerated all the different considerations—those based on the constitution of the organs, as well as those based on the other symptoms which, as just mentioned, occur both before and after the gullet has been exposed—we have thus sufficiently proved that the inner coat exists for the purpose of attraction and the outer for the purpose of propulsion.

Now the original task we set before ourselves was to demonstrate that the *retentive* faculty exists in every one of the organs, just as in the previous book we proved the existence of the *attractive*, and, over and above this, the *alterative* faculty. Thus, in the natural course of our argument, we have demonstrated these four faculties existing in the stomach—the attractive faculty in connection with swallowing, the retentive with digestion, the expulsive with vomiting and with the descent of digested food into the small intestine—and digestion itself we have shown to be a process of *alteration*.

IX

Οὔκουν ἔτ᾽ ἀπορήσομεν οὐδὲ περὶ τοῦ σπληνός,
εἰ ἕλκει μὲν τὸ οἰκεῖον, ἀποκρίνει δὲ τὸ ἀλλότριον,
ἀλλοιοῦν δὲ καὶ κατέχειν, ὅσον ἂν ἐπισπάσηται,
πέφυκεν, οὐδὲ περὶ ἥπατος ἢ φλεβὸς ἢ ἀρτηρίας
178 ἢ καρδίας ἢ τῶν || ἄλλων τινός· ἀναγκαῖαι γὰρ
ἐδείχθησαν αἱ τέτταρες αὗται δυνάμεις ἅπαντι
μορίῳ τῷ μέλλοντι θρέψεσθαι καὶ διὰ τοῦτ᾽ αὐτὰς
ὑπηρέτιδας εἶναι θρέψεως ἔφαμεν· ὡς γὰρ τὸ τῶν
ἀνθρώπων ἀποπάτημα τοῖς κυσὶν ἥδιστον, οὕτω
καὶ τὰ τοῦ ἥπατος περιττώματα τὸ μὲν τῷ
σπληνί, τὸ δὲ τῇ χοληδόχῳ κύστει, τὸ δὲ τοῖς
νεφροῖς οἰκεῖον.

X

Καὶ λέγειν ἔτι περὶ τῆς τούτων γενέσεως οὐκ
ἂν ἐθέλοιμι μεθ᾽ Ἱπποκράτην καὶ Πλάτωνα καὶ
Ἀριστοτέλην καὶ Διοκλέα καὶ Πραξαγόραν καὶ
Φιλότιμον· οὐδὲ γὰρ οὐδὲ περὶ τῶν δυνάμεων
εἶπον ἄν, εἴ τις τῶν ἔμπροσθεν ἀκριβῶς ἐξειργά-
σατο τὸν ὑπὲρ αὐτῶν λόγον.

Ἐπεὶ δ᾽ οἱ μὲν παλαιοὶ καλῶς ὑπὲρ αὐτῶν
ἀποφηνάμενοι παρέλιπον ἀγωνίσασθαι τῷ λόγῳ,
μηδ᾽ ὑπονοήσαντες ἔσεσθαί τινας εἰς τοσοῦτον
ἀναισχύντους σοφιστάς, ὡς ἀντιλέγειν ἐπιχειρῆ-
σαι τοῖς ἐναργέσιν, οἱ νεώτεροι δὲ τὸ μέν τι

---

[1] cf. p. 205.

IX

Concerning the spleen, also, we shall therefore have
no further doubts [1] as to whether it attracts what is
proper to it, rejects what is foreign, and has a
natural power of altering and retaining all that it
attracts ; nor shall we be in any doubt as to the liver,
veins, arteries, heart, or any other organ.   For these
four faculties have been shown to be necessary for
every part which is to be nourished ; this is why we
have called these faculties the *handmaids of nutrition.*
For just as human faeces are most pleasing to dogs,
so the residual matters from the liver are, some of
them, proper to the spleen,[2] others to the gall-bladder,
and others to the kidneys.

X

I should not have cared to say anything further as
to the origin of these [surplus subtances] after Hippocrates, Plato, Aristotle, Diocles, Praxagoras, and
Philotimus, nor indeed should I even have said
anything about the *faculties,* if any of our predecessors
had worked out this subject thoroughly.

While, however, the statements which the Ancients
made on these points were correct, they yet omitted
to defend their arguments with logical proofs ; of
course they never suspected that there could be
sophists so shameless as to try to contradict obvious
facts.   More recent physicians, again, have been

---

[2] Thus Galen elsewhere calls the spleen a mere *emunctory*
(ἐκμαγεῖον) of the liver.   *cf.* p. 214, note 1.

νικηθέντες ὑπὸ τῶν σοφισμάτων ἐπείσθησαν
αὐτοῖς, τὸ δέ τι καὶ ἀντιλέγειν ἐπιχειρήσαντες
ἀποδεῖν μοι πολὺ τῆς τῶν παλαιῶν ἔδοξαν δυνά-
179 μεως, ‖ διὰ τοῦθ', ὡς ἂν ἐκείνων αὐτῶν, εἴπερ ἔτ'
ἦν τις, ἀγωνίσασθαί μοι δοκεῖ πρὸς τοὺς ἀνατρέ-
ποντας τῆς τέχνης τὰ κάλλιστα, καὶ αὐτὸς οὕτως
ἐπειράθην συνθεῖναι τοὺς λόγους.

Ὅτι δ' ἢ οὐδὲν ἢ παντάπασιν ἀνύσω τι σμικρόν, οὐκ ἀγνοῶ· πάμπολλα γὰρ εὑρίσκω τελέως
μὲν ἀποδεδειγμένα τοῖς παλαιοῖς, οὔτε δὲ συνετὰ
τοῖς πολλοῖς τῶν νῦν δι' ἀμαθίαν ἀλλ' οὐδ'
ἐπιχειρούμενα γιγνώσκεσθαι διὰ ῥαθυμίαν, οὔτ',
εἰ καὶ γνωσθείη τινί, δικαίως ἐξεταζόμενα.

Χρὴ γὰρ τὸν μέλλοντα γνώσεσθαί τι τῶν πολλῶν ἄμεινον εὐθὺς μὲν καὶ τῇ φύσει καὶ τῇ πρώτῃ
διδασκαλίᾳ πολὺ τῶν ἄλλων διενεγκεῖν· ἐπειδὰν
δὲ γένηται μειράκιον, ἀληθείας τινὰ σχεῖν ἐρω-
τικὴν μανίαν, ὥσπερ ἐνθουσιῶντα καὶ μήθ' ἡμέρας
μήτε νυκτὸς διαλείπειν σπεύδοντά τε καὶ συντε-
ταμένον ἐκμαθεῖν, ὅσα τοῖς ἐνδοξοτάτοις εἴρηται
τῶν παλαιῶν· ἐπειδὰν δ' ἐκμάθῃ, κρίνειν αὐτὰ
καὶ βασανίζειν χρόνῳ παμπόλλῳ καὶ σκοπεῖν,
πόσα μὲν ὁμολογεῖ τοῖς ἐναργῶς φαινομένοις,
180 πόσα δὲ διαφέρεται, ‖ καὶ οὕτω τὰ μὲν αἱρεῖσθαι,
τὰ δ' ἀποστρέφεσθαι. τῷ μὲν δὴ τοιούτῳ πάνυ
σφόδρα χρησίμους ἤλπικα τοὺς ἡμετέρους ἔσε-

partly conquered by the sophistries of these fellows and have given credence to them; whilst others who attempted to argue with them appear to me to lack to a great extent the power of the Ancients. For this reason I have attempted to put together my arguments in the way in which it seems to me the Ancients, had any of them been still alive, would have done, in opposition to those who would overturn the finest doctrines of our art.

I am not, however, unaware that I shall achieve either nothing at all or else very little. For I find that a great many things which have been conclusively demonstrated by the Ancients are unintelligible to the bulk of the Moderns owing to their ignorance—nay, that, by reason of their laziness, they will not even make an attempt to comprehend them; and even if any of them have understood them, they have not given them impartial examination.

The fact is that he whose purpose is to know anything better than the multitude do must far surpass all others both as regards his nature and his early training. And when he reaches early adolescence he must become possessed with an ardent love for truth, like one inspired; neither day nor night may he cease to urge and strain himself in order to learn thoroughly all that has been said by the most illustrious of the Ancients. And when he has learnt this, then for a prolonged period he must test and prove it, observing what part of it is in agreement, and what in disagreement with obvious fact; thus he will choose this and turn away from that. To such an one my hope has been that my treatise would prove of the very greatest assistance. . . .

σθαι λόγους· εἶεν δ' ἂν ὀλίγοι παντάπασιν οὗτοι·
τοῖς δ' ἄλλοις οὕτω γενήσεται τὸ γράμμα περιττ-
τόν, ὡς εἰ καὶ μῦθον ὄνῳ τις λέγοι.

XI

Συμπεραντέον οὖν ἡμῖν τὸν λόγον ἕνεκα τῶν
τῆς ἀληθείας ἐφιεμένων ὅσα λείπει κατ' αὐτὸν
ἔτι προσθεῖσιν. ὡς γὰρ ἡ γαστὴρ ἕλκει μὲν
ἐναργῶς καὶ κατασπᾷ τὰ σιτία τοῖς σφόδρα
πεινώδεσι, πρὶν ἀκριβῶς ἐν τῷ στόματι λειωθῆναι, δυσχεραίνει δὲ καὶ ἀπωθεῖται τοῖς ἀποσί-
τοις τε καὶ πρὸς ἀνάγκην ἐσθίουσιν, οὕτω καὶ
τῶν ἄλλων ὀργάνων ἕκαστον ἀμφοτέρας ἔχει τὰς
δυνάμεις, τήν τε τῶν οἰκείων ἑλκτικὴν καὶ τὴν
τῶν ἀλλοτρίων ἀποκριτικήν. καὶ διὰ τοῦτο, κἂν
ἐξ ἑνὸς ᾖ χιτῶνος ὄργανόν τι συνεστώς, ὥσπερ
καὶ αἱ κύστεις ἀμφότεραι καὶ αἱ μῆτραι καὶ αἱ
φλέβες, ἀμφότερα τῶν ἰνῶν ἔχει τὰ γένη, τῶν
εὐθειῶν τε καὶ τῶν ἐγκαρσίων.

181 Καὶ μέν γε καὶ τρίτον τι || γένος ἰνῶν ἐστι
<τῶν> λοξῶν, ἔλαττον πολὺ τῷ πλήθει τῶν
προειρημένων δύο γενῶν. εὑρίσκεται δ' ἐν μὲν
τοῖς ἐκ δυοῖν χιτώνων συνεστηκόσιν ὀργάνοις
ἐν θατέρῳ μόνῳ ταῖς εὐθείαις ἰσὶν ἀναμεμιγμένον,
ἐν δὲ τοῖς ἐξ ἑνὸς ἅμα τοῖς ἄλλοις δύο γένεσι.
συνεπιλαμβάνουσι δ' αὗται μέγιστον τῇ τῆς καθεκτικῆς ὀνομασθείσης δυνάμεως ἐνεργείᾳ· δεῖται
γὰρ ἐν τούτῳ τῷ χρόνῳ πανταχόθεν ἐσφίγχθαι
καὶ περιτετάσθαι τοῖς ἐνυπάρχουσι τὸ μόριον, ἡ.

---

[1] cf. p. 269.

Still, such people may be expected to be quite few in number, while, as for the others, this book will be as superfluous to them as a tale told to an ass.

XI

For the sake, then, of those who are aiming at truth, we must complete this treatise by adding what is still wanting in it. Now, in people who are very hungry, the stomach obviously attracts or draws down the food before it has been thoroughly softened in the mouth, whilst in those who have no appetite or who are being forced to eat, the stomach is displeased and rejects the food.[1] And in a similar way each of the other organs possesses both faculties —that of attracting what is proper to it, and that of rejecting what is foreign. Thus, even if there be any organ which consists of only one coat (such as the two bladders,[2] the uterus, and the veins), it yet possesses both kinds of fibres, the longitudinal and the transverse.

But further, there are fibres of a third kind—the *oblique*—which are much fewer in number than the two kinds already spoken of. In the organs consisting of two coats this kind of fibre is found in the one coat only, mixed with the longitudinal fibres; but in the organs composed of one coat it is found along with the other two kinds. Now, these are of the greatest help to the action of the faculty which we have named *retentive*. For during this period the part needs to be tightly contracted and stretched over its contents at every point—the

[2] The urinary bladders of pigs (such as Galen dissected) are thin, and appear to have only one coat.

μὲν γαστὴρ ἐν τῷ τῆς πέψεως, αἱ μῆτραι δ' ἐν τῷ τῆς κυήσεως χρόνῳ παντί.

Ταῦτ' ἄρα καὶ ὁ τῆς φλεβὸς χιτὼν εἷς ὢν ἐκ πολυειδῶν ἰνῶν ἐγένετο καὶ τῶν τῆς ἀρτηρίας ὁ μὲν ἔξωθεν ἐκ τῶν στρογγύλων, ὁ δ' ἔσωθεν ἐκ μὲν τῶν εὐθειῶν πλείστων, ὀλίγων δέ τινων σὺν αὐταῖς καὶ τῶν λοξῶν, ὥστε τὰς μὲν φλέβας ταῖς μήτραις καὶ ταῖς κύστεσιν ἐοικέναι κατά γε τὴν τῶν ἰνῶν σύνθεσιν, εἰ καὶ τῷ πάχει λείπονται, τὰς δ' ἀρτηρίας τῇ γαστρί. μόνα δὲ πάντων ὀργάνων ἐκ δυοῖν θ' ἅμα καὶ ἀμφοτέρων ἐγκαρσίας ἐχόντων τὰς ἶνας ἐγένετο τὰ ἔντερα. τὸ δ' ὅτι
182 βέλτιον ἦν ‖ τῶν τ' ἄλλων ἑκάστῳ τοιούτῳ τὴν φύσιν ὑπάρχειν, οἱόνπερ καὶ νῦν ἐστι, τοῖς τ' ἐντέροις ἐκ δυοῖν ὁμοίων χιτώνων συγκεῖσθαι, τῆς περὶ χρείας μορίων πραγματείας ἐστίν. οὔκουν νῦν χρὴ ποθεῖν ἀκούειν περὶ τῶν τοιούτων, ὥσπερ οὐδὲ διὰ τί περὶ τοῦ πλήθους τῶν χιτώνων ἑκάστου τῶν ὀργάνων διαπεφώνηται τοῖς ἀνατομικοῖς ἀνδράσιν. ὑπὲρ μὲν γὰρ τούτων αὐτάρκως ἐν τοῖς περὶ τῆς ἀνατομικῆς διαφωνίας εἴρηται· περὶ δὲ τοῦ διότι τοιοῦτον ἕκαστον ἐγένετο τῶν ὀργάνων, ἐν τοῖς περὶ χρείας μορίων εἰρήσεται.

XII

Νυνὶ δ' οὐδέτερον τούτων πρόκειται λέγειν, ἀλλὰ τὰς φυσικὰς δυνάμεις μόνας ἀποδεικνύειν ἐν ἑκάστῳ τῶν ὀργάνων τέτταρας ὑπαρχούσας. ἐπὶ τοῦτ' οὖν πάλιν ἐπανελθόντες ἀναμνήσωμέν τε

---

[1] cf. p. 243.
[2] My suggestion is that Galen refers to (1) the *mucous*

stomach during the whole period of digestion,[1] and the uterus during that of gestation.

Thus too, the coat of a vein, being single, consists of various kinds of fibres; whilst the outer coat of an artery consists of circular fibres, and its inner coat mostly of longitudinal fibres, but with a few oblique ones also amongst them. Veins thus resemble the uterus or the bladder as regards the arrangement of their fibres, even though they are deficient in thickness; similarly arteries resemble the stomach. Alone of all organs the intestines consist of two coats of which both have their fibres transverse.[2] Now the proof that it was *for the best* that all the organs should be naturally such as they are (that, for instance, the intestines should be composed of two coats) belongs to the subject of the *use of parts*[3]; thus we must not now desire to hear about matters of this kind nor why the anatomists are at variance regarding the number of coats in each organ. For these questions have been sufficiently discussed in the treatise "On Disagreement in Anatomy." And the problem as to why each organ has such and such a character will be discussed in the treatise "On the Use of Parts."

XII

It is not, however, our business to discuss either of these questions here, but to consider duly the *natural faculties*, which, to the number of four, exist in each organ. Returning then, to this point, let us

coat, with its *valvulae conniventes*, and (2) the *muscular* coat, of which the chief layer is made up of circular fibres. *cf.* p. 262, note 1.   [3] Or *utility*.   283

τῶν ἔμπροσθεν εἰρημένων ἐπιθῶμέν τε κεφαλὴν
ἤδη τῷ λόγῳ παντὶ τὸ λεῖπον ἔτι προσθέντες.
ἐπειδὴ γὰρ ἕκαστον τῶν ἐν τῷ ζῴῳ μορίων ἕλκειν
εἰς ἑαυτὸ τὸν οἰκεῖον χυμὸν ἀποδέδεικται καὶ
πρώτη σχεδὸν αὕτη τῶν φυσικῶν ἐστι δυνάμεων,
183 ἐφεξῆς ‖ ἐκείνῳ γνωστέον, ὡς οὐ πρότερον ἀποτρίβεται τὴν ἐλχθεῖσαν <τροφὴν> ἤτοι σύμπασαν
ἢ καί τι περίττωμα αὐτῆς, πρὶν ἂν εἰς ἐναντίαν
μεταπέσῃ διάθεσιν ἢ αὐτὸ τὸ ὄργανον ἢ καὶ τῶν
περιεχομένων ἐν αὐτῷ τὰ πλεῖστα. ἡ μὲν οὖν
γαστήρ, ἐπειδὰν μὲν ἱκανῶς ἐμπλησθῇ τῶν σιτίων
καὶ τὸ χρηστότατον αὐτῶν εἰς τοὺς ἑαυτῆς χιτῶνας ἐναπόθηται βδάλλουσα, τηνικαῦτ' ἤδη τὸ
λοιπὸν ἀποτρίβεται καθάπερ ἄχθος ἀλλότριον·
αἱ κύστεις δ', ἐπειδὰν ἕκαστον τῶν ἐλχθέντων ἢ
τῷ πλήθει διατεῖνον ἢ τῇ ποιότητι δάκνον ἀνιαρὸν
γένηται.

Τῷ δ' αὐτῷ τρόπῳ καὶ αἱ μῆτραι· ἤτοι γάρ,
ἐπειδὰν μηκέτι φέρωσι διατεινόμεναι, τὸ λυποῦν
ἀποθέσθαι σπεύδουσιν ἢ τῇ ποιότητι δακνόμεναι
τῶν ἐκχυθέντων εἰς αὐτὰς ὑγρῶν. ἑκάτερον δὲ
τῶν εἰρημένων γίγνεται μὲν καὶ βιαίως ἔστιν ὅτε
καὶ ἀμβλώσκουσι τηνικαῦτα, γίγνεται δ' ὡς τὰ
πολλὰ καὶ προσηκόντως, ὅπερ οὐκ ἀμβλώσκειν
ἀλλ' ἀποκυΐσκειν τε καὶ τίκτειν ὀνομάζεται. τοῖς
μὲν οὖν ἀμβλωθριδίοις φαρμάκοις ἤ τισιν ἄλλοις
184 παθήμασι διαφθεί‖ρουσι τὸ ἔμβρυον ἤ τινας τῶν
ὑμένων αὐτοῦ ῥηγνύουσιν αἱ ἀμβλώσεις ἕπονται,
οὕτω δὲ κἀπειδὰν ἀνιαθῶσί ποθ' αἱ μῆτραι κακῶς
ἔχουσαι τῇ διατάσει, ταῖς δὲ τῶν ἐμβρύων αὐτῶν
κινήσεσι ταῖς σφοδροτάταις οἱ τόκοι, καθάπερ
καὶ τοῦθ' Ἱπποκράτει καλῶς εἴρηται. κοινὸν δ'

recall what has already been said, and set a crown to the whole subject by adding what is still wanting. For when every part of the animal has been shewn to draw into itself the juice which is proper to it (this being practically *the first of the natural faculties*), the next point to realise is that the part does not get rid either of this attracted nutriment as a whole, or even of any superfluous portion of it, until either the organ itself, or the major part of its contents also have their condition reversed. Thus, when the stomach is sufficiently filled with the food and has absorbed and stored away the most useful part of it in its own coats, it then rejects the rest like an alien burden. The same happens to the bladders, when the matter attracted into them begins to give trouble either because it distends them through its quantity or irritates them by its quality.

And this also happens in the case of the uterus ; for it is either because it can no longer bear to be stretched that it strives to relieve itself of its annoyance, or else because it is irritated by the quality of the fluids poured out into it. Now both of these conditions sometimes occur with actual violence, and then *miscarriage* takes place. But for the most part they happen in a normal way, this being then called not miscarriage but *delivery* or *parturition*. Now abortifacient drugs or certain other conditions which destroy the embryo or rupture certain of its membranes are followed by abortion, and similarly also when the uterus is in pain from being in a bad state of tension ; and, as has been well said by Hippocrates, excessive movement on the part of the embryo itself brings on labour. Now

ἁπασῶν τῶν διαθέσεων ἡ ἀνία καὶ ταύτης αἴτιον
τριττὸν ἢ ὄγκος περιττὸς ἤ τι βάρος ἢ δῆξις·
ὄγκος μέν, ἐπειδὰν μηκέτι φέρωσι διατεινόμεναι,
βάρος δ᾿, ἐπειδὰν ὑπὲρ τὴν ῥώμην αὐτῶν ᾖ τὸ
περιεχόμενον, δῆξις δ᾿, ἐπειδὰν ἤτοι τὰ πρότερον
ἐν τοῖς ὑμέσιν ὑγρὰ στεγόμενα ῥαγέντων αὐτῶν
εἰς αὐτὰς ἐκχυθῇ τὰς μήτρας ἢ καὶ σύμπαν
ἀποφθαρὲν τὸ κύημα σηπόμενόν τε καὶ διαλυό-
μενον εἰς μοχθηροὺς ἰχῶρας οὕτως ἐρεθίζῃ τε καὶ
δάκνῃ τὸν χιτῶνα τῶν ὑστερῶν.

Ἀνάλογον οὖν ἐν ἅπασι τοῖς ὀργάνοις ἕκαστα
τῶν τ᾿ ἔργων αὐτῶν τῶν φυσικῶν καὶ μέντοι τῶν
παθημάτων τε καὶ νοσημάτων φαίνεται γιγνόμενα, τὰ μὲν ἐναργῶς καὶ σαφῶς οὕτως, ὡς
ἀποδείξεως δεῖσθαι μηδέν, τὰ δ᾿ ἧττον μὲν ἐναρ-
185 γῶς, οὐ μὴν ἄγνωστά γε παντάπασι τοῖς ‖ ἐθέλουσι προσέχειν τὸν νοῦν.

Ἐπὶ μὲν οὖν τῆς γαστρὸς αἵ τε δήξεις ἐναργεῖς,
διότι πλείστης αἰσθήσεως μετέχει, τά τ᾿ ἄλλα
παθήματα τά τε ναυτίαν ἐμποιοῦντα καὶ οἱ
καλούμενοι καρδιωγμοὶ σαφῶς ἐνδείκνυνται τὴν
ἀποκριτικήν τε καὶ ἀπωστικὴν τῶν ἀλλοτρίων
δύναμιν, οὕτω δὲ κἀπὶ τῶν ὑστερῶν τε καὶ τῆς
κύστεως τῆς τὸ οὖρον ὑποδεχομένης· ἐναργῶς γὰρ
οὖν καὶ αὕτη φαίνεται μέχρι τοσούτου τὸ ὑγρὸν
ὑποδεχομένη τε καὶ ἀθροίζουσα, ἄχρις ἂν ἤτοι
πρὸς τοῦ πλήθους αὐτοῦ διατεινομένη μηκέτι
φέρῃ τὴν ἀνίαν ἢ πρὸς τῆς ποιότητος δακνομένη·
χρονίζον γὰρ ἕκαστον τῶν περιττωμάτων ἐν τῷ
σώματι σήπεται δηλονότι, τὸ μὲν ἐλάττονι, τὸ δὲ
πλείονι χρόνῳ, καὶ οὕτω δακνῶδές τε καὶ δριμὺ
καὶ ἀνιαρὸν τοῖς περιέχουσι γίγνεται. οὐ μὴν

pain is common to all these conditions, and of this there are three possible causes—either excessive bulk, or weight, or irritation; bulk when the uterus can no longer support the stretching, weight when the contents surpass its strength, and irritation when the fluids which had previously been pent up in the membranes, flow out, on the rupture of these, into the uterus itself, or else when the whole foetus perishes, putrefies, and is resolved into pernicious ichors, and so irritates and bites the coat of the uterus.

In all organs, then, both their natural effects and their disorders and maladies plainly take place on analogous lines,[1] some so clearly and manifestly as to need no demonstration, and others less plainly, although not entirely unrecognizable to those who are willing to pay attention.

Thus, to take the case of the stomach: the irritation is evident here because this organ possesses most sensibility, and among its other affections those producing nausea and the so-called heartburn clearly demonstrate the eliminative faculty which expels foreign matter. So also in the case of the uterus and the urinary bladder; this latter also may be plainly observed to receive and accumulate fluid until it is so stretched by the amount of this as to be incapable of enduring the pain; or it may be the quality of the urine which irritates it; for every superfluous substance which lingers in the body must obviously putrefy, some in a shorter, and some in a longer time, and thus it becomes pungent, acrid, and burdensome to the organ which contains it. This

[1] Relationship between physiology and pathology again emphasized. *cf.* p. 188, note 2.

ἐπί γε τῆς ἐπὶ τῷ ἥπατι κύστεως ὁμοίως. ἔχει·
ᾧ δῆλον, ὅτι νεύρων ἥκιστα μετέχει. χρὴ δὲ
κἀνταῦθα τόν γε φυσικὸν ἄνδρα τὸ ἀνάλογον
ἐξευρίσκειν. εἰ γὰρ ἕλκειν τε τὸν οἰκεῖον ἀπε-
δείχθη χυμόν, ὡς φαίνεσθαι πολλάκις μεστήν,
186 ἀποκρί∥νειν τε τὸν αὐτὸν τοῦτον οὐκ εἰς μακράν,
ἀναγκαῖόν ἐστιν αὐτὴν ἢ διὰ τὸ πλῆθος βαρυνο-
μένην ἢ τῆς ποιότητος μεταβαλλούσης ἐπὶ τὸ
δακνῶδές τε καὶ δριμὺ τῆς ἀποκρίσεως ἐφίεσθαι.
οὐ γὰρ δὴ τὰ μὲν σιτία τὴν ἀρχαίαν ὑπαλλάττει
ποιότητα ταχέως οὕτως, ὥστ᾽, ἐπειδὰν ἐμπέσῃ
τοῖς λεπτοῖς ἐντέροις, εὐθὺς εἶναι κόπρον, ἡ χολὴ
δ᾽ οὐ πολὺ μᾶλλον ἢ τὸ οὖρον, ἐπειδὰν ἅπαξ
ἐκπέσῃ τῶν φλεβῶν, ἐξαλλάττει τὴν ποιότητα,
τάχιστα μεταβάλλοντα καὶ σηπόμενα. καὶ μὴν
εἴπερ ἐπί τε τῶν κατὰ τὰς ὑστέρας καὶ τὴν
κοιλίαν καὶ τὰ ἔντερα καὶ προσέτι τὴν τὸ οὖρον
ὑποδεχομένην κύστιν ἐναργῶς φαίνεται διάτασίς
τις ἢ δῆξις ἢ ἄχθος ἐπεγεῖρον ἕκαστον τῶν
ὀργάνων εἰς ἀπόκρισιν, οὐδὲν χαλεπὸν κἀπὶ
τῆς χοληδόχου κύστεως ταὐτὸ τοῦτ᾽ ἐννοεῖν ἐπί
τε τῶν ἄλλων ἁπάντων ὀργάνων, ἐξ ὧν δηλονότι
καὶ αἱ ἀρτηρίαι καὶ αἱ φλέβες εἰσίν.

XIII

Οὐ μὴν οὐδὲ τὸ διὰ τοῦ αὐτοῦ πόρου τήν θ᾽
ὁλκὴν γίγνεσθαι καὶ τὴν ἀπόκρισιν ἐν διαφέ-
187 ρουσι ∥ χρόνοις οὐδὲν ἔτι χαλεπὸν ἐξευρεῖν, εἴ γε
καὶ τῆς γαστρὸς ὁ στόμαχος οὐ μόνον ἐδέσματα

does not apply, however, in the case of the bladder alongside the liver, whence it is clear that it possesses fewer nerves than do the other organs. Here too, however, at least the physiologist[1] must discover an analogy. For since it was shown that the gall-bladder attracts its own special juice, so as to be often found full, and that it discharges it soon after, this desire to discharge must be either due to the fact that it is burdened by the quantity or that the bile has changed in quality to pungent and acrid. For while food does not change its original quality so fast that it is already ordure as soon as it falls into the small intestine, on the other hand the bile even more readily than the urine becomes altered in quality as soon as ever it leaves the veins, and rapidly undergoes change and putrefaction. Now, if there be clear evidence in relation to the uterus, stomach, and intestines, as well as to the urinary bladder, that there is either some distention, irritation, or burden inciting each of these organs to elimination, there is no difficulty in imagining this in the case of the gall-bladder also, as well as in the other organs,—to which obviously the arteries and veins also belong.

XIII

Nor is there any further difficulty in ascertaining that it is through the same channel that both attraction and discharge take place at different times. For obviously the inlet to the stomach does not merely

[1] Or physicist—the investigator of the Physis or Nature. *cf.* p. 196, note 2. Note here the use of analogical reasoning. *cf.* p. 113, note 2.

καὶ πόματα παράγων εἰς αὐτήν, ἀλλὰ κἂν ταῖς
ναυτίαις τὴν ἐναντίαν ὑπηρεσίαν ὑπηρετῶν ἐναργῶς φαίνεται, καὶ τῆς ἐπὶ τῷ ἥπατι κύστεως
ὁ αὐχὴν εἰς ὃν ἅμα μὲν πληροῖ δι' αὐτοῦ τὴν
κύστιν, ἅμα δ' ἐκκενοῖ, καὶ τῶν μητρῶν ὁ
στόμαχος ὡσαύτως ὁδός ἐστιν εἴσω μὲν τοῦ
σπέρματος, ἔξω δὲ τοῦ κυήματος.

Ἀλλὰ κἀνταῦθα πάλιν ἡ μὲν ἐκκριτικὴ δύναμις
ἐναργής, οὐ μὴν ὁμοίως γ' αὐτῇ σαφὴς τοῖς
πολλοῖς ἡ ἑλκτική· ἀλλ' Ἱπποκράτης μὲν ἀρρώστου μήτρας αἰτιώμενος αὐχένα φησί· "Οὐ γὰρ
δύναται αὐτέης ὁ στόμαχος εἰρύσαι τὴν γονήν."

Ἐρασίστρατος δὲ καὶ Ἀσκληπιάδης εἰς τοσοῦτον ἥκουσι σοφίας, ὥστ' οὐ μόνον τὴν κοιλίαν καὶ
τὰς μήτρας ἀποστεροῦσι τῆς τοιαύτης δυνάμεως
ἀλλὰ καὶ τὴν ἐπὶ τῷ ἥπατι κύστιν ἅμα τοῖς
νεφροῖς. καίτοι γ' ὅτι μηδ' εἰπεῖν δυνατὸν ἕτερον
αἴτιον ἢ οὔρων ἢ χολῆς διακρίσεως, ἐν τῷ πρώτῳ
δέδεικται λόγῳ.

Καὶ μήτραν οὖν καὶ γαστέρα καὶ τὴν ἐπὶ
188 τῷ ἥπατι κύστιν δι' ἑνὸς καὶ ταὐτοῦ στο‖μάχου
τήν θ' ὁλκὴν καὶ τὴν ἀπόκρισιν εὑρίσκοντες
ποιουμένας μηκέτι θαυμάζωμεν, εἰ καὶ διὰ τῶν
φλεβῶν ἡ φύσις ἐκκρίνει πολλάκις εἰς τὴν
γαστέρα περιττώματα. τούτου δ' ἔτι μᾶλλον
οὐ χρὴ θαυμάζειν, εἰ, δι' ὧν εἰς ἧπαρ ἀνεδόθη
φλεβῶν ἐκ γαστρός, αὖθις εἰς αὐτὴν ἐξ ἥπατος
ἐν ταῖς μακροτέραις ἀσιτίαις ἕλκεσθαί τις δύναται τροφή. τὸ γὰρ τοῖς τοιούτοις ἀπιστεῖν

---

[1] cf. p. 95.   [2] I. xiii. ; II. ii.
[3] Galen's idea is that if reversal of the direction of flow

290

conduct food and drink into this organ, but in the condition of nausea it performs the opposite service. Further, the neck of the bladder which is beside the liver, albeit single, both fills and empties the bladder. Similarly the canal of the uterus affords an entrance to the semen and an exit to the foetus.

But in this latter case, again, whilst the eliminative faculty is evident, the attractive faculty is not so obvious to most people. It is, however, the cervix which Hippocrates blames for inertia of the uterus when he says :—" Its orifice has no power of attracting semen." [1]

Erasistratus, however, and Asclepiades reached such heights of wisdom that they deprived not merely the stomach and the womb of this faculty but also the bladder by the liver, and the kidneys as well. I have, however, pointed out in the first book that it is impossible to assign any other cause for the secretion of urine or bile. [2]

Now, when we find that the uterus, the stomach and the bladder by the liver carry out attraction and expulsion through one and the same duct, we need no longer feel surprised that Nature should also frequently discharge waste-substances into the stomach through the veins. Still less need we be astonished if a certain amount of the food should, during long fasts, be drawn back from the liver into the stomach through the same veins [3] by which it was yielded up to the liver during absorption of nutriment. [4] To disbelieve such things

---

can occur in the *primae viae* (in vomiting), it may also be expected to occur in the *secundae viae* or absorptive channels.

[4] For this "delivery," "up-yield," or *anadosis, v.* p. 13, note 5.

ὅμοιόν ἐστι δήπου τῷ μηκέτι πιστεύειν μηδ'
ὅτι τὰ καθαίροντα φάρμακα διὰ τῶν αὐτῶν
στομάτων ἐξ ὅλου τοῦ σώματος εἰς τὴν γαστέρα
τοὺς οἰκείους ἐπισπᾶται χυμούς, δι' ὧν ἔμπροσθεν ἡ ἀνάδοσις ἐγένετο, ἀλλ' ἕτερα μὲν ζητεῖν
ἀναδόσεως, ἕτερα δὲ καθάρσεως στόματα. καὶ
μὴν εἴπερ ἓν καὶ ταὐτὸ στόμα διτταῖς ὑπηρετεῖ
δυνάμεσιν, ἐν διαφόροις χρόνοις εἰς τἀναντία
τὴν ὁλκὴν ποιουμέναις, ἔμπροσθεν μὲν τῇ κατὰ
τὸ ἧπαρ, ἐν δὲ τῷ τῆς καθάρσεως καιρῷ τῇ
τοῦ φαρμάκου, τί θαυμαστόν ἐστι διττὴν ὑπηρεσίαν τε καὶ χρείαν εἶναι ταῖς φλεψὶ ταῖς
ἐν τῷ μέσῳ τεταγμέναις ἥπατός τε καὶ τῶν
κατὰ τὴν κοιλίαν, ὥσθ', ὁπότε μὲν ἐν τούτοις
ἄφθονος εἴη περιεχομένη τροφή, διὰ τῶν εἰρη-
189 μένων εἰς ‖ ἧπαρ ἀναφέρεσθαι φλεβῶν, ὁπότε δ'
εἴη κενὰ καὶ δεόμενα τρέφεσθαι, διὰ τῶν αὐτῶν
αὖθις ἐξ ἥπατος ἕλκεσθαι;

Πᾶν γὰρ ἐκ παντὸς ἕλκειν φαίνεται καὶ παντὶ
μεταδιδόναι καὶ μία τις εἶναι σύρροια καὶ σύμ-
πνοια πάντων, καθάπερ καὶ τοῦθ' ὁ θειότατος
Ἱπποκράτης εἶπεν. ἕλκει μὲν οὖν τὸ ἰσχυρότερον,
ἐκκενοῦται δὲ τὸ ἀσθενέστερον.

Ἰσχυρότερον δὲ καὶ ἀσθενέστερον ἕτερον ἑτέρου
μόριον ἢ ἁπλῶς καὶ φύσει καὶ κοινῇ πᾶσίν ἐστιν
ἢ ἰδίως τῷδέ τινι γίγνεται. φύσει μὲν καὶ κοινῇ
πᾶσιν ἀνθρώποις θ' ἅμα καὶ ζῴοις ἡ μὲν καρδία
τοῦ ἥπατος, τὸ δ' ἧπαρ τῶν ἐντέρων τε καὶ τῆς
γαστρός, αἱ δ' ἀρτηρίαι τῶν φλεβῶν ἑλκύσαι τε
τὸ χρήσιμον ἑαυταῖς ἀποκρῖναί τε τὸ μὴ τοιοῦτον

---

[1] The mesenteric veins.

would of course be like refusing to believe that
purgative drugs draw their appropriate humours from
all over the body by the same stomata through
which absorption previously takes place, and to look
for separate stomata for absorption and purgation
respectively. As a matter of fact one and the same
stoma subserves two distinct faculties, and these
exercise their pull at different times in opposite
directions—first it subserves the pull of the liver
and, during catharsis, that of the drug. What is
there surprising, then, in the fact that the veins
situated between the liver and the region of the
stomach [1] fulfil a double service or purpose? Thus,
when there is abundance of nutriment contained in
the food-canal, it is carried up to the liver by the
veins mentioned ; and when the canal is empty and
in need of nutriment, this is again attracted from the
liver by the same veins.

For everything appears to attract from and to go
shares with everything else, and, as the most divine
Hippocrates has said, there would seem to be a consensus in the movements of fluids and vapours.[2]
Thus the stronger draws and the weaker is evacuated.

Now, one part is weaker or stronger than another
either absolutely, by nature, and in all cases, or else
it becomes so in such and such a particular instance.
Thus, by nature and in all men alike, the heart is
stronger than the liver at attracting what is serviceable to it and rejecting what is not so ; similarly the
liver is stronger than the intestines and stomach, and

[2] Linacre renders : "Una omnium confluxio ac conspira-
tio"; and he adds the marginal note "Totum corpus nostrum
est conspirabile et confluxile per meatus communes." *cf.*
p. 48.

ἰσχυρότεραι. καθ᾽ ἕκαστον δ᾽ ἡμῶν ἰδίως ἐν μὲν τῷδε τῷ καιρῷ τὸ ἧπαρ ἰσχυρότερον ἕλκειν, ἡ γαστὴρ δ᾽ ἐν τῷδε. πολλῆς μὲν γὰρ ἐν τῇ κοιλίᾳ περιεχομένης τροφῆς καὶ σφοδρῶς ὀρεγομένου τε καὶ χρῄζοντος τοῦ ἥπατος, πάντως ἰσχυρότερον ἕλκει τὸ σπλάγχνον· ἔμπαλιν δὲ τοῦ μὲν ἥπατος 190 ἐμπεπλησμένου τε καὶ δια‖τεταμένου, τῆς γαστρὸς δ᾽ ὀρεγομένης καὶ κενῆς ὑπαρχούσης ἡ τῆς ὁλκῆς ἰσχὺς εἰς ἐκείνην μεθίσταται.

Ὡς γάρ, εἰ κἂν ταῖς χερσί τινα σιτία κατέχοντες ἀλλήλων ἁρπάζοιμεν, εἰ μὲν ὁμοίως εἴημεν δεόμενοι, περιγίγνεσθαι τὸν ἰσχυρότερον εἰκός, εἰ δ᾽ οὗτος μὲν ἐμπεπλησμένος εἴη καὶ διὰ τοῦτ᾽ ἀμελῶς κατέχων τὰ περιττὰ ἢ καί τινι μεταδοῦναι ποθῶν, ὁ δ᾽ ἀσθενέστερος ὀρέγοιτο δεινῶς, οὐδὲν ἂν εἴη κώλυμα τοῦ μὴ πάντα λαβεῖν αὐτόν, οὕτω καὶ ἡ γαστὴρ ἐκ τοῦ ἥπατος ἐπισπᾶται ῥαδίως, ὅταν αὐτὴ μὲν ἱκανῶς ὀρέγηται τροφῆς, ἐμπεπλησμένον δ᾽ ᾖ τὸ σπλάγχνον. καὶ τοῦ γε μὴ πεινῆν ἐνίοτε τὸ ζῷον ἡ περιουσία τῆς ἐν ἥπατι τροφῆς αἰτία· κρείττονα γὰρ ἔχουσα καὶ ἑτοιμοτέραν ἡ γαστὴρ τροφὴν οὐδὲν δεῖται τῆς ἔξωθεν· εἰ δέ γέ ποτε δέοιτο μέν, ἀποροίη δέ, πληροῦται περιττωμάτων. ἰχῶρες δέ τινές εἰσι ταῦτα χολώδεις τε καὶ φλεγματώδεις καὶ ὀρρώδεις, οὓς μόνους ἑλκούσῃ μεθίησιν αὐτῇ τὸ ἧπαρ, ὅταν ποτὲ καὶ αὐτὴ δέηται τροφῆς.

Ὥσπερ οὖν ἐξ ἀλλήλων ἕλκει τὰ μόρια ‖ 191 τροφήν, οὕτω καὶ ἀποτίθεταί ποτ᾽ εἰς ἄλληλα

the arteries than the veins. In each of us personally, however, the liver has stronger drawing power
at one time, and the stomach at another. For when
there is much nutriment contained in the alimentary
canal and the appetite and craving of the liver is violent, then the viscus[1] exerts far the strongest traction.
Again, when the liver is full and distended and the
stomach empty and in need, then the force of the
traction shifts to the latter.

Suppose we had some food in our hands and were
snatching it from one another; if we were equally
in want, the stronger would be likely to prevail, but
if he had satisfied his appetite, and was holding what
was over carelessly, or was anxious to share it with
somebody, and if the weaker was excessively desirous
of it, there would be nothing to prevent the latter
from getting it all. In a similar manner the stomach
easily attracts nutriment from the liver when it [the
stomach] has a sufficiently strong craving for it,
and the appetite of the viscus is satisfied. And
sometimes the surplusage of nutriment in the liver
is a reason why the animal is not hungry; for when
the stomach has better and more available food it
requires nothing from extraneous sources, but if ever
it is in need and is at a loss how to supply the need,
it becomes filled with waste-matters; these are
certain biliary, phlegmatic [mucous] and serous fluids,
and are the only substances that the liver yields in
response to the traction of the stomach, on the
occasions when the latter too is in want of nutriment.

Now, just as the parts draw food from each other,
so also they sometimes deposit their excess substances

---

[1] The alimentary canal, as not being edible, is not considered a *splanchnon* or viscus.

τὸ περιττὸν καὶ ὥσπερ ἑλκόντων ἐπλεονέκτει τὸ
ἰσχυρότερον, οὕτω καὶ ἀποτιθεμένων καὶ τῶν γε
καλουμένων ῥευμάτων ἥδε ἡ πρόφασις. ἕκαστον
γὰρ τῶν μορίων ἔχει τινὰ τόνον σύμφυτον, ᾧ
διωθεῖται τὸ περιττόν. ὅταν οὖν ἓν ἐξ αὐτῶν
ἀρρωστότερον γένηται κατὰ δή τινα διάθεσιν, ἐξ
ἁπάντων εἰς ἐκεῖνο συρρεῖν ἀνάγκη τὰ περιττώματα. τὸ μὲν γὰρ ἰσχυρότατον ἐναποτίθεται
τοῖς πλησίον ἅπασιν, ἐκείνων δ᾽ αὖ πάλιν
ἕκαστον εἰς ἕτερ᾽ ἄττα τῶν ἀσθενεστέρων, εἶτ᾽
αὖθις ἐκείνων ἕκαστον εἰς ἄλλα καὶ τοῦτ᾽ ἐπὶ
πλεῖστον γίγνεται, μέχρι περ ἂν ἐξ ἁπάντων
ἐλαυνόμενον τὸ περίττωμα καθ᾽ ἕν τι μείνῃ τῶν
ἀσθενεστάτων· ἐντεῦθεν γὰρ οὐκέτ᾽ εἰς ἄλλο
δύναται μεταρρεῖν, ὡς ἂν μήτε δεχομένου τινὸς
αὐτὸ τῶν ἰσχυροτέρων μήτ᾽ ἀπώσασθαι δυναμένου
τοῦ πεπονθότος.

Ἀλλὰ περὶ μὲν τῶν παθῶν τῆς γενέσεως καὶ
τῆς ἰάσεως αὖθις ἡμῶν ἐπιδεικνύντων ἱκανὰ κἀξ
ἐκείνων ἔσται λαβεῖν μαρτύρια τῶν ἐν τῷδε τῷ
192 λόγῳ παντὶ ‖ δεδειγμένων ὀρθῶς. ὃ δ᾽ ἐν τῷ
παρόντι δεῖξαι προὔκειτο, πάλιν ἀναλάβωμεν, ὡς
οὐδὲν θαυμαστὸν ἐξ ἥπατος ἥκειν τινὰ τροφὴν
ἐντέροις τε καὶ γαστρὶ διὰ τῶν αὐτῶν φλεβῶν,
δι᾽ ὧν ἔμπροσθεν ἐξ ἐκείνων εἰς ἧπαρ ἀνεδίδοτο.
καὶ πολλοῖς ἀθρόως τε καὶ τελέως ἀποστᾶσιν
ἰσχυρῶν γυμνασίων ἤ τι κῶλον ἀποκοπεῖσιν
αἵματος διὰ τῶν ἐντέρων γίγνεται κένωσις ἔκ
τινων περιόδων, ὥς που καὶ Ἱπποκράτης ἔλεγεν,
οὐδὲν μὲν ἄλλο λυποῦσα, καθαίρουσα δ᾽ ὀξέως τὸ
πᾶν σῶμα καὶ τὰς πλησμονὰς ἐκκενοῦσα, διὰ τῶν

in each other, and just as the stronger prevailed when the two were exercising traction, so it is also when they are depositing; this is the cause of the so-called fluxions,[1] for every part has a definite inborn tension, by virtue of which it expels its superfluities, and, therefore, when one of these parts,—owing, of course, to some special condition—becomes weaker, there will necessarily be a confluence into it of the superfluities from all the other parts. The strongest part deposits its surplus matter in all the parts near it; these again in other parts which are weaker; these next into yet others; and this goes on for a long time, until the superfluity, being driven from one part into another, comes to rest in one of the weakest of all; it cannot flow from this into another part, because none of the stronger ones will receive it, while the affected part is unable to drive it away.

When, however, we come to deal again with the origin and cure of disease, it will be possible to find there also abundant proofs of all that we have correctly indicated in this book. For the present, however, let us resume again the task that lay before us, *i.e.* to show that there is nothing surprising in nutriment coming from the liver to the intestines and stomach by way of the very veins through which it had previously been yielded up from these organs into the liver. And in many people who have suddenly and completely given up active exercise, or who have had a limb cut off, there occurs at certain periods an evacuation of blood by way of the intestines—as Hippocrates has also pointed out somewhere. This causes no further trouble but sharply purges the whole body and evacuates the plethoras;

---

[1] Lit. *rheums*; hence our term *rheumatism*.

αὐτῶν δήπου φλεβῶν τῆς φορᾶς τῶν περιττῶν ἐπιτελουμένης, δι' ὧν ἔμπροσθεν ἡ ἀνάδοσις ἐγίγνετο.

Πολλάκις δ' ἐν νόσοις ἡ φύσις διὰ μὲν τῶν αὐτῶν δήπου φλεβῶν τὸ πᾶν ἐκκαθαίρει ζῷον, οὐ μὴν αἱματώδης γ' ἡ κένωσις αὐτοῖς, ἀλλὰ κατὰ τὸν λυποῦντα γίγνεται χυμόν. οὕτω δὲ κἂν ταῖς χολέραις ἐκκενοῦται τὸ πᾶν σῶμα διὰ τῶν εἰς ἔντερά τε καὶ γαστέρα καθηκουσῶν φλεβῶν.

Τὸ δ' οἴεσθαι μίαν εἶναι ταῖς ὕλαις φορὰν
193 τελέως ἀγνοοῦντός ἐστι τὰς φυσικὰς ‖ δυνάμεις τάς τ' ἄλλας καὶ τὴν ἐκκριτικὴν ἐναντίαν οὖσαν τῇ ἑλκτικῇ· ταῖς γὰρ ἐναντίαις δυνάμεσιν ἐναντίας κινήσεις τε καὶ φορὰς τῶν ὑλῶν ἀναγκαῖον ἀκολουθεῖν. ἕκαστον γὰρ τῶν μορίων, ὅταν ἑλκύσῃ τὸν οἰκεῖον χυμόν, ἔπειτα κατάσχῃ καὶ ἀπολαύσῃ, τὸ περιττὸν ἅπαν ἀποθέσθαι σπεύδει, καθότι μάλιστα δύναται τάχιστά θ' ἅμα καὶ κάλλιστα, κατὰ τὴν τοῦ περιττοῦ ῥοπήν.

Ὅθεν ἡ γαστὴρ τὰ μὲν ἐπιπολάζοντα τῶν περιττωμάτων ἐμέτοις ἐκκαθαίρει, τὰ δ' ὑφιστάμενα διαρροίαις. καὶ τό γε ναυτιῶδες γίγνεσθαι τὸ ζῷον τοῦτ' ἔστιν ὁρμῆσαι τὴν γαστέρα κενωθῆναι δι' ἐμέτου. οὕτω δὲ δή τι βίαιον καὶ σφοδρὸν ἡ ἐκκριτικὴ δύναμις ἔχει, ὥστ' ἐν τοῖς εἰλεοῖς, ὅταν ἀποκλεισθῇ τελέως ἡ κάτω διέξοδος, ἐμεῖται κόπρος. καίτοι πρὶν διελθεῖν τό τε λεπτὸν ἔντερον ἅπαν καὶ τὴν νῆστιν καὶ τὸν πυλωρὸν καὶ τὴν γαστέρα καὶ τὸν οἰσοφάγον οὐχ οἷόν τε διὰ τοῦ στόματος ἐκπεσεῖν οὐδενὶ τοιούτῳ περιττώματι. τί δὴ θαυμαστόν, εἰ κἀκ τῆς ἐσχάτης

---

[1] Here Galen apparently indicates that vital functions are

the passage of the superfluities is effected, of course, through the same veins by which absorption took place.

Frequently also in disease Nature purges the animal through these same veins—although in this case the discharge is not sanguineous, but corresponds to the humour which is at fault. Thus in *cholera* the entire body is evacuated by way of the veins leading to the intestines and stomach.

To imagine that matter of different kinds is carried in one direction only would characterise a man who was entirely ignorant of all the natural faculties, and particularly of the eliminative faculty, which is the opposite of the attractive. For opposite movements of matter, active and passive, must necessarily follow opposite faculties ; that is to say, every part, after it has attracted its special nutrient juice and has retained and taken the benefit of it hastens to get rid of all the surplusage as quickly and effectively as possible, and this it does in accordance with the mechanical tendency of this surplus matter.[1]

Hence the stomach clears away by vomiting those superfluities which come to the surface of its contents,[2] whilst the sediment it clears away by diarrhœa. And when the animal becomes sick, this means that the stomach is striving to be evacuated by vomiting. And the expulsive faculty has in it so violent and forcible an element that in cases of *ileus* [volvulus], when the lower exit is completely closed, vomiting of faeces occurs ; yet such surplus matter could not be emitted from the mouth without having first traversed the whole of the small intestine, the jejunum, the pylorus, the stomach, and the oesophagus. What is there to wonder at, then, if something

at least partly explicable in terms of mechanical law. *cf.* Introduction, p. xxviii.   [2] *cf.* pp. 211, 247.

GALEN

ἐπιφανείας τῆς κατὰ τὸ δέρμα μέχρι τῶν ἐντέρων
194 τε καὶ τῆς γαστρὸς ἀφικνοῖτό τι ‖ μεταλαμβανό-
μενον, ὡς καὶ τοῦθ' Ἱπποκράτης ἡμᾶς ἐδίδαξεν,
οὐ πνεῦμα μόνον ἢ περίττωμα φάσκων ἀλλὰ καὶ
τὴν τροφὴν αὐτὴν ἐκ τῆς ἐσχάτης ἐπιφανείας
αὖθις ἐπὶ τὴν ἀρχήν, ὅθεν ἀνηνέχθη, καταφέρε-
σθαι. ἐλάχισται γὰρ ῥοπαὶ κινήσεων τὴν
ἐκκριτικὴν ταύτην οἰακίζουσι δύναμιν, ὡς ἂν διὰ
τῶν ἐγκαρσίων μὲν ἰνῶν γιγνομένην, ὠκύτατα δὲ
διαδιδομένην ἀπὸ τῆς κινησάσης ἀρχῆς ἐπὶ τὰ
καταντικρὺ πέρατα. οὔκουν ἀπεικὸς οὐδ' ἀδύνατον ἀήθει ποτὲ ψύξει τὸ πρὸς τῷ δέρματι
μόριον ἐξαίφνης πιληθὲν ἅμα μὲν ἀρρωστότερον
αὐτὸ γενόμενον, ἅμα δ' οἷον ἄχθος τι μᾶλλον ἢ
παρασκευὴν θρέψεως ἔχον τὴν ἔμπροσθεν ἀλύπως
αὐτῷ παρεσπαρμένην ὑγρότητα καὶ διὰ τοῦτ'
ἀπωθεῖσθαι σπεῦδον, ἅμα δὲ τῆς ἔξω φορᾶς
ἀποκεκλεισμένης τῇ πυκνώσει, πρὸς τὴν λοιπὴν
ἐπιστραφῆναι καὶ οὕτω βιασάμενον εἰς τὸ
παρακείμενον αὐτῷ μόριον ἀθρόως ἀπώσασθαι
τὸ περιττόν, ἐκεῖνο δ' αὖ πάλιν εἰς τὸ μετ' αὐτό, ‖
195 καὶ τοῦτο μὴ παύσασθαι γιγνόμενον, ἄχρις ἂν ἡ
μετάληψις ἐπὶ τὰ ἐντὸς πέρατα τῶν φλεβῶν
τελευτήσῃ.

Αἱ μὲν δὴ τοιαῦται κινήσεις θᾶττον ἀπο-
παύονται, αἱ δ' ἀπὸ τῶν ἔνδοθεν διερεθιζόντων,
ὡς ἔν τε τοῖς καθαίρουσι φαρμάκοις καὶ ταῖς
χολέραις ἰσχυρότεραί τε πολὺ καὶ μονιμώτεραι
γίγνονται καὶ διαμένουσιν, ἔστ' ἂν καὶ ἡ περὶ
τοῖς στόμασι τῶν ἀγγείων διάθεσις, ἢ τὸ πλησίον

---

[1] See p. 298, note 1.

should also be transferred from the extreme skin-surface and so reach the intestines and stomach ? This also was pointed out to us by Hippocrates, who maintained that not merely pneuma or excess-matter, but actual nutriment is brought down from the outer surface to the original place from which it was taken up. For the slightest mechanical movements [1] determine this expulsive faculty, which apparently acts through the transverse fibres, and which is very rapidly transmitted from the source of motion to the opposite extremities. It is, therefore, neither unlikely nor impossible that, when the part adjoining the skin becomes suddenly oppressed by an unwonted cold, it should at once be weakened and should find that the liquid previously deposited beside it without discomfort had now become more of a burden than a source of nutrition, and should therefore strive to put it away. Finally, seeing that the passage outwards was shut off by the condensation [of tissue], it would turn to the remaining exit and would thus forcibly expel all the waste-matter at once into the adjacent part; this would do the same to the part following it ; and the process would not cease until the transference finally terminated at the inner ends of the veins.[2]

Now, movements like these come to an end fairly soon, but those resulting from internal irritants (*e.g.,* in the administration of purgative drugs or in cholera) become much stronger and more lasting ; they persist as long as the condition of things [3] about the mouths of the veins continues, that is, so long as

[2] The ends of the veins in the alimentary canal from which absorption or *anadosis* had originally taken place.

[3] *Diathesis.*

ἕλκουσα, παραμένῃ. αὕτη μὲν γὰρ τὸ συνεχὲς
ἐκκενοῖ μόριον, ἐκεῖνο δ' αὖ τὸ μετ' αὐτὸ καὶ τοῦτ'
οὐ παύεται μέχρι τῆς ἐσχάτης ἐπιφανείας, ὥστε
διαδιδόντων τῶν ἐφεξῆς ἀεὶ μορίων ἑτέρων ἑτέροις
τὸ πρῶτον πάθος ὠκύτατα διικνεῖσθαι μέχρι τῶν
ἐσχάτων. οὕτως οὖν ἔχει κἀπὶ τῶν εἰλεῶν. αὐτὸ
μὲν γὰρ τὸ φλεγμαῖνον ἔντερον οὔτε τοῦ βάρους
οὔτε τῆς δριμύτητος ἀνέχεται τῶν περιττωμάτων
καὶ διὰ τοῦτ' ἐκκρίνειν αὐτὰ σπεύδει καὶ ἀπωθεῖσθαι πορρωτάτω. κωλυόμενον δὲ κάτω ποιεῖσθαι τὴν δίωσιν, ὅταν ἐνταυθοῖ ποτε τὸ σφοδρό-
τατον ᾖ τῆς φλεγμονῆς, εἰς τὰ πλησιάζοντα τῶν
ὑπερκειμένων ἐντέρων ἀπωθεῖται. καὶ οὕτως ἤδη
196 κατὰ ‖ τὸ συνεχὲς τὴν ῥοπὴν τῆς ἐκκριτικῆς
δυνάμεως ἄνω ποιησαμένης ἄχρι τοῦ στόματος
ἐπανέρχεται τὰ περιττώματα.

Ταῦτα μὲν οὖν δὴ κἀν τοῖς τῶν νοσημάτων
λογισμοῖς ἐπὶ πλέον εἰρήσεται. τὸ δ' ἐκ παντὸς
εἰς πᾶν φέρεσθαί τι καὶ μεταλαμβάνεσθαι καὶ
μίαν ἁπάντων εἶναι σύμπνοιάν τε καὶ σύρροιαν,
ὡς Ἱπποκράτης ἔλεγεν, ἤδη μοι δοκῶ δεδεῖχθαι
σαφῶς καὶ μηκέτ' ἄν τινα, μηδ' εἰ βραδὺς αὐτῷ
νοῦς ἐνείη, περὶ τῶν τοιούτων ἀπορῆσαι μηδενός,
οἷον ὅπως ἡ γαστὴρ ἢ τὰ ἔντερα τρέφεται καὶ
τίνα τρόπον ἐκ τῆς ἐσχάτης ἐπιφανείας εἴσω τι
διικνεῖται. πάντων γὰρ τῶν μορίων ἕλκειν μὲν
τὸ προσῆκόν τε καὶ φίλιον, ἀποκρίνειν δὲ τὸ
βαρῦνον ἢ δάκνον ἐχόντων δύναμιν οὐδὲν θαυμαστὸν ἐναντίας συνεχῶς γίγνεσθαι κινήσεις ἐν

these continue to attract what is adjacent. For this condition [1] causes evacuation of the contiguous part, and that again of the part next to it, and this never stops until the extreme surface is reached; thus, as each part keeps passing on matter to its neighbour, the original affection [2] very quickly arrives at the extreme termination. Now this is also the case in *ileus*; the inflamed intestine is unable to support either the weight or the acridity of the waste substances and so does its best to excrete them, in fact to drive them as far away as possible. And, being prevented from effecting an expulsion downwards when the severest part of the inflammation is there, it expels the matter into the adjoining part of the intestines situated above. Thus the tendency of the eliminative faculty is step by step upwards, until the superfluities reach the mouth.

Now this will be also spoken of at greater length in my treatise on disease. For the present, however, I think I have shewn clearly that there is a universal conveyance or transference from one thing into another, and that, as Hippocrates used to say, there exists in everything a consensus in the movement of air and fluids. And I do not think that anyone, however slow his intellect, will now be at a loss to understand any of these points,—how, for instance, the stomach or intestines get nourished, or in what manner anything makes its way inwards from the outer surface of the body. Seeing that all parts have the faculty of attracting what is suitable or well-disposed and of eliminating what is troublesome or irritating, it is not surprising that opposite movements should occur in them consecutively—as may

[1] *Diathesis.* [2] *Pathos.*

αὐτοῖς, ὥσπερ ἐπί τε τῆς καρδίας ὁρᾶται σαφῶς
καὶ τῶν ἀρτηριῶν ἁπασῶν καὶ τοῦ θώρακος καὶ
τοῦ πνεύμονος. ἐπὶ μέν γε τούτων ἁπάντων
μόνον οὐ καθ' ἑκάστην καιροῦ ῥοπὴν τὰς ἐναντίας
κινήσεις θ' ἅμα τῶν ὀργάνων καὶ φορὰς τῶν
197 ὑλῶν ‖ ἐναργῶς ἔστιν ἰδεῖν γιγνομένας. εἶτ' ἐπὶ
μὲν τῆς τραχείας ἀρτηρίας οὐκ ἀπορεῖς ἐναλλὰξ
ποτὲ μὲν εἴσω παραγούσης εἰς τὸν πνεύμονα τὸ
πνεῦμα, ποτὲ δ' ἔξω, καὶ τῶν κατὰ τὰς ῥῖνας
πόρων καὶ ὅλου τοῦ στόματος ὡσαύτως οὐδ' εἶναί
σοι δοκεῖ θαυμαστὸν οὐδὲ παράδοξον, εἰ, δι' οὗ
μικρῷ πρόσθεν εἴσω παρεκομίζετο τὸ πνεῦμα, διὰ
τούτου νῦν ἐκπέμπεται, περὶ δὲ τῶν ἐξ ἥπατος
εἰς ἔντερά τε καὶ γαστέρα καθηκουσῶν φλεβῶν
ἀπορεῖς καί σοι θαυμαστὸν εἶναι φαίνεται, διὰ
τῶν αὐτῶν ἀναδίδοσθαί θ' ἅμα τὴν τροφὴν εἰς
ἧπαρ ἕλκεσθαί τ' ἐξ ἐκείνου πάλιν εἰς γαστέρα;
διόρισαι δὴ τὸ ἅμα τοῦτο ποτέρως λέγεις. εἰ μὲν
γὰρ κατὰ τὸν αὐτὸν χρόνον, οὐδ' ἡμεῖς τοῦτό γέ
φαμεν. ὥσπερ γὰρ εἰσπνέομεν ἐν ἑτέρῳ χρόνῳ
καὶ αὖθις πάλιν ἐν ἑτέρῳ ἀντεκπνέομεν, οὕτω καὶ
τροφὴν ἐν ἑτέρῳ μὲν χρόνῳ τὸ ἧπαρ ἐκ τῆς
γαστρός, ἐν ἑτέρῳ δ' ἡ γαστὴρ ἐκ τοῦ ἥπατος
ἐπισπᾶται. εἰ δ' ὅτι καθ' ἓν καὶ ταὐτὸ ζῷον ἓν
ὄργανον ἐναντίαις φοραῖς ὑλῶν ὑπηρετεῖ, τοῦτό
σοι βούλεται δηλοῦν τὸ ἅμα καὶ τοῦτό σε ταράτ-
198 τει, τήν τ' ‖ εἰσπνοὴν ἰδὲ καὶ τὴν ἐκπνοήν. πάντως που καὶ αὗται διὰ μὲν τῶν αὐτῶν ὀργάνων
γίγνονται, τρόπῳ δὲ κινήσεώς τε καὶ φορᾶς τῶν
ὑλῶν διαφέρουσιν.

---

[1] He means, not only under the stress of special circumstances, but also normally.

be clearly seen in the case of the heart, in the various arteries, in the thorax, and lungs. In all these[1] the active movements of the organs and therewith the passive movements of [their contained] matters may be seen taking place almost every second in opposite directions. Now, you are not astonished when the trachea-artery[2] alternately draws air into the lungs and gives it out, and when the nostrils and the whole mouth act similarly; nor do you think it strange or paradoxical that the air is dismissed through the very channel by which it was admitted just before. Do you, then, feel a difficulty in the case of the veins which pass down from the liver into the stomach and intestines, and do you think it strange that nutriment should at once be yielded up to the liver and drawn back from it into the stomach by the same veins? You must define what you mean by this expression " at once." If you mean " at the same time " this is not what we ourselves say; for just as we take in a breath at one moment and give it out again at another, so at one time the liver draws nutriment from the stomach, and at another the stomach from the liver. But if your expression " at once " means that in one and the same animal a single organ subserves the transport of matter in opposite directions, and if it is this which disturbs you, consider inspiration and expiration. For of course these also take place through the same organs, albeit they differ in their manner of movement, and in the way in which the matter is conveyed through them.

[2] Lit. "rough artery." The air-passages as well as the arteries proper were supposed by the Greeks to carry air (pneuma); diastole of arteries was, like expansion of the chest, a movement for drawing in air. cf. p. 317, note 1.

Ὁ πνεύμων μὲν οὖν καὶ ὁ θώραξ καὶ ἀρτηρίαι
αἱ τραχεῖαι καὶ αἱ λεῖαι καὶ καρδία καὶ στόμα
καὶ ῥῖνες ἐν ἐλαχίσταις χρόνου ῥοπαῖς εἰς ἐναντίας κινήσεις αὐτά τε μεταβάλλει καὶ τὰς ὕλας
μεθίστησιν. αἱ δ' ἐξ ἥπατος εἰς ἔντερα καὶ γαστέρα καθήκουσαι φλέβες οὐκ ἐν οὕτω βραχέσι
χρόνου μορίοις ἀλλ' ἐν πολλαῖς ἡμέραις ἅπαξ
ἐνίοτε τὴν ἐναντίαν κινοῦνται κίνησιν.

Ἔχει γὰρ ὧδε τὸ σύμπαν. ἕκαστον τῶν ὀργάνων εἰς ἑαυτὸ τὴν πλησιάζουσαν ἐπισπᾶται
τροφὴν ἐκβοσκόμενον αὐτῆς ἅπασαν τὴν χρηστὴν
νοτίδα, μέχρις ἂν ἱκανῶς κορεσθῇ, καὶ ταύτην,
ὡς καὶ πρόσθεν ἐδείκνυμεν, ἐναποτίθεται ἑαυτῷ
καὶ μετὰ ταῦτα προσφύει τε καὶ ὁμοιοῖ, τουτ-
έστι τρέφεται. διώρισται γὰρ ἱκανῶς ἔμπροσθεν ἕτερόν τι τῆς θρέψεως ἐξ ἀνάγκης αὐτῆς
προηγούμενον ἡ πρόσφυσις ὑπάρχειν, ἐκείνης δ'
199 ἔτι πρότερον ἡ πρόσθεσις. ὥσπερ οὖν ‖ τοῖς
ζῴοις αὐτοῖς ὅρος ἐστὶ τῆς ἐδωδῆς τὸ πληρῶσαι
τὴν γαστέρα, κατὰ τὸν αὐτὸν τρόπον ἑκάστῳ
τῶν μορίων ὅρος ἐστὶ τῆς προσθέσεως ἡ πλήρωσις
τῆς οἰκείας ὑγρότητος. ἐπεὶ τοίνυν ἅπαν μόριον
τῇ γαστρὶ ὁμοίως ὀρέγεται τρέφεσθαι, καὶ περι-
πτύσσεται τῇ τροφῇ καὶ οὕτω σφίγγει πανταχόθεν αὐτὴν ὡς ἡ γαστήρ. ἕπεται δ' ἐξ ἀνάγκης
τούτῳ, καθάπερ καὶ πρόσθεν ἐρρήθη, τὸ πέττεσθαι
τοῖς σιτίοις, τῆς γαστρὸς οὐ διὰ τοῦτο περι-
στελλομένης αὐτοῖς, ἵν' ἐπιτήδεια τοῖς ἄλλοις
ἐργάσηται μορίοις· οὕτω γὰρ ἂν οὐκέτι φυσικὸν

---

[1] cf. p. 39, chap. xi.
[2] Lit. orexis.

Now the lungs, the thorax, the arteries rough and smooth, the heart, the mouth, and the nostrils reverse their movements at very short intervals and change the direction of the matters they contain. On the other hand, the veins which pass down from the liver to the intestines and stomach reverse the direction of their movements not at such short intervals, but sometimes once in many days.

The whole matter, in fact, is as follows :—Each of the organs draws into itself the nutriment alongside it, and devours all the useful fluid in it, until it is thoroughly satisfied ; this nutriment, as I have already shown, it stores up in itself, afterwards making it adhere and then assimilating it—that is, it becomes nourished by it. For it has been demonstrated with sufficient clearness already [1] that there is something which necessarily precedes actual nutrition, namely *adhesion*, and that before this again comes *presentation*. Thus as in the case of the *animals* themselves the end of eating is that the stomach should be filled, similarly in the case of each of the *parts*, the end of presentation is the filling of this part with its appropriate liquid. Since, therefore, every part has, like the stomach, a *craving* [2] to be nourished, it too envelops its nutriment and clasps it all round as the stomach does. And this [action of the stomach], as has been already said, is necessarily followed by the digestion of the food, although it is not to make it suitable for the other parts that the stomach contracts upon it ; if it did so, it would no longer be a physiological organ,[3] but an animal possessing reason

[3] Lit. a "physical" organ ; that is, a mere instrument or organon of the Physis,—not one of the Psyche or conscious personality. *cf.* semen, p. 132, note 1.

ὄργανον ἀλλὰ ζῷόν τι γίγνοιτο λογισμόν τε κα
νοῦν ἔχον, ὡς αἱρεῖσθαι τὸ βέλτιον.

Ἀλλ' αὕτη μὲν περιστέλλεται τῷ τὸ πᾶ
σῶμα δύναμιν ἑλκτικήν τινα καὶ ἀπολαυστικὴ
κεκτῆσθαι τῶν οἰκείων ποιοτήτων, ὡς ἔμπροσθε
ἐδείκνυτο· συμβαίνει δ' ἐν τούτῳ τοῖς σιτίοι
ἀλλοιοῦσθαι. καὶ μέντοι καὶ πληρωθεῖσα τῇ
ἐξ αὐτῶν ὑγρότητος καὶ κορεσθεῖσα βάρος ἡγεῖτα
τὸ λοιπὸν αὐτά. τὸ περιττὸν οὖν εὐθὺς ἀπο
200 τρίβεταί τε καὶ ὠθεῖ κάτω πρὸς ‖ ἕτερον ἔργον
αὐτὴ τρεπομένη, τὴν πρόσφυσιν. ἐν δὲ τούτῳ
τῷ χρόνῳ διερχομένη τὸ ἔντερον ἅπαν ἡ τροφὴ
διὰ τῶν εἰς αὐτὸ καθηκόντων ἀγγείων ἀναρπά
ζεται, πλείστη μὲν εἰς τὰς φλέβας, ὀλίγη δέ τι
εἰς τὰς ἀρτηρίας, ὡς μικρὸν ὕστερον ἀποδείξομεν.
ἐν τούτῳ δ' αὖ τῷ χρόνῳ καὶ τοῖς τῶν ἐντέρων
χιτῶσι προστίθεται.

Καί μοι τεμὼν ἤδη τῷ λογισμῷ τὴν τῆς τροφῆς
οἰκονομίαν ἅπασαν εἰς τρεῖς μοίρας χρόνων, ἐ
μὲν τῇ πρώτῃ νόει μένουσάν θ' ἅμα κατὰ τὴ
κοιλίαν αὐτὴν καὶ πεττομένην καὶ προστιθεμένη
εἰς κόρον τῇ γαστρὶ καί τι καὶ τῷ ἥπατι παρ'
αὐτῆς ἀναφερόμενον.

Ἐν δὲ τῇ δευτέρᾳ διερχομένην τά τ' ἔντερα
καὶ προστιθεμένην εἰς κόρον αὐτοῖς τε τούτοις καὶ
τῷ ἥπατι καί τι βραχὺ μέρος αὐτῆς πάντη τοῦ
σώματος φερόμενον· ἐν δὲ δὴ τούτῳ τῷ καιρῷ
τὸ προστεθὲν ἐν τῷ πρώτῳ χρόνῳ προσφύεσθαι
νόει τῇ γαστρί.

Κατὰ δὲ τὴν τρίτην μοῖραν τοῦ χρόνου τρέ-

---

[1] cf. p. 317, note 2; p. 319, chap. xv.

and intelligence, with the power of choosing the better [of two alternatives].

But while the stomach contracts for the reason that the whole body possesses a power of attracting and of utilising appropriate qualities, as has already been explained, it also happens that, in this process, the food undergoes alteration; further, when filled and saturated with the fluid pabulum from the food, it thereafter looks on the food as a burden; thus it at once gets rid of the excess—that is to say, drives it downwards—itself turning to another task, namely that of causing adhesion. And during this time, while the nutriment is passing along the whole length of the *intestine*, it is caught up by the vessels which pass into the intestine; as we shall shortly demonstrate,[1] most of it is seized by the veins, but a little also by the arteries; at this stage also it becomes *presented* to the coats of the intestines.

Now imagine the whole economy of nutrition divided into three periods. Suppose that in the first period the nutriment remains in the stomach and is digested and presented to the stomach until satiety is reached, also that some of it is taken up from the stomach to the liver.[2]

During the second period it passes along the intestines and becomes presented both to them and to the liver—again until the stage of satiety—while a small part of it is carried all over the body.[2] During this period, also imagine that what was presented to the stomach in the first period becomes now adherent to it.

During the third period the stomach has reached

[2] Note that absorption takes place from the stomach as well as the intestines. *cf.* p. 118, note 1.

φεσθαι μὲν ἤδη τὴν κοιλίαν ὁμοιώσασαν ἑαυτῇ
τελέως τὰ προσφύντα, πρόσφυσιν δὲ τοῖς ἐντέροις καὶ τῷ ἥπατι γίγνεσθαι τῶν προστεθέντων,
201 ἀνά‖δοσιν δὲ πάντη τοῦ σώματος καὶ πρόσθεσιν.
εἰ μὲν οὖν ἐπὶ τούτοις εὐθέως τὸ ζῷον λαμβάνοι
τροφήν, ἐν ᾧ πάλιν ἡ γαστὴρ χρόνῳ πέττει τε
ταύτην καὶ ἀπολαύει προστιθεῖσα πᾶν ἐξ αὐτῆς
τὸ χρηστὸν τοῖς ἑαυτῆς χιτῶσι, τὰ μὲν ἔντερα
τελέως ὁμοιώσει τὸν προσφύντα χυμόν, ὡσαύτως
δὲ καὶ τὸ ἥπαρ. ἐν ὅλῳ δὲ τῷ σώματι πρόσφυσις
τῶν προστεθέντων τῆς τροφῆς ἔσται μορίων.
εἰ δ᾽ ἄσιτος ἀναγκάζοιτο μένειν ἡ γαστὴρ ἐν
τούτῳ τῷ χρόνῳ, παρὰ τῶν ἐν μεσεντερίῳ τε καὶ
ἥπατι φλεβῶν ἕλξει τὴν τροφήν· οὐ γὰρ ἐξ αὐτοῦ
γε τοῦ σώματος τοῦ ἥπατος. λέγω δὲ σῶμα τοῦ
ἥπατος αὐτήν τε τὴν ἰδίαν αὐτοῦ σάρκα πρώτην
καὶ μάλιστα, μετὰ δὲ τήνδε καὶ τῶν ἀγγείων
ἕκαστον τῶν κατ᾽ αὐτό. τὸν μὲν γὰρ ἐν ἑκάστῳ
τῶν μορίων ἤδη περιεχόμενον χυμὸν οὐκέτ᾽
εὔλογον ἀντισπᾶν ἑτέρῳ μορίῳ καὶ μάλισθ᾽ ὅταν
ἤδη πρόσφυσις ἢ ἐξομοίωσις αὐτοῦ γίγνηται. τὸν
δ᾽ ἐν ταῖς εὐρυχωρίαις τῶν φλεβῶν τὸ μᾶλλον
ἰσχύον θ᾽ ἅμα καὶ δεόμενον ἀντισπᾷ μόριον.
202 Οὕτως οὖν καὶ ἡ γαστὴρ ἐν ‖ ᾧ χρόνῳ δεῖται
μὲν αὐτὴ τροφῆς, ἐσθίει δ᾽ οὐδέπω τὸ ζῷον, ἐν
τούτῳ τῶν κατὰ τὸ ἥπαρ ἐξαρπάζει φλεβῶν.
ἐπεὶ δὲ καὶ τὸν σπλῆνα διὰ τῶν ἔμπροσθεν
ἐδείκνυμεν ὅσον ἐν ἥπατι παχύτερον ἕλκοντα

---

[1] That is, among the ultimate tissues or cells.

the stage of receiving nourishment; it now entirely
assimilates everything that had become adherent to
it: at the same time in the intestines and liver there
takes place adhesion of what had been before
presented, while dispersal [anadosis] is taking place
to all parts of the body,[1] as also presentation.
Now, if the animal takes food immediately after
these [three stages] then, during the time that the
stomach is again digesting and getting the benefit
of this by presenting all the useful part of it to
its own coats, the intestines will be engaged in final
assimilation of the juices which have adhered to
them, and so also will the liver: while in the various
parts of the body there will be taking place adhesion
of the portions of nutriment presented. And if the
stomach is forced to remain without food during this
time, it will draw its nutriment from the veins in
the mesentery and liver; for it will not do so from
the actual body of the liver (by *body of the liver* I
mean first and foremost its flesh proper, and after
this all the vessels contained in it), for it is irrational
to suppose that one part would draw away from
another part the juice already contained in it,
especially when adhesion and final assimilation of
that juice were already taking place; the juice,
however, that is in the cavity of the veins will be
abstracted by the part which is stronger and more in
need.

It is in this way, therefore, that the stomach,
when it is in need of nourishment and the animal
has nothing to eat, seizes it from the veins in the
liver. Also in the case of the spleen we have shown
in a former passage [2] how it draws all material from

[2] Pp. 205-9.

κατεργάζεσθαί τε καὶ μεταβάλλειν ἐπὶ τὸ χρη-
στότερον, οὐδὲν οὐδ' ἐνταῦθα θαυμαστὸν ἕλκεσθαί
τι κἀκ τοῦ σπληνὸς εἰς ἕκαστον τῶν κοινωνούν-
των αὐτῷ κατὰ τὰς φλέβας ὀργάνων, οἷον εἰς
ἐπίπλοον καὶ μεσεντέριον καὶ λεπτὸν ἔντερον καὶ
κῶλον καὶ αὐτὴν τὴν γαστέρα· κατὰ δὲ τὸν αὐτὸν
τρόπον ἐξερεύγεσθαι μὲν εἰς τὴν γαστέρα τὸ
περίττωμα καθ' ἕτερον χρόνον, αὐτὸν δ' αὖθις
ἐκ τῆς γαστρὸς ἕλκειν τι τῆς οἰκείας τροφῆς ἐν
ἑτέρῳ καιρῷ.

Καθόλου δ' εἰπεῖν, ὃ καὶ πρόσθεν ἤδη λέλεκται,
πᾶν ἐκ παντὸς ἕλκειν τε καὶ πέμπειν ἐγχωρεῖ
κατὰ διαφέροντας χρόνους, ὁμοιοτάτου γιγνομένου
τοῦ συμβαίνοντος, ὡς εἰ καὶ ζῷα νοήσαις πολλὰ
τροφὴν ἄφθονον ἐν κοινῷ κατακειμένην, εἰς ὅσον
βούλεται, προσφερόμενα. καθ' ὃν γὰρ ἤδη πέ-
παυται χρόνον ἕτερα, κατὰ τοῦτον εἰκὸς ἐσθίειν
203 ἕτερα, καὶ μέλλειν γε τὰ μὲν || παύεσθαι, τὰ δ'
ἄρχεσθαι, καί τινα μὲν συνεσθίοντα, τὰ δ' ἀνὰ
μέρος ἐσθίοντα καὶ ναὶ μὰ Δία γε τὸ ἕτερον ἁρπά-
ζειν θατέρου πολλάκις, εἰ τὸ μὲν ἕτερον ἐπιδέοιτο,
τῷ δ' ἀφθόνως παρακέοιτο. καὶ οὕτως οὐδὲν
θαυμαστὸν οὔτ' ἐκ τῆς ἐσχάτης ἐπιφανείας εἴσω
τι πάλιν ὑποστρέφειν οὔτε διὰ τῶν αὐτῶν ἀγγείων ἐξ ἥπατός τε καὶ σπληνὸς εἰς κοιλίαν
ἀνενεχθῆναί τι, δι' ὧν ἐκ ταύτης εἰς ἐκεῖνα
πρότερον ἀνηνέχθη.

Κατὰ μὲν γὰρ τὰς ἀρτηρίας ἱκανῶς ἐναργὲς τὸ
τοιοῦτον, ὥσπερ καὶ κατὰ τὴν καρδίαν τε καὶ τὸν
θώρακα καὶ τὸν πνεύμονα. τούτων γὰρ ἁπάντων
διαστελλομένων τε καὶ συστελλομένων ἐναλλὰξ
ἀναγκαῖον, ἐξ ὧν εἱλκύσθη τι πρότερον, εἰς ταῦθ'

the liver that tends to be thick, and by working it up converts it into more useful matter. There is nothing surprising, therefore, if, in the present instance also, some of this should be drawn from the spleen into such organs as communicate with it by veins, *e.g.* the omentum, mesentery, small intestine, colon, and the stomach itself. Nor is it surprising that the spleen should disgorge its surplus matters into the stomach at one time, while at another time it should draw some of its appropriate nutriment from the stomach.

For, as has already been said, speaking generally, everything has the power at different times of attracting from and of adding to everything else. What happens is just as if you might imagine a number of animals helping themselves at will to a plentiful common stock of food; some will naturally be eating when others have stopped, some will be on the point of stopping when others are beginning, some eating together, and others in succession. Yes, by Zeus! and one will often be plundering another, if he be in need while the other has an abundant supply ready to hand. Thus it is in no way surprising that matter should make its way back from the outer surface of the body to the interior, or should be carried from the liver and spleen into the stomach by the same vessels by which it was carried in the reverse direction.

In the case of the arteries[1] this is clear enough, as also in the case of heart, thorax, and lungs; for, since all of these dilate and contract alternately, it must needs be that matter is subsequently discharged back into the parts from which it was

[1] By this term, of course, the air-passages are also meant; *cf.* p. 305.

ὕστερον ἐκπέμπεσθαι. καὶ ταύτην ἄρα τὴν ἀνάγκην ἡ φύσις προγιγνώσκουσα τοῖς ἐν τῇ καρδίᾳ
στόμασι τῶν ἀγγείων ὑμένας ἐπέφυσε κωλύ-
σοντας εἰς τοὐπίσω φέρεσθαι τὰς ὕλας. ἀλλ᾽
ὅπως μὲν τοῦτο γίγνεται καὶ καθ᾽ ὅντινα τρόπον,
ἐν τοῖς περὶ χρείας μορίων εἰρήσεται δεικνύντων
ἡμῶν τά τ᾽ ἄλλα καὶ ὡς ἀδύνατον οὕτως ἀκριβῶς
204 κλείεσθαι τὰ στόματα τῶν ἀγγείων, ὡς ‖ μηδὲν
παλινδρομεῖν. εἰς μὲν γὰρ τὴν ἀρτηρίαν τὴν
φλεβώδη, καὶ γὰρ καὶ τοῦτ᾽ ἐν ἐκείνοις δειχθή-
σεται, πολὺ πλέον ἢ διὰ τῶν ἄλλων στομάτων
εἰς τοὐπίσω πάλιν ἀναγκαῖον ἐπανέρχεσθαι. τὸ
δ᾽ εἰς τὰ παρόντα χρήσιμον, ὡς οὐκ ἐνδέχεταί τι
τῶν αἰσθητὴν καὶ μεγάλην ἐχόντων εὐρύτητα
μὴ οὐκ ἤτοι διαστελλόμενον ἕλκειν ἐξ ἁπάντων
τῶν πλησίον ἢ ἐκθλίβειν αὖθις εἰς ταῦτα συστελλόμενον ἔκ τε τῶν ἤδη προειρημένων ἐν τῷδε
τῷ λόγῳ σαφὲς ἂν εἴη κἀξ ὧν Ἐρασίστρατός τε
καὶ ἡμεῖς ἑτέρωθι περὶ τῆς πρὸς τὸ κενούμενον
ἀκολουθίας ἐδείξαμεν.

XIV

Ἀλλὰ μὴν καὶ ὡς ἐν ἑκάστῃ τῶν ἀρτηριῶν
ἐστί τις δύναμις ἐκ τῆς καρδίας ἐπιρρέουσα, καθ᾽
ἣν διαστέλλονταί τε καὶ συστέλλονται, δέδεικται
δι᾽ ἑτέρων.

Εἴπερ οὖν συνθείης ἄμφω τό τε ταύτην εἶναι
τὴν κίνησιν αὐταῖς τό τε πᾶν τὸ διαστελλόμενον

---

[1] cf. p. 34, note 1.   [2] cf. p. 121, note 4.
[3] Pulmonary vein, or rather, left auricle. Galen means a
reflux through the mitral orifice; the left auricle was looked

previously drawn. Now Nature foresaw this ne-cessity,[1] and provided the cardiac openings of the vessels with membranous attachments,[2] to prevent their contents from being carried backwards. How and in what manner this takes place will be stated in my work "On the Use of Parts," where among other things I show that it is impossible for the openings of the vessels to be closed so accurately that nothing at all can run back. Thus it is in-evitable that the reflux into the *venous artery*[3] (as will also be made clear in the work mentioned) should be much greater than through the other openings. But what it is important for our present purpose to recognise is that every thing possessing a large and appreciable cavity must, when it dilates, abstract matter from all its neighbours, and, when it contracts, must squeeze matter back into them. This should all be clear from what has already been said in this treatise and from what Erasistratus and I myself have demonstrated elsewhere respecting the tendency of a vacuum to become refilled.[4]

XIV

AND further, it has been shown in other treatises that all the arteries possess a power which derives from the heart, and by virtue of which they dilate and contract.

Put together, therefore, the two facts—that the arteries have this motion, and that everything, when

on rather as the termination of the pulmonary veins than as a part of the heart. *cf.* p. 323, note 4. He speaks here of a kind of "physiological" mitral incompetence.

[4] *Horror vacui.*

ἕλκειν ἐκ τῶν πλησίον εἰς ἑαυτό, θαυμαστὸν
οὐδέν σοι φανεῖται τὰς ἀρτηρίας, ὅσαι μὲν εἰς τὸ
δέρμα περαίνουσιν αὐτῶν, ἐπισπᾶσθαι τὸν ἔξωθεν
ἀέρα διαστελλομένας, ὅσαι δὲ κατά τι πρὸς τὰς ‖
205 φλέβας ἀνεστόμωνται, τὸ λεπτότατον ἐν αὐταῖς
καὶ ἀτμωδέστατον ἐπισπᾶσθαι τοῦ αἵματος, ὅσαι
δ' ἐγγὺς τῆς καρδίας εἰσίν, ἐξ αὐτῆς ἐκείνης ποιεῖσθαι τὴν ὁλκήν. ἐν γὰρ τῇ πρὸς τὸ κενούμενον
ἀκολουθίᾳ τὸ κουφότατόν τε καὶ λεπτότατον
ἕπεται πρῶτον τοῦ βαρυτέρου τε καὶ παχυτέρου·
κουφότατον δ' ἐστὶ καὶ λεπτότατον ἁπάντων τῶν
κατὰ τὸ σῶμα πρῶτον μὲν τὸ πνεῦμα, δεύτερον
δ' ὁ ἀτμός, ἐπὶ τούτῳ δὲ τρίτον, ὅσον ἂν ἀκριβῶς
ᾖ κατειργασμένον τε καὶ λελεπτυσμένον αἷμα.

Ταυτ' οὖν εἰς ἑαυτὰς ἕλκουσιν αἱ ἀρτηρίαι
πανταχόθεν, αἱ μὲν εἰς τὸ δέρμα καθήκουσαι τὸν
ἔξωθεν ἀέρα· πλησίον τε γὰρ αὐταῖς οὗτός ἐστι
καὶ κουφότατος ἐν τοῖς μάλιστα· τῶν δ' ἄλλων
ἡ μὲν ἐπὶ τὸν τράχηλον ἐκ τῆς καρδίας ἀνιοῦσα
καὶ ἡ κατὰ ῥάχιν, ἤδη δὲ καὶ ὅσαι τούτων ἐγγὺς
ἐξ αὐτῆς μάλιστα τῆς καρδίας· ὅσαι δὲ καὶ τῆς
καρδίας πορρωτέρω καὶ τοῦ δέρματος, ἕλκειν
ταύταις ἀναγκαῖον ἐκ τῶν φλεβῶν τὸ κουφότατον τοῦ αἵματος· ὥστε καὶ τῶν εἰς τὴν γαστέρα
τε καὶ τὰ ἔντερα καθηκουσῶν ἀρτηριῶν τὴν
ὁλκὴν ἐν τῷ διαστέλλεσθαι γίγνεσθαι παρά τε
206 τῆς ‖ καρδίας αὐτῆς καὶ τῶν παρακειμένων αὐτῇ
φλεβῶν παμπόλλων οὐσῶν. οὐ γὰρ δὴ ἔκ γε τῶν
ἐντέρων καὶ τῆς κοιλίας τροφὴν οὕτω παχεῖάν τε
καὶ βαρεῖαν ἐν ἑαυτοῖς ἐχόντων δύνανταί τι
μεταλαμβάνειν, ὅ τι καὶ ἄξιον λόγου, φθάνουσαι
πληροῦσθαι τοῖς κουφοτέροις. οὐδὲ γὰρ εἰ καθεὶς

it dilates, draws neighbouring matter into itself—and
you will find nothing strange in the fact that those
arteries which reach the skin draw in the outer air
when they dilate, while those which anastomose at
any point with the veins attract the thinnest and
most vaporous part of the blood which these contain,
and as for those arteries which are near the heart,
it is on the heart itself that they exert their traction.
For, by virtue of the tendency by which a vacuum
becomes refilled, the lightest and thinnest part obeys
the tendency before that which is heavier and
thicker. Now the lightest and thinnest of anything
in the body is firstly pneuma, secondly vapour, and
in the third place that part of the blood which has
been accurately elaborated and refined.

These, then, are what the arteries draw into
themselves on every side; those arteries which
reach the skin draw in the outer air[1] (this being
near them and one of the lightest of things); as
to the other arteries, those which pass up from the
heart into the neck, and that which lies along the
spine, as also such arteries as are near these—draw
mostly from the heart itself; and those which are
further from the heart and skin necessarily draw the
lightest part of the blood out of the veins. So also
the traction exercised by the diastole of the arteries
which go to the stomach and intestines takes place
at the expense of the heart itself and the numerous
veins in its neighbourhood ; for these arteries cannot
get anything worth speaking of from the thick
heavy nutriment contained in the intestines and
stomach,[2] since they first become filled with lighter
elements. For if you let down a tube into a vessel

---

[1] cf. p. 305, note 2.   [2] cf. p. 308, note 1.

αὐλίσκον εἰς ἀγγεῖον ὕδατός τε καὶ ψάμμου πλῆρες
ἐπισπάσαιο τῷ στόματι τὸν ἐκ τοῦ αὐλίσκου
ἀέρα, δύναιτ᾿ ἂν ἀκολουθῆσαί σοι πρὸ τοῦ ὕδατος
ἡ ψάμμος· ἀεὶ γὰρ ἐν τῇ πρὸς τὸ κενούμενον
ἀκολουθίᾳ τὸ κουφότερον ἔπεται πρότερον.

XV

Οὔκουν χρὴ θαυμάζειν, εἰ παντελῶς ὀλίγον ἐκ
τῆς κοιλίας, ὅσον ἂν ἀκριβῶς ᾖ κατειργασμένον,
εἰς τὰς ἀρτηρίας παραγίγνεται φθανούσας πληροῦσθαι τῶν κουφοτέρων, ἀλλ᾿ ἐκεῖνο γιγνώσκειν,
ὡς δύ᾿ ἐστὸν ὁλκῆς εἴδη, τὸ μὲν τῇ πρὸς τὸ
κενούμενον ἀκολουθίᾳ, τὸ δ᾿ οἰκειότητι ποιότητος
γιγνόμενον· ἑτέρως μὲν γὰρ εἰς τὰς φύσας ὁ ἀήρ,
ἑτέρως δ᾿ ὁ σίδηρος ὑπὸ τῆς ἡρακλείας ἐπισπᾶται
λίθου· καὶ ὡς ἡ μὲν πρὸς τὸ κενούμενον ἀκο-
207 λουθία ‖ τὸ κουφότερον ἕλκει πρότερον, ἡ δὲ
κατὰ τὴν τῆς ποιότητος οἰκειότητα πολλάκις,
εἰ οὕτως ἔτυχε, τὸ βαρύτερον, ἂν τῇ φύσει συγ-
γενέστερον ὑπάρχῃ. καὶ τοίνυν καὶ ταῖς ἀρτηρίαις τε καὶ τῇ καρδίᾳ, ὡς μὲν κοίλοις τε καὶ
διαστέλλεσθαι δυναμένοις ὀργάνοις, ἀεὶ τὸ κουφότερον ἀκολουθεῖ πρότερον, ὡς δὲ τρέφεσθαι
δεομένοις, εἰς αὐτοὺς τοὺς χιτῶνας, οἳ δὴ τὰ
σώματα τῶν ὀργάνων εἰσίν, ἕλκεται τὸ οἰκεῖον.
ὅσον ἂν οὖν εἰς τὴν κοιλότητα διαστελλομένων
αὐτῶν αἵματος μεταληφθῇ, τούτου τὸ οἰκειότατον

---

[1] The "mechanical" principle of *horror vacui* contrasted
with the "physical" or semi-physiological principle of
specific attraction. *Appropriateness* here might almost be
rendered *affinity* or *kinship*. *cf.* note 2, *infra*.

full of water and sand, and suck the air out of the tube with your mouth, the sand cannot come up to you before the water, for in accordance with the principle of the refilling of a vacuum the lighter matter is always the first to succeed to the evacuation.

XV

IT is not to be wondered at, therefore, that only a very little [nutrient matter] such, namely, as has been accurately elaborated—gets from the stomach into the arteries, since these first become filled with lighter matter. We must understand that *there are two kinds of attraction*, that by which a vacuum becomes refilled and that caused by appropriateness of quality;[1] air is drawn into bellows in one way, and iron by the lodestone in another. And we must also understand that the traction which results from evacuation acts primarily on what is light, whilst that from appropriateness of quality acts frequently, it may be, on what is heavier (if this should be naturally more nearly related[2]). Therefore, in the case of the heart and the arteries, it is in so far as they are hollow organs, capable of diastole, that they always attract the lighter matter first, while, in so far as they require nourishment, it is actually into their *coats* (which are the real *bodies* of these organs) that the appropriate matter is drawn.[3] Of the blood, then, which is taken into their cavities when they dilate, that part which is most proper to them and

[2] "Related," "akin." *cf.* p. 36, note 2.
[3] The coats exercise the *vital* traction, the cavities the merely *mechanical*. *cf.* p. 165, note 2.

τε καὶ μάλιστα τρέφειν δυνάμενον οἱ χιτῶνες
αὐτοὶ τῶν ἀγγείων ἐπισπῶνται.

Τοῦ δ' ἐκ τῶν φλεβῶν εἰς τὰς ἀρτηρίας μετα-
λαμβάνεσθαί τι πρὸς τοῖς εἰρημένοις ἱκανὸν καὶ
τοῦτό γε τεκμήριον. εἰ πολλὰς καὶ μεγάλας
ἀρτηρίας διατεμὼν ἀποκτεῖναι τὸ ζῷον βουληθείης, εὑρήσεις αὐτοῦ τὰς φλέβας ὁμοίως ταῖς
ἀρτηρίαις ἐκκενουμένας, οὐκ ἂν τούτου ποτὲ
γενομένου χωρὶς τῶν πρὸς ἀλλήλας αὐταῖς
ἀναστομώσεων. ὡσαύτως δὲ καὶ κατ' αὐτὴν τὴν
καρδίαν ἐκ τῆς δεξιᾶς κοιλίας εἰς τὴν ἀριστερὰν
208 ἕλκεται τὸ λεπτό‖τατον ἔχοντός τινα τρήματα
τοῦ μέσου διαφράγματος αὐτῶν, ἃ μέχρι μὲν
πλείστου δυνατόν ἐστιν ἰδεῖν, οἷον βοθύνους τινὰς
ἐξ εὐρυτέρου στόματος ἀεὶ καὶ μᾶλλον εἰς στενό-
τερον προϊόντας. οὐ μὴν αὐτά γε τὰ ἔσχατα
πέρατα δυνατὸν ἔτι θεάσασθαι διά τε σμικρότητα
καὶ ὅτι τεθνεῶτος ἤδη τοῦ ζῴου κατέψυκταί τε
καὶ πεπύκνωται πάντα. ἀλλ' ὁ λόγος κἀνταῦθα
πρῶτον μὲν ἐκ τοῦ μηδὲν ὑπὸ τῆς φύσεως
γίγνεσθαι μάτην ὁρμώμενος ἐξευρίσκει τὰς
ἀναστομώσεις ταύτας τῶν κοιλιῶν τῆς καρδίας·
οὐ γὰρ δὴ εἰκῇ γε καὶ ὡς ἔτυχεν οἱ ἐς στενὸν οὕτω
τελευτῶντες ἐγένοντο βόθυνοι.

Δεύτερον δὲ κἀκ τοῦ δυοῖν ὄντοιν στομάτοιν ἐν
τῇ δεξιᾷ τῆς καρδίας κοιλίᾳ τοῦ μὲν εἰσάγοντος
τὸ αἷμα, τοῦ δ' ἐξάγοντος πολὺ μεῖζον εἶναι τὸ
εἰσάγον. ὡς γὰρ οὐ παντὸς τοῦ αἵματος, ὅσον ἡ
κοίλη φλὲψ δίδωσι τῇ καρδίᾳ, πάλιν ἐξ ἐκείνης

---

[1] Chap. xiv.
[2] These *fossae* were probably the recesses between the
columnae carneae. [3] On *logos* cf. p. 226, note 2.

most able to afford nourishment is attracted by their actual coats.

Now, apart from what has been said,[1] the following is sufficient proof that something is taken over from the veins into the arteries. If you will kill an animal by cutting through a number of its large arteries, you will find the veins becoming empty along with the arteries: now, this could never occur if there were not anastomoses between them. Similarly, also, in the heart itself, the thinnest portion of the blood is drawn from the right ventricle into the left, owing to there being perforations in the septum between them: these can be seen for a great part [of their length]; they are like a kind of fossae [pits] with wide mouths, and they get constantly narrower; it is not possible, however, actually to observe their extreme terminations, owing both to the smallness of these and to the fact that when the animal is dead all the parts are chilled and shrunken.[2] Here, too, however, our argument,[3] starting from the principle that nothing is done by Nature in vain, discovers these anastomoses between the ventricles of the heart; for it could not be at random and by chance that there occurred fossae ending thus in narrow terminations.

And secondly [the presence of these anastomoses has been assumed] from the fact that, of the two orifices in the right ventricle, the one conducting blood in and the other out, the former[4] is much the larger. For, the fact that the insertion of the vena cava into the heart[5] is larger than the

---

[4] He means the tricuspid orifice. *cf.* p. 121, note 4.

[5] The right auricle was looked on less as a part of the heart than as an expansion or "insertion" of the vena cava.

GALEN

ἐκπεμπομένου τῷ πνεύμονι, μείζων ἐστὶν ἡ ἀπὸ
τῆς κοίλης εἰς αὐτὴν ἔμφυσις τῆς ἐμφυομένης εἰς
209 τὸν πνεύμονα φλεβός. οὐδὲ ‖ γὰρ τοῦτ᾽ ἔστιν
εἰπεῖν, ὡς ἐδαπανήθη τι τοῦ αἵματος εἰς τὴν αὐτοῦ
τοῦ σώματος τῆς καρδίας θρέψιν. ἑτέρα γάρ
ἐστι φλὲψ ἡ εἰς ἐκεῖνο κατασχιζομένη μήτε τὴν
γένεσιν ἐκ τῆς καρδίας αὐτῆς μήτε τὴν τοῦ
αἵματος ἔχουσα μετάληψιν. εἰ δὲ καὶ δαπανᾶταί
τι, ἀλλ᾽ οὐ τοσοῦτόν γε μείων ἐστὶν ἡ εἰς τὸν
πνεύμονα φλὲψ ἄγουσα τῆς εἰς τὴν καρδίαν
ἐμφυομένης, ὅσον εἰκὸς εἰς τὴν τροφὴν ἀνηλῶσθαι
τῆς καρδίας, ἀλλὰ πλέον πολλῷ. δῆλον οὖν,
ὡς εἰς τὴν ἀριστεράν τι μεταλαμβάνεται κοιλίαν.

Καὶ γὰρ οὖν καὶ τῶν κατ᾽ ἐκείνην ἀγγείων δυοῖν
ὄντων ἔλαττόν ἐστι πολλῷ τὸ ἐκ τοῦ πνεύμονος εἰς
αὐτὴν εἰσάγον τὸ πνεῦμα τῆς ἐκφυομένης ἀρτηρίας
τῆς μεγάλης, ἀφ᾽ ἧς αἱ κατὰ τὸ σῶμα σύμπασαι
πεφύκασιν, ὡς ἂν μὴ μόνον ἐκ τοῦ πνεύμονος πνεῦμα μεταλαμβανούσης αὐτῆς, ἀλλὰ κἀκ τῆς δεξιᾶς
κοιλίας αἷμα διὰ τῶν εἰρημένων ἀναστομώσεων.

Ὅτι δ᾽ ἄμεινον ἦν τοῖς τοῦ σώματος μορίοις
τοῖς μὲν ὑπὸ καθαροῦ καὶ λεπτοῦ καὶ ἀτμώδους
αἵματος τρέφεσθαι, τοῖς δ᾽ ὑπὸ παχέος καὶ
θολεροῦ καὶ ὡς οὐδ᾽ ἐνταῦθά τι παρεώραται τῇ
210 φύσει, τῆς ‖ περὶ χρείας μορίων πραγματείας
ἐστίν, ὥστ᾽ οὐ χρὴ νῦν ὑπὲρ τούτων ἔτι λέγειν,

---

[1] This "vein" (really the pulmonary artery) was supposed
to be the channel by which the lungs received nutriment
from the right heart. cf. p. 121, note 3.

[2] The coronary vein.

[3] Galen's conclusion, of course, is, so far, correct, but he
has substituted an imaginary direct communication between
the ventricles for the actual and more roundabout pulmonary

vein which is inserted into the lungs[1] suggests that not all the blood which the vena cava gives to the heart is driven away again from the heart to the lungs. Nor can it be said that any of the blood is expended in the nourishment of the actual body of the heart, since there is another vein[2] which breaks up in it and which does not take its origin nor get its share of blood from the heart itself. And even if a certain amount is so expended, still the vein leading to the lungs is not to such a slight extent smaller than that inserted into the heart as to make it likely that the blood is used as nutriment for the heart: the disparity is much too great for such an explanation. It is, therefore, clear that something *is* taken over into the left ventricle.[3]

Moreover, of the two vessels connected with it, that which brings pneuma into it from the lungs[4] is much smaller than the great outgrowing artery[5] from which the arteries all over the body originate; this would suggest that it not merely gets pneuma from the lungs, but that it also gets blood from the right ventricle through the anastomoses mentioned.

Now it belongs to the treatise "On the Use of Parts" to show that it was best that some parts of the body should be nourished by pure, thin, and vaporous blood, and others by thick, turbid blood, and that in this matter also Nature has overlooked nothing. Thus it is not desirable that these matters should be further discussed. Having mentioned,

circulation, of whose existence he apparently had no idea. His views were eventually corrected by the Renascence anatomists. *cf.* Introduction, pp. xxii.–xxiii.

[4] He means the left auricle, considered as the termination of the pulmonary " arteries "; *cf.* p. 314, note 3.

[5] The aorta, its orifice being circular, appears bigger than the slit-like mitral orifice.

ἀλλ' ὑπομνήσαντας, ὡς δύο ἐστὸν ὁλκῆς εἴδη, τῶν μὲν εὐρείαις ὁδοῖς ἐν τῷ διαστέλλεσθαι τῇ πρὸς τὸ κενούμενον ἀκολουθίᾳ τὴν ἕλξιν ποιουμένων, τῶν δ' οἰκειότητι ποιότητος, ἐφεξῆς λέγειν, ὡς τὰ μὲν πρότερα καὶ πόρρωθεν ἕλκειν τι δύναται, τὰ δὲ δεύτερα ἐκ τῶν ἐγγυτάτω μόνων. αὐλίσκον μὲν γὰρ ὅτι μήκιστον εἰς ὕδωρ ἔνεστι καθέντα ῥᾳδίως ἀνασπᾶν εἰς τὸ στόμα δι' αὐτοῦ τὸ ὑγρόν· οὐ μὴν εἴ γ' ἐπὶ πλέον ἀπαγάγοις τῆς ἡρακλείας λίθου τὸν σίδηρον ἢ τοὺς πυροὺς τοῦ κεραμίου — καὶ γὰρ καὶ τοιοῦτόν τι πρόσθεν ἐλέγετο παράδειγμα — δύναιτ' ἂν ἔτι γενέσθαι τις ὁλκή.

Σαφέστατα δ' ἂν αὐτὸ μάθοις ἐπὶ τῶν ἐν τοῖς κήποις ὀχετῶν· ἐκ τούτων γὰρ εἰς μὲν τὰ παρακείμενα καὶ πλησίον ἅπαντα διαδίδοταί τις ἰκμάς, εἰς δὲ τὰ πορρωτέρω προσελθεῖν οὐκέτι δύναται, καὶ διὰ τοῦτ' ἀναγκάζονται πολλοῖς ὀχετοῖς μικροῖς ἀπὸ τοῦ μεγάλου τετμημένοις εἰς ἕκαστον μέρος τοῦ κήπου τὴν ἐπίρρυσιν τοῦ ὕδατος ἐπι-
211 τεχνᾶσθαι· καὶ τηλικαῦτά γε τὰ ‖ μεταξὺ διαστήματα τούτων τῶν μικρῶν ὀχετῶν ποιοῦσιν, ἡλίκα μάλιστα νομίζουσιν ἀρκεῖν εἰς τὸ ἱκανῶς ἀπολαύειν ἕλκοντα τῆς ἑκατέρωθεν αὐτοῖς ἐπιρρεούσης ὑγρότητος. οὕτως οὖν ἔχει κἂν τοῖς τῶν ζῴων σώμασιν. ὀχετοὶ πολλοὶ κατὰ πάντα τὰ μέλη διεσπαρμένοι παράγουσιν αὐτοῖς αἷμα καθάπερ ἐν κήποις ὑδρείαν τινά. καὶ τούτων τῶν ὀχετῶν τὰ μεταξὺ διαστήματα θαυμαστῶς ὑπὸ τῆς φύσεως εὐθὺς ἐξ ἀρχῆς διατέτακται πρὸς τὸ μήτ' ἐνδεῶς χορηγεῖσθαι τοῖς μεταξὺ μορίοις ἕλκουσιν εἰς ἑαυτὰ τὸ αἷμα μήτε κατακλύζεσθαι

however, that there are two kinds of attraction, certain bodies exerting attraction along wide channels during diastole (by virtue of the principle by which a vacuum becomes refilled) and others exerting it by virtue of their appropriateness of quality, we must next remark that the former bodies can attract even from a distance, while the latter can only do so from among things which are quite close to them ; the very longest tube let down into water can easily draw up the liquid into the mouth, but if you withdraw iron to a distance from the lodestone or corn from the jar (an instance of this kind has in fact been already given[1]) no further attraction can take place.

This you can observe most clearly in connection with *garden conduits*. For a certain amount of moisture is distributed from these into every part lying close at hand but it cannot reach those lying further off : therefore one has to arrange the flow of water into all parts of the garden by cutting a number of small channels leading from the large one. The intervening spaces between these small channels are made of such a size as will, presumably, best allow them [the spaces] to satisfy their needs by drawing from the liquid which flows to them from every side. So also is it in the bodies of animals. Numerous conduits distributed through the various limbs bring them pure blood, much like the garden water-supply, and, further, the intervals between these conduits have been wonderfully arranged by Nature from the outset so that the intervening parts should be plentifully provided for when absorbing blood, and that they should never

[1] p. 87.

GALEN

ποτ' αὐτὰ πλήθει περιττῆς ὑγρότητος ἀκαίρως
ἐπιρρεούσης.

Ὁ γὰρ δὴ τρόπος τῆς θρέψεως αὐτῶν τοιόσδε
τίς ἐστι. τοῦ συνεχοῦς ἑαυτῷ σώματος, οἱόνπερ
τὸ ἁπλοῦν ἀγγεῖον Ἐρασίστρατος ὑποτίθεται, τὰ
μὲν ἐπιπολῆς μέρη πρῶτα τῆς ὁμιλούσης ἀπολαύει τροφῆς· ἐκ δὲ τούτων αὖ μεταλαμβάνει
κατὰ τὸ συνεχὲς ἕλκοντα τὰ τούτων ἑξῆς, εἶτ' ἐξ
ἐκείνων αὖθις ἕτερα καὶ τοῦτ' οὐ παύεται γιγνόμενον, ἄχρις ἂν εἰς ἅπαντ' αὐτοῦ διαδοθῇ τὰ
μόρια τῆς τρεφούσης οὐσίας ἡ ποιότης. ὅσα δὲ
212 τῶν μορίων ἐπὶ πλέον ‖ ἀλλοιουμένου δεῖται τοῦ
μέλλοντος αὐτὰ θρέψειν χυμοῦ, τούτοις ὥσπερ τι
ταμιεῖον ἡ φύσις παρεσκεύασεν ἤτοι κοιλίας ἢ
σήραγγας ἤ τι ταῖς σήραγξιν ἀνάλογον. αἱ μὲν
γὰρ σάρκες αἵ τε τῶν σπλάγχνων ἁπάντων αἵ τε
τῶν μυῶν ἐξ αἵματος αὐτοῦ τρέφονται βραχεῖαν
ἀλλοίωσιν δεξαμένου. τὰ δ' ὀστᾶ παμπόλλης ἐν
τῷ μεταξὺ δεῖται τῆς μεταβολῆς, ἵνα τραφῇ, καὶ
ἔστιν οἱόνπερ τὸ αἷμα ταῖς σαρξί, τοιοῦτος ὁ
μυελὸς τοῖς ὀστοῖς ἐν μὲν τοῖς μικροῖς τε καὶ
ἀκοιλίοις κατὰ τὰς σήραγγας αὐτῶν διεσπαρ-
μένος, ἐν δὲ τοῖς μείζοσί τε καὶ κοιλίας ἔχουσιν ἐν
ἐκείναις ἠθροισμένος.

Ὡς γὰρ καὶ διὰ τοῦ πρώτου γράμματος ἐδείκνυτο, τοῖς μὲν ὁμοίαν ἔχουσι τὴν οὐσίαν εἰς
ἄλληλα μεταβάλλειν ἐγχωρεῖ, τοῖς δὲ πάμπολυ
διεστῶσιν ἀμήχανον ἀλλήλοις ὁμοιωθῆναι χωρὶς
τῶν ἐν μέσῳ μεταβολῶν. τοιοῦτόν τι καὶ τοῖς

---

[1] Or we may render it "corpuscle"; Galen practically
means the cell. cf. p. 153, note 2.

be deluged by a quantity of superfluous fluid running in at unsuitable times.

For the way in which they obtain nourishment is somewhat as follows. In the body[1] which is continuous throughout, such as Erasistratus supposes his *simple vessel* to be, it is the superficial parts which are the first to make use of the nutriment with which they are brought into contact; then the parts coming next draw their share from these by virtue of their contiguity; and again others from these; and this does not stop until the quality of the nutrient substance has been distributed among all parts of the corpuscle in question. And for such parts as need the humour which is destined to nourish them to be altered still further, Nature has provided a kind of storehouse, either in the form of a central cavity or else as separate caverns,[2] or something analogous to caverns. Thus the flesh of the viscera and of the muscles is nourished from the blood directly, this having undergone merely a slight alteration; the bones, however, in order to be nourished, require very great change, and what blood is to flesh marrow is to bone; in the case of the small bones, which do not possess central cavities, this marrow is distributed in their caverns, whereas in the larger bones which do contain central cavities the marrow is all concentrated in these.

For, as was pointed out in the first book,[3] things having a similar substance can easily change into one another, whereas it is impossible for those which are very different to be assimilated to one another without intermediate stages. Such a·one in respect to

[2] *cf.* the term "cavernous tissue."
[3] I. x.

χόνδροις ἐστὶ τὸ περικεχυμένον μυξῶδες καὶ τοῖς
συνδέσμοις καὶ τοῖς ὑμέσι καὶ τοῖς νεύροις τὸ
παρεσπαρμένον ἐν αὐτοῖς ὑγρὸν γλίσχρον· ἕκα-
213 στον γὰρ ‖ τούτων ἐξ ἰνῶν σύγκειται πολλῶν,
αἵπερ ὁμοιομερεῖς τ' εἰσὶ καὶ ὄντως αἰσθητὰ
στοιχεῖα. κατὰ δὲ τὰς μεταξὺ χώρας αὐτῶν ὁ
οἰκειότατος εἰς θρέψιν παρέσπαρται χυμός, ὃν
εἵλκυσαν μὲν ἐκ τῶν φλεβῶν τοῦ αἵματος, ὅσον
οἷόν τ' ἦν ἐκλεξάμεναι τὸν ἐπιτηδειότατον, ἐξ-
ομοιοῦσι δὲ κατὰ βραχὺ καὶ μεταβάλλουσιν εἰς
τὴν ἑαυτῶν οὐσίαν.

Ἅπαντ' οὖν ταῦτα καὶ ἀλλήλοις ὁμολογεῖ καὶ
τοῖς ἔμπροσθεν ἀποδεδειγμένοις ἱκανῶς μαρτυρεῖ
καὶ οὐ χρὴ μηκύνειν ἔτι τὸν λόγον· ἐκ γὰρ τῶν
εἰρημένων ἔνεστιν ἑκάστῳ τὰ κατὰ μέρος ἅπαντα
καθ' ὅντινα γίγνεται τρόπον ἐξευρίσκειν ἑτοίμως,
ὥσπερ καὶ διὰ τί πολλοῖς κωθωνιζομένοις πάμπολυ τάχιστα· μὲν ἀναδίδοται τὸ ποθέν, οὐρεῖται
δ' ὀλίγου δεῖν ἅπαν ἐντὸς οὐ πολλοῦ χρόνου. καὶ
γὰρ κἀνταῦθα τῇ τε τῆς ποιότητος οἰκειότητι καὶ
τῇ τῆς ὑγρότητος λεπτότητι καὶ τῇ τῶν ἀγγείων
τε καὶ τῶν κατ' αὐτὰ στομάτων εὐρύτητι καὶ τῇ
τῆς ἑλκτικῆς δυνάμεως εὐρωστίᾳ τὸ τάχος συντελεῖται τῆς ἀναδόσεως, τῶν μὲν πλησίον τῆς
κοιλίας τεταγμένων μορίων οἰκειότητι ποιότη-
214 τος ‖ ἑαυτῶν ἕνεκα ἑλκόντων τὸ πόμα, τῶν δ'

cartilage is the myxoid substance which surrounds it, and in respect to ligaments, membranes, and nerves the viscous liquid dispersed inside them ; for each of these consists of numerous fibres, which are homogeneous[1]—in fact, actual *sensible elements*; and in the intervals between these fibres is dispersed the humour most suited for nutrition; this they have drawn from the blood in the veins, choosing the most appropriate possible, and now they are assimilating it step by step and changing it into their own substance.

All these considerations, then, agree with one another, and bear sufficient witness to the truth of what has been already demonstrated ; there is thus no need to prolong the discussion further. For, from what has been said, anyone can readily discover in what way all the particular [vital activities] come about. For instance, we could in this way ascertain why it is that in the case of many people who are partaking freely of wine, the fluid which they have drunk is rapidly absorbed[2] through the body and almost the whole of it is passed by the kidneys within a very short time. For here, too, the rapidity with which the fluid is absorbed depends on appropriateness of quality, on the thinness of the fluid, on the width of the vessels and their mouths, and on the efficiency of the attractive faculty. The parts situated near the alimentary canal, by virtue of their appropriateness of quality, draw in the imbibed food for their own purposes, then the parts next to them

---

[1] Lit. *homoeomerous*, *i.e.* "the same all through," of similar structure throughout, the *elements* of living matter. *cf.* p. 20, note 3, and *cf.* also the " cell " of Erasistratus, p. 153.
[2] " Delivered," " dispersed " ; *cf.* p. 13, note 5.

ἑξῆς τούτοις ἐξαρπαζόντων καὶ αὐτῶν εἰς ἑαυτὰ
κἄπειτα τῶν ἐφεξῆς πάλιν ἐκ τούτων μεταλαμ-
βανόντων, ἄχρις ἂν εἰς τὴν κοίλην ἀφίκηται
φλέβα, τοὐντεῦθεν δ᾽ ἤδη τῶν νεφρῶν τὸ οἰκεῖον
ἐπισπωμένων. ὥστ᾽ οὐδὲν θαυμαστὸν οἶνον μὲν
ὕδατος ἀναλαμβάνεσθαι θᾶττον οἰκειότητι ποιότητος, αὐτὸν δὲ τὸν οἶνον τὸν μὲν λευκὸν καὶ
καθαρὸν ἑτοίμως ἀναδίδοσθαι διὰ λεπτότητα, τὸν
δ᾽ αὖ μέλανα καὶ θολερὸν ἴσχεσθαί τε κατὰ τὴν
ὁδὸν καὶ βραδύνειν ὑπὸ πάχους.

Εἴη δ᾽ ἂν ταῦτα καὶ τῶν ὑπὲρ τῶν ἀρτηριῶν
ἔμπροσθεν εἰρημένων οὐ σμικρὰ μαρτύρια. παν-
ταχοῦ γὰρ ὅσον οἰκεῖόν τε καὶ λεπτὸν αἷμα τοῦ
μὴ τοιούτου ῥᾷον ἔπεται τοῖς ἕλκουσιν. ἀτμὸν
οὖν ἕλκουσαι καὶ πνεῦμα καὶ λεπτὸν αἷμα κατὰ
τὰς διαστάσεις αἱ ἀρτηρίαι τῶν κατὰ τὴν κοιλίαν
καὶ τὰ ἔντερα περιεχομένων χυμῶν ἢ οὐδ᾽ ὅλως ἢ
παντάπασιν ἐπισπῶνται βραχύ.

in their turn snatch it away, then those next again take it from these, until it reaches the vena cava, whence finally the kidneys attract that part of it which is proper to them. Thus it is in no way surprising that wine is taken up more rapidly than water, owing to its appropriateness of quality, and, further, that the white clear kind of wine is absorbed more rapidly owing to its thinness, while black turbid wine is checked on the way and retarded because of its thickness.

These facts, also, will afford abundant proof of what has already been said about the arteries ; everywhere, in fact, such blood as is both specifically appropriate and at the same time thin in consistency answers more readily to their traction than does blood which is not so ; this is why the arteries which, in their diastole, absorb vapour, pneuma, and thin blood attract either none at all or very little of the juices contained in the stomach and intestines.

INDEX AND GLOSSARY

INDEX AND GLOSSARY

(The numbers refer to the pages of the present edition; fuller references will be found in the footnotes.)

INDEX AND GLOSSARY

INDEX AND GLOSSARY

337

INDEX AND GLOSSARY

INDEX AND GLOSSARY